Enterprise Networks and Logistics for Agile Manufacturing

Lihui Wang · S.C. Lenny Koh
Editors

Enterprise Networks and Logistics for Agile Manufacturing

Springer

Prof. Lihui Wang
University of Skövde
Virtual Systems Research Centre
Intelligent Automation
PO Box 408
541 28 Skövde
Sweden
lihui.wang@his.se

Prof. S.C. Lenny Koh
Sheffield University
Management School
Logistics and Supply Chain Management
(LSCM) Research Centre
9 Mappin Street
Sheffield S1 4DT
UK
s.c.l.koh@sheffield.ac.uk

ISBN 978-1-4471-5701-4 ISBN 978-1-84996-244-5 (eBook)
DOI 10.1007/978-1-84996-244-5
Springer London Dordrecht Heidelberg New York

British Library Cataloguing in Publication Data
A catalogue record for this book is available from the British Library

Cover illustration: Lihui Wang

Cover design: eStudioCalamar, Figueres/Berlin

Printed on acid-free paper

Springer is part of Springer Science+Business Media (www.springer.com)

Preface

Manufacturing has been one of the key areas that support and influence a nation's economy since the eighteenth century. As the primary driving force behind economic growth, manufacturing serves as the foundation of and contributes to other industries, with products ranging from heavy-duty machinery to hi-tech home electronics. In past centuries, manufacturing has contributed significantly to modern civilisation and created the momentum that drives today's economy. Despite various revolutionary changes and innovations in the twentieth century that contributed to manufacturing advancement, we are constantly facing new challenges in the global marketplace.

Today, agile manufacturing has gained prominence due to recent business decentralisation and outsourcing. Manufacturing companies are competing in a dynamic marketplace that demands a short response time to changing markets and agility in production. In the twenty-first century, manufacturing is gradually shifting to a distributed environment with increasing dynamism. In order to win orders, locally or globally, customer satisfaction is treated as priority. This has led to mass customisation and ever more agile manufacturing processes, from the shop floor to every level of the manufacturing supply chain. At the same time, outsourcing has forged a multi-tier supplier structure with numerous small-to-medium-sized enterprises (SMEs) involved, where enterprise networks are formed and broken dynamically in order to deal with issues of logistics and supply chain management, effectively and efficiently. Moreover, environmental concerns have forced companies to address the recycling and re-manufacturing of end-of-life products, and this has created problems for both the reverse supply chain and reverse logistics. These issues constantly challenge manufacturing companies, and create a lot of uncertainty in agile manufacturing. Engineers across organisations often find themselves in situations that demand advanced planning and management capability when dealing with daily operations related to enterprise networks and logistics.

Targeting the uncertainty issues in agile manufacturing, over the past decade, research efforts have focused on improving the flexibility, adaptability, productivity, agility and leagility of manufacturing, particularly in supply chain management and logistics of decentralised enterprise networks. Various Web-based and artificial intelligence (AI) based tools have been developed to deal with these issues, and many research projects have been devoted to improving the throughput and efficiency of agile manufacturing. Thanks to recent advancements in information technology, research in supply chain management and logistics has progressed to a new level in adaptive decision making and trouble shooting, in order to address the problems encountered in today's enterprise network environment with increasing globalisation and outsourcing. While research and development efforts have resulted in a large volume of publications and impacted both present and future practices in agile manufacturing, there still exists a gap in the literature for a focused collection of works dedicated to enterprise networks and logistics. To bridge this gap and present the state-of-the-art to a broad readership, from academic researchers to practicing engineers, is the primary motivation behind this book.

As a general overview, Chapter 1 begins with a clear definition of enterprise network, logistics, supply chain, supply network and value chains, and explains the contexts within which they differ. Based on a comparative analysis of the existing literature, this chapter provides a discussion on decentralised decision making and presents both the current status and potential future trends in enterprise networks and logistics within the context of agile manufacturing. The discussion of decentralised decision making is extended in Chapter 2. Particularly, it reviews the research and practices of the industrial networks of the future. This chapter also identifies the fundamental challenges of preparing for the industrial networks of 2020 and beyond. Chapter 3 then introduces a unique perspective showing where agile manufacturing can position itself in complex supply networks. Through a Co-OPERATE project, it aims to develop a Web-based system for improved coordination of manufacturing planning and control activities across a supply network.

Recognising the importance of structure versus operation of an organisation, Chapter 4 focuses its attention around enterprise architecture in order to determine how an organisation can most effectively achieve its current and future objectives. Assuming that a portion of the value of an enterprise architecture initiative is in the form of embedded options (or real options), this chapter proposes the use of real options that allow flexibility for architects to change plans, so that uncertainties can be resolved over time. In light of the current popularity of information and communication technologies (ICT), Chapter 5 reports on ICT standardisation, aiming at ensuring interoperability between the various systems of an enterprise network.

Chapter 6 highlights ways of collaborative demand planning, particularly when information is shared in the downstream supply chain between manufacturer and retailer. It regards information sharing concerning demand signals within supply chains as one of the keys to responding to retail demands with greater agility.

In the area of supply selection, Chapter 7 depicts an empirical analysis of value creation and supplier selection. This chapter also examines the criteria used in the suppliers' selection process and thereby in the supply chain. Continuing this theme, Chapter 8 utilises a fuzzy AHP (analytic hierarchy process) approach to address the supplier selection problem. When faced with incomplete information from experts, the fuzzy set theory is found to be useful to handle uncertainties.

These discussions are extended in Chapter 9 to include a sustainable green supply chain platform in a globally integrated supply chain network. Based on preliminary analyses, this chapter offers some suggestions to help manufacturers and logistics service providers to restructure their supply chain strategies.

The primary goal of a supply chain is to meet the varying demand of customers where coordination among the customers is paramount. Realising this, Chapter 10 proposes a multi-agent self-healing approach that can assist in selecting outsourcing partners, and developing effective coordination among themselves and between manufacturing units. The agent-based approach is extended in Chapter 11 to cover simulation-based optimisation for supply chain management, and considers the entities (*e.g.* supplier, manufacturer, distributor and retailer) in a supply chain as intelligent agents in a simulation. This chapter also gives an outline on how these agents pursue their local objectives as well as how they react and interact with each other to achieve a more holistic outcome.

In addition to forward supply chains, reverse supply chains are becoming equally important, owing to increasing environmental concerns. Chapter 12 identifies the major barriers of a battery recycling system as an example, and shows how the interaction among those barriers hinders the recycling activities along its reverse supply chain. The issue of the reverse supply chain is further discussed in Chapter 13, looking at the optimal design of reverse logistics and closed-loop supply chain networks.

In a decentralised environment, global logistics services have increased dramatically and become extremely complex and dynamic. The logistics industry is changing in a variety of ways, including mergers to form integrated transportation service providers, outsourcing, and the increased use of information technology. Chapter 14 provides an overview of this evolution and looks at important trends in the logistics services industry. In this sector, routing and scheduling of delivery vehicles often involves complex decision making. Chapter 15 addresses the problem of multiple-vehicle pick-up and delivery, with time windows and heterogeneous capacitated vehicles, using simulated annealing as well as a simple and fast meta-heuristic.

Chapter 16 proposes the use of conventional simulation tools to model and visualise the coordinating behaviours of a networked distributed system. This can be a great assistance in accelerating system development, especially when it is large in size and complex in nature.

Finally, Chapter 17 discusses the implication of robustness and capability indices in the optimisation process of an airline's fleet. It introduces a technique capable of effectively addressing contradicting outcomes and minimising potential losses.

All together, the seventeen chapters provide an overview of some recent R&D achievements in supply chain design, supplier selection, vehicle routing, and system visualisation. With the rapid advancement of ICT, particularly Internet- and Web-based, we believe that this will continue to be a very active research field for years.

The editors would like to take this opportunity express their deep appreciation to all the authors for their significant contributions to this book. Their commitment, enthusiasm, and technical expertise are what made this book possible. We are also grateful to the publisher for supporting this project, and would especially like to thank Anthony Doyle, Senior Editor for Engineering, and Claire Protherough, Senior Editorial Assistant, for their constructive assistance and earnest cooperation, both with the publishing venture in general and the editorial details. We hope that readers find this book informative and useful.

Skövde, Sweden — Lihui Wang
Sheffield, United Kingdom — S.C. Lenny Koh
December 2009

Contents

List of Contributors

Blandine Ageron
Department of Supply Chain and Information Systems
University of Grenoble
26901 Valence Cedex 9
France

S. Arunachalam
School of Computing and Technology
University of East London
Essex
UK

Tehseen Aslam
Virtual Systems Research Centre
University of Skövde
PO Box 408, 541 28 Skövde
Sweden

David Bennett
Operations & Information Management Group
Aston University
Birmingham B4 7ET
UK

Z.M. Bi
Department of Engineering
Indiana Purdue University Fort Wayne
Fort Wayne, IN 46805-1499
USA

Elias G. Carayannis
School of Business
George Washington University
Washington, DC 20052
USA

F.T.S. Chan
Department of Industrial and Systems Engineering
The Hong Kong Polytechnic University
Hung Hom, Hong Kong
China

Federica Cucchiella
Department of Electrical and Information Engineering
University of L'Aquila
Monteluco di Roio, 67040 L'Aquila
Italy

Rob Dekkers
University of the West of Scotland
Paisley PA1 2BE
United Kingdom

K. Ganesh
Global Business Services –
Global Delivery
IBM India Private Ltd.
Bandra Kula Complex, Mumbai, 400051
India

Massimo Gastaldi
Department of Electrical and Information Engineering
Faculty of Engineering
University of L'Aquila
Monteluco di Roio, 67040 L'Aquila
Italy

Mark Goh
NUS Business School
National University of Singapore
Singapore 117574

A. Noorul Haq
Department of Production Engineering
National Institute of Technology
Tiruchirappalli, 620 015
India

Eleftherios Iakovou
Industrial Management Division
Department of Mechanical Engineering
Aristotle University of Thessaloniki
54124 Thessaloniki
Greece

Cengiz Kahraman
Department of Industrial Engineering
Istanbul Technical University
34367 Macka, Istanbul
Turkey

İhsan Kaya
Department of Industrial Engineering
Istanbul Technical University
34367 Macka, Istanbul
Turkey

S.C. Lenny Koh
Management School
The University of Sheffield
9 Mappin Street, Sheffield S1 4DT
UK

Leo D. Kounis
Department of Aviation Technology
Halkis Polytechnic
34 400 Psachna Evias
KEA, Research Department
State Aircraft Factory
Hellinikon, Athens
Greece

V. Kumar
Department of Management
Exeter Business School
University of Exeter
Exeter, EX4 4PU
United Kingdom

Sushmera Manikandan
The Logistics Institute – Asia Pacific
National University of Singapore
Singapore 117574

Aristides Matopoulos
Department of Business Administration
and Economics
International Faculty of the University of
Sheffield
54626 Thessaloniki
Greece

N. Mishra
School of Computer Science and
Information Technology
University of Nottingham
Nottingham, NG8 1BB
UK

S.P. Nachiappan
Department of Mechanical Engineering
Thiagarajar College of Engineering
Madurai
India

Amos Ng
Virtual Systems Research Centre
University of Skövde
PO Box 408, 541 28 Skövde
Sweden

Yiannis Nikolaidis
Department of Technology Management
University of Macedonia
59200 Naousa
Greece

David van Over
Faculty of Business and Management
University of Wollongong in Dubai
Knowledge Village, Dubai, 20183
UAE

Eleni-Maria Papadopoulou
Department of Applied Informatics
University of Macedonia
156 Egnatia Street, 540 06, Thessaloniki
Greece

Karine Evrard Samuel
Centre of Studies and Research in
Management
University of Grenoble
38040 Grenoble Cedex 9
France

P. Sasikumar
Department of Production Engineering
National Institute of Technology
Tiruchirappalli, 620 015
India

P. Sivakumar
Vickram College of Engineering
Madurai-Anna University
Tiruchirappalli
India

Robert de Souza
The Logistics Institute – Asia Pacific
National University of Singapore
Singapore 117574

Alain Spalanzani
University of Grenoble
51, rue B. de Laffemas – BP 29
26901 Valence Cedex 9
France

Balan Sundarakani
Faculty of Business and Management
University of Wollongong in Dubai
Knowledge Village, Dubai, 20183
UAE

Lihui Wang
Virtual Systems Research Centre
University of Skövde
Sweden

Anastasios Xanthopoulos
Department of Mechanical Engineering
Aristotle University of Thessaloniki
54124 Thessaloniki
Greece

Henry Xu
UQ Business School
The University of Queensland
St Lucia, Queensland, 4072
Australia

1

Overview of Enterprise Networks and Logistics for Agile Manufacturing

S.C. Lenny Koh[1] and Lihui Wang[2]

[1] Logistics and Supply Chain Management (LSCM) Research Centre
Management School, The University of Sheffield
9 Mappin Street, Sheffield S1 4DT, UK
Email: s.c.l.koh@sheffield.ac.uk

[2] Virtual Systems Research Centre
University of Skövde
PO Box 408, 541 28 Skövde, Sweden
Email: lihui.wang@his.se

Abstract

The demand for research and development of enterprise networks and logistics has been on an upward trajectory over the last decades. With a need for more innovative and responsive enterprise network structure, technology and supply chain to deal with an ever-changing and highly competitive market, the agility of processes, organisations and their supply chain, particularly in a manufacturing environment, need to be re-examined. This chapter provides an overview of the current status and potential future trends in this area. More specifically, this will be analysed within the context of agile manufacturing.

1.1 Introduction

The terms of enterprise network, logistics, supply chain, supply network and value chain are often used interchangeably and interpreted synonymously in the literature. The terms carry different meanings, depending on how these terms are interpreted and in what context they are being used.

Taking a normalised perspective from the literature, this chapter begins with a clear definition of their variations and explains the contexts within which they differ. We will then overview and critically analyse enterprise networks and logistics in the context of agile manufacturing. Previous literature in these related fields will be drawn on to provide a baseline for comparative analytics driving the discussions between current and future projections of enterprise network and logistics for agile manufacturing.

1.2 Logistics

Authors often use the term supply chain management synonymously with the term logistics. Logistics is actually a sub-set of supply chain management. Logistics refers to the distribution and movement of materials, components, parts, products and services from one node to another, up and down the supply chain. Logistics involves deciding upon various transportation modes, for example, air, rail, road and sea, to manage the movement and distribution of the above. From an organisational perspective, logistics could also be categorised into inbound and outbound logistics. Inbound logistics deals with managing the inward flow of materials, components, parts, products and services from suppliers or third party logistics to the organisation. Outbound logistics deals with managing the outward flow of materials, components, parts, products and services from the organisation to customers or third party logistics. Many organisations, in diverse industries, do not want to manage their own logistics operation, and use third party services in this area. Fourth party logistics has also emerged providing another layer of services to third party logistics. When the demand on third party logistics is very high and triggers insufficient capacity (*e.g.* fleet and so on) to manage the delivery, fourth party logistics will be used to meet the demand. Both inbound and outbound logistics requires good relationship management with suppliers and customers. The relationship with tier suppliers is paramount and the same applies to tier 1 customers. A tier 1 customer could be a distributor or retailer and this provides a large market size for the product or service. Hence, management of the supply chain is very important in ensuring that the right quality and the right quantity are delivered and received at the right time.

Reverse logistics is equally important given the nature for rework and re-distribution of products in order to satisfy various environmental requirements. When designing a logistics operation, one must consider the element of reverse logistics and how this could be designed into or designed out of the process. Designing reverse logistics into the operation includes considerations such as the methods by which the product could be returned directly to manufacturers. Designing reverse logistics out of the operation includes consideration such as the methods by which good product design eliminates the needs for return (*e.g.* decomposable materials).

1.3 Supply Chain Management

Supply chain management, taking logistics as a sub-set, integrates with all other important elements such as suppliers, manufacturers, distributors, retailers and customers in a holistic whole to ensure that the entire supply chain is integrated upstream and downstream. Supply chain management activities include sourcing, procurement, manufacturing and logistics. In a supply chain, in addition to managing the flow of materials, components, parts, products and services, managing information/knowledge, cash and intellectual capital flow are equally important. Building a long-term partnership with suppliers rather than an arms-length relationship is paramount in a supply chain.

Supply chains compete, not organisations. It is fundamental that organisations should re-examine their supply network and, if necessary, restructure the supply chain in order to compete with other supply chains. An enterprise network is the basis of a supply network. An enterprise network is a group of organisations working together for a common goal. The notion of an enterprise network interlinks with the work in cluster, enterprise system and extended enterprise. An enterprise network could be formed formally or informally. A formally structured enterprise network, such as a consortium, provides buying power for the group of organisations in the enterprise network. An informally structured enterprise network exists in a more virtual manner, which comes together and dissolves depending on specific opportunistic alliances and joint ventures. Unlike a cluster, the formation of an enterprise network could be independent of sector. A cluster, whether formal or informal, is normally structured around a sector, for example, the cheese and wine cluster in south east Europe. In contrast, an enterprise network is formed around the supply chain of the organisation; for example, there is an enterprise network around the ODM/OEM (original design/equipment manufacturer) suppliers to ACER and Phillips. When an enterprise network matures over time, it provides an opportunity to enable the supplier to work more closely with the manufacturer. This scenario will lead to potentially three outcomes: (1) a supply network, (2) a value chain, and (3) an integrated supply chain. Supply network formation creates a mutually beneficial environment with a common supply base to enable organisations to flexibly source the required products or services from the supply network. When value is added to the process in this supply network, for example, outsourcing of some processes to suppliers, a value chain is created. This enables an even closer collaboration between the suppliers and manufacturers and creates an environment for innovation. When the relationship between the supplier and the manufacturer has reached a further maturity point, it creates an opportunity to enable the supplier to have a physical presence at the manufacturer's plant, providing the highly responsive and agile processes required to fulfil demand. This leads to an integrated supply chain, where the supplier's supply chain is integrated with the manufacturer's supply chain. In this scenario, the supplier is still owned by the supplier and not the manufacturer, which makes it different to vertical integration. The automotive industry is pioneering the notion of integrated supply chains and the shipping industry is also looking at how the integrated supply chain model could be adapted to suit demand in the shipping industry given the need to re-examine their infrastructure. The notion of integrated supply chain was derived from Dell's supply chain model, but with an extension to consider ways in which it could be adapted to different industries' supply chains and ways in which relevant information systems are required to enable seamless exchange and sharing of information and resources.

1.4 Agile Manufacturing – Towards Leagile Manufacturing and Supply Chain?

Agile manufacturing environment requires responsive-to-demand facility and lean production. An agile manufacturing environment creates processes, tools, and knowledge base to enable the organisation to respond quickly to customer needs and market changes whilst still controlling costs and quality. Agile manufacturing cannot be achieved without facilitation by appropriate manufacturing and information technologies, and, more importantly, the appropriate integration of these technologies along the supply chain, including responsive manufacturing system, flexible manufacturing system, virtual manufacturing system, ultra rapid prototyping, process modelling, Computer Aided Manufacturing (CAM), Enterprise Resource Planning (ERP), mobile manufacturing services, on-line stock control system, satellite controlled networked maintenance, repair and overhaul database, Customer Relationship Management (CRM), Supplier Relationship Management (SRM), RFID, e-commerce, e-business and so on. These are crucial technologies required to enable seamless exchange and sharing of information, and provide a responsive manufacturing capacity required.

One of the biggest challenges facing organisations today is dealing with volatility in demand. Due to high demand volatility, there is no one strategy that can be adopted and this has led to the need for organisations to adopt a multiple chain strategy. This helps them to quickly respond to the both in terms of changed variety and volume.

One way to identify the type of supply chain strategies that will best suit the organisation is to position the products in an organisations portfolio according to their supply and demand characteristics. 'Supply characteristics' means the amount of time that it takes to replenish the stock. 'Demand characteristics', on the other hand, deals with how well the organisation can predict the demand for goods and services. To achieve both of these objectives satisfactorily, an organisation must re-examine how responsive and how agile their systems are.

Figure 1.1 suggests four generic strategies that can be adopted to meet demand and these are dependent on the combination of supply and demand characteristics for each product.

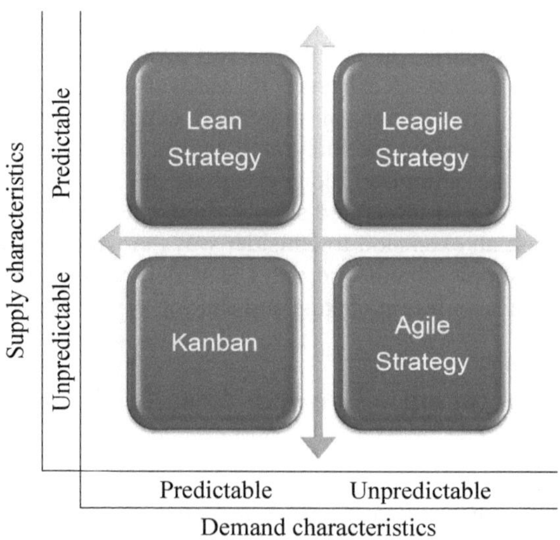

Figure 1.1. Generic supply chain strategies [1.1]

1.4.1 Lean Strategy

Womack and Jones [1.2] developed the *lean enterprise* concept and later expanded it into the wider concept of *lean thinking*. Leanness is about doing more with less. It owes its origins to the Toyota Production System (TPS) [1.3], where the concern was the reduction of waste (or *muda* in Japanese) within the factory environment. The focus of lean thinking is to eliminate all type of waste such as reduction of inventories, lot-size, supplier base and elimination of paperwork so that a level schedule can be established. However, the problem with lean thinking is that it originated in the Japanese automobile industry of the 1970s, whereas now we are in a different era of manufacturing, with lower demand, higher variety and higher uncertainty in the supply chain. Christopher [1.1] states that 'lean' works best in high volume, low variety and predictable environments. This led to the development of the agile concept.

1.4.2 Agile Strategy

Hiebelar *et al.* [1.4] introduced the agile strategy with the aim to satisfy demand by taking minimal lead times. 'Agility' is primarily concerned with responsiveness and the ability to match supply and demand in unpredictable markets where the demand for variety is very high. The distinguishing feature of agile supply chain is that it is 'market sensitive'. The idea of manufacturing flexibility was later extended by Nagel and Dove [1.5] into a wider framework and the concept of agility as a supply chain paradigm was born. However, Harrison *et al.* [1.6] realised that for agility to work, information flow within the supply chain partners is necessary, and stated that it could only happen with the use of information technology. This will then minimise the lost sales and also reduce the cost of stocking inventory.

1.4.3 Leagile Strategy

The top-right quadrant in Figure 1.1 represents a situation where the lead times are long and demand is unpredictable. In such situation, the first priority is to decrease the lead times since the variability of demand is totally uncertain and beyond the control of the organisation. However, if lead time cannot be reduced, then the next option is to seek to create a hybrid lean/agile solution. Various researchers suggest that the lean and agile approaches can be integrated to form a 'leagile' strategy. Christopher and Towill [1.7] formed the following three distinct lean–agile hybrids:

- *Pareto rule*
  This recognises that 80% of an organisation's revenue is generated from 20% of its products. Goldsby and Garcia-Dastugue [1.8] suggest that if 20% of the production is managed in a lean manner given that demand is stable, the remaining 80% can be managed in an agile manner.
- *Base and surplus demand*
  This is founded on the principle of base and surplus demand, which assumes that most organisations experience a base level of demand that can be

managed by a lean strategy, and the remaining demand that is periodical or seasonal can be managed by an agile strategy.

- *Postponement*
  Postponement strategy is founded on the principle of postponement, which requires the supply chain to be 'de-coupled' through holding strategic inventory in some generic or unfurnished form, with final configuration being completed rapidly once the real demand is known. Bucklin [1.9] states that the risk and uncertainty costs mainly occur due to the differentiation in products in the supply chain and that the postponement strategy will help to reduce or fully eliminate this cost by postponing certain activities until the actual demand arises.

Leagile supply chain systems have several advantages:

- they increase the organisation's ability to adjust products to specific customer wishes;
- inventory can be held at a generic level, resulting in lower stock-keeping and hence reducing the holding, transportation and obsolescence costs;
- keeping the inventory in a generic form gives greater flexibility, as the same inventory can be used to produce variety of end products;
- forecasting is easier at the generic level than at the level of the finished item;
- finally, the ability to customise products locally means that a higher level of variety may be offered at a lower total cost, enabling strategies of 'mass-customisation' to be pursued.

Taking the analysis of the leagile strategy and agile manufacturing together, the literature above suggests the extension of agile manufacturing towards a leagile manufacturing and leagile supply chain direction.

1.5 Cases from Logistics Sectors

This section summarises the cases published in Koh and Tan [1.10] and extends the narratives by considering the leagility of their supply chains as a result of changes made to their logistics operations and enterprise network. Due to confidentiality requests, both Company A Logistics and Company B Logistics prefer to remain anonymous in any publications.

1.5.1 Foreign 3PL: Company A Logistics and Maersk Logistics

Technological use, including the application of e-commerce in Company A Logistics and Maersk Logistics, is advanced or even in leading position in the industry. For example, Company A Logistics spent around US$200 million in IT development and is maintaining the technological leader position of the 3PL logistic industry in the world. The general manager of Company A Logistics pointed out that the current concerns of e-commerce in Company A Logistics are not to develop new e-commerce technologies, but to apply all existing functions to the China market.

This finding suggests the importance of the diffusion of technology across the supply chain. Once it is proven to provide significant improvement in one site of the chain, the organisation is keen to extend that across the chain.

Maersk Logistics' parent companies have invested heavily in developing new technologies (*e.g.* some e-commerce functions such as M*Power Web Report Builder, M*Power Web Search, M*Power Web Shipper, Startrack, e-SOP, *etc.*). This supports the finding from Langley *et al.* [1.11] that competition at the technological level is one of most important future trends, hence that providing more reliable and comprehensive services to customers in 3PL industry through the use of e-commerce could be regarded as providing the critical competitive advantage. The use and application of e-commerce must be supported by reliable technologies. Therefore, the development of new technologies makes it possible to supply better and more reliable services than competitors. This finding suggests the thirst of the organisation to search for new and innovative technologies to enable information exchange and sharing across the supply chain. Given that Maersk Logistics is a key player in the logistic sector, it is not surprising to note this demand. With the introduction of leagile manufacturing and supply chain, the projection of the future trends in the application of the leagile supply chain in this sector is promising.

1.5.2 Domestic 3PL: Longfei Logistics and Company B Logistics

The use of technologies and e-commerce in Longfei Logistics and Company B Logistics tends to be behind their foreign competitors due to the lack of sufficient funds and capabilities to develop leading technologies. The sources of e-commerce applications in these two domestic 3PL providers are mainly through two channels, namely, purchase from external vendors or cooperation with their partners. Longfei Logistics purchased all its logistics software as off-the-shelf packages and Company B Logistics purchased significant parts of their software using the same method. They do not own or use any advanced technologies such as track and trace systems, EDI with customers or JIT services. However, on some occasions they could provide those services to their customers by cooperating with partners who have the relevant technologies. For example, Company B Logistics share the warehousing systems of Maersk Logistics, and Longfei Logistics provide part of their goods tracking by using their partners' capabilities.

These technological strategies have many disadvantages. For example, they may never catch up with the new technology development and may never become a technology leader in the industry. Sometimes, they may be somewhat controlled by the partners. Besides, using partners' capabilities and/or purchasing from external vendors may cause an increase in cost, and thus diminish their competitiveness.

Despite these disadvantages, the two domestic 3PL providers are found to be willing to pursue their current strategies for practical reasons; it is almost impossible for them to catch up with the technology level of Company A Logistics and Maersk Logistics, whose parent companies invest heavily in R&D to keep their leading positions in new technology development. Longfei Logistics and Company B Logistics benefit from the cheaper solutions since they could acquire new technologies quickly and only purchase those technologies that they need or cooperate with those partners who could provide them with such technological advantages. Their technological strategies are more geared towards customer needs rather than developing new or advanced technologies which need huge amount of investment and may not be used in the near future.

The cases from the domestic 3PL sector discussed above illustrate different angle in terms of their degree of diffusion and adoption of enterprise network and logistics for leagility, as compared to the previously reported cases from the foreign 3PL sector. However, these cases suggest an interesting point, which is that they all rely on information technology to provide the competitiveness and responsiveness required. Due to the nature of this sector (*i.e.* not manufacturing), we cannot extract the manufacturing conditions from the above four cases. However, extrapolation of the findings suggests that pressure from demand (in manufacturing organisations) for such a movement indicates that the responsiveness of an organisation does relate to demand. This implies that higher leagility in the supply chain facilitated by innovative manufacturing and information technologies are essential to compete with other supply chains.

1.6 Supply Chain Transformation

Understanding a supply chain requires understanding the ways in which the organisations in the supply chain operate. Abundant research has examined organisational-level intervention, for example, lean production, agile manufacturing and so on. Research on the supply chain domain has been illumined considerably over the last decade spawning from the globalisation debate. The research on the supply chain itself has also evolved, and this section connects its evolution with enterprise network, logistics and agile manufacturing.

For a manufacturing organisation to sustain its competitiveness, it is important that the organisation re-examines its supply chain structure (including evaluation of the enterprise network and logistics operations). A supply chain structure can be represented in the following ways and this represents the transformation (periodically) now and into the future. The supply chain transformation starts from the classical linear supply chain, which represents the baseline of a normal buyer and supplier scenario. Integrating this from the traditional economic model and the purchasing techniques in the supply chain literature, it represents an arms-length relationship where there is minimal partnership, sharing of information and joint development between supplier and customer. The classical linear supply chain is the most basic form.

This type of supply chain then evolves to a more dynamic and responsive form of supply chain, which captures the importance of lean production, agile manufacturing and leagile strategy. In the dynamic and responsive supply chain, it encapsulates all the discussions above in this chapter, depending on the diffusion and use of manufacturing and information technologies to respond with maximum leagility. The relationship-based characteristic starts to emerge in this form, but further consolidated and solidified via the collaborative and relationship-based supply chain. It is over an acceptable period of time and numerous projects and collaborations between the suppliers, manufacturers and customers that then further extend those relationships to a solid collaborative nature. This involves consistent joint venture and joint development, sharing of information as well as resources. The integrated supply chain falls into this formation.

Simply competing between supply chains with the normal economic indicators, such as price, market share and so on, is no longer adequate. The market with increased awareness of green consumerisms, the legislation with tightened taxation and financial penalty, the industry with intense competition for a lower-carbon product and service, the manufacturer with demand on green purchasing and standards in place, such as ISO14000, WEEE, RoHS and so on, have all driven the transformation of supply chain to a new level. This new level of supply chain is termed the green and low-carbon supply chain, and encapsulates the notion of the triple bottom-line objectives, *i.e.* economic, environmental, and social. This implies that the KPI (key performance indicator) and priorities in organisations and supply chains need to be reshuffled in order to reflect this direction. There are massive challenges in creating a green and low-carbon organisation, let alone a green and low-carbon supply chain. Hence, an increased effort has been invested in finding innovative ways to lower CO_2 from a supply chain perspective, which also provides a positive response to social and economic objectives. This challenge is currently facing many industries and supply chains. Given also the importance of ensuring sustainability in how we respond to the changes, a balanced and next-generation supply chain form will emerge. The rapid transformation of supply chain formations does not start or stop periodically (discrete), it overlaps with classical and future forms (continuous) and it hybridises many characteristics from various forms.

1.7 Conclusions

This chapter provides an overview of the upward trajectory trend over the last few decades in enterprise networks and logistics, and how this shapes and influences the development of manufacturing and supply chain management. A detailed discussion supported by four industrial cases rationalising the need for more innovative and responsive enterprise network structure, technology and supply chain to deal with the ever-changing and highly-competitive market characteristics are presented. Agility of processes, organisations and its supply chain, particularly in the manufacturing environment, need to be re-examined. The analysis suggests that agile manufacturing is inadequate and we must look at leagile manufacturing and leagile supply chain in order to compete effectively with other supply chains. An overview of the current status and potential future trends in this area is provided, suggesting also supply chain transformation and how these are shaping future research in this area.

References

[1.1] Christopher, M., 2005, *Logistics and Supply Chain Management: Creating Value-Adding Networks*, Financial Times Prentice Hall, Upper Saddle River, NJ.

[1.2] Womack, J.P. and Jones, D.T., 1996, "Beyond Toyota: how to root out waste and pursue perfection," *Harvard Business Review*, **74**(5), pp. 140–153.

[1.3] Monden, Y., 1983, *Toyota Production System*, Institute of Industrial Engineers, Norcross, GA.
[1.4] Hiebelar, R., Kelly, T. and Katteman, C., 1998, *Best Practices Building Your Business with Customer Focussed Solutions*, Simon and Schuster, New York.
[1.5] Nagel, R.N. and Dove, R., 1991, *21st Century Manufacturing Enterprise Strategy: An Industry Led View*, Diane Publishing Company, Darby, PA.
[1.6] Harrison, A., Van Hoek, R. and Christopher, M., 1999, "Creating the agile supply chain," *School of Management Working Paper*, Cranfield University, Cranfield.
[1.7] Christopher, M. and Towill, D., 2001, "An integrated model for the design of agile supply chains," *International Journal of Physical Distribution and Logistics Management*, **31**(4), pp. 235–246.
[1.8] Goldsby, T.J. and Garcia-Dastugue, S.J., 2003, "The manufacturing flow management process," *International Journal of Logistics Management*, **14**(2), pp. 33–45.
[1.9] Bucklin, L.P., 1965, "Postponement, speculation and the structure of distribution channels," *Journal of Marketing Research*, **2**, pp. 26–31.
[1.10] Koh, S.C.L. and Tan, Z., 2005, "Using e-commerce to gain a competitive advantage in 3PL enterprises in China," *International Journal of Logistics Systems and Management*, **1**(2), pp. 187–210.
[1.11] Langley, J.L. Jr., Allen, G.R. and Tyndall, G.R., 2001, *Third-Party Logistics Study: Results and Findings of the 2001 Sixth Annual Study*, Georgia Institute of Technology, Atlanta, GA.

2

A Review of Research and Practice for the Industrial Networks of the Future

Rob Dekkers[1] and David Bennett[2]

[1] University of the West of Scotland, Paisley PA1 2BE, UK
Email: rob.dekkers@uws.ac.uk

[2] Aston University, Birmingham B4 7ET, UK
Email: d.j.bennett@aston.ac.uk

Abstract
Academic researchers have followed closely the interest of companies in establishing industrial networks by studying aspects such as social interaction and contractual relationships. But what patterns underlie the emergence of industrial networks and what support should research provide for practitioners? First, it appears that manufacturing is becoming a commodity rather than a unique capability, which accounts especially for low-technology approaches in downstream parts of the network, for example, in assembly operations. Second, the increased tendency towards specialisation has forced other, upstream, parts of industrial networks to introduce advanced manufacturing technologies for niche markets. Third, the capital market for investments in capacity, and the trade in manufacturing as a commodity, dominates resource allocation to a larger extent than was previously the case. Fourth, there is becoming a continuous move towards more loosely connected entities that comprise manufacturing networks. Finally, in these networks, concepts for supply chain management should address collaboration and information technology that supports decentralised decision-making, in particular to address sustainable and green supply chains. More traditional concepts, such as the keiretsu and chaibol networks of some Asian economies, do not sufficiently support the demands now being placed on networks. Research should address these five fundamental challenges to prepare for the industrial networks of 2020 and beyond.

2.1 Introduction

In recent years, practitioners and researchers have started to look increasingly at companies as part of networks within which they operate. The emergence of manufacturing networks is often associated with the possibilities offered by information technology and data-communication for collaboration and coordination, the globalisation of markets and the increasing tendency of companies to specialise, *e.g.* [2.1]. These possibilities provide firms with easier access to the capabilities and resources of others, moving them further away from the traditional logic behind the make-or-buy decision; even though this particular manufacturing decision still attracts attention from researchers to develop appropriate models, *e.g.* [2.2–2.4]. Additionally, the world of management has seen an abundance of theories that might have been adequate to deal with the contemporary challenges for some enterprises, but not for many others [2.5, 2.6]. The notion of core competencies and the concept of *lean production* serve as examples of such theories that address questions relating to supply chain management in the context of industrial networks; but it could be questioned whether they really deal with the characteristics of networked organisations. Capello [2.7] (p. 496) supports this statement by noting that not enough is known about the failure of networks. In this chapter, we argue that industrial networks require the adaptation of existing theories to fit their particular characteristics as well as the development of grounded theories based on the unique characteristics of industrial collaboration.

2.1.1 Brief History of Industrial Networks

Although the study of industrial networks has attracted recent attention among researchers, there was already an awareness of the implications associated with the particular characteristics of networked organisations [2.8, 2.9]. In particular, academic interest has centred on two periods in the past. The first of these is in the 1970s and 1980s, when attention was focused on Japanese manufacturing concepts and techniques, including just-in-time (JIT), co-production and 'keiretsu' networks. The second period starts in the 1990s, after the bursting of Japan's 'bubble' economy, as a consequence of the drive for even lower cost, greater efficiency, and responsiveness to customer demands. This resulted in a more formal recognition of the networked organisation as a follow-up to the paradigm of core competencies and the consequent escalation in outsourcing. Mayntz [2.10] acknowledges networks as capable of solving complex tasks and exceeding the capability of individual firms. The earlier overview by Miles and Snow [2.11] illustrated the move from the simpler paradigms to more complicated forms of network-based organisations that subsequently have been witnessed in recent years (see Table 2.1) and consequently have attracted academic deliberation.

The establishment and emergence of industrial networks is closely related to the subject of manufacturing strategy. Since Skinner's seminal work in 1969 [2.12], manufacturing has been recognised as a fundamental cornerstone for achieving corporate competitive advantage. Although it recognises the traditional and limited perspective of considering low cost and high efficiency as dominant objectives within manufacturing strategy, this earlier work of Skinner is still rooted in the tradition that economies of scale provide competitive opportunities (see pp. 260–265 in [2.13]). That tradition gave rise to the monolithic company driven by forward and backward integration [2.14], which implied an emphasis on the coordination of operations. Only later, in 1986, does Skinner consider the role of smaller-scale units that may now be regarded as elements of an industrial network [2.15], while subsequently questioning the traditional effort towards productivity improvement through making large capital investments in manufacturing [2.16]. According to Sturgeon [2.17] (pp. 8–10), American firms – compared with most Asian and many European companies – have generally placed manufacturing in a low position on the

Table 2.1. Evolution of organisation forms [2.11]. This indicates the evolution of organisation forms that are both internally and externally consistent. Miles and Snow [2.11] state in their paper that a minimal fit is necessary for survival, and that tight fit associates with corporate excellence, and early fit provides a competitive advantage. Therefore, dynamic networks (industrial networks) require both internal fits and external fits, giving early adopters a competitive advantage.

Period	Product-market strategy	Organisation structure	Inventor or early user	Core activating and control mechanisms
1800–	Single product or service. Local/regional markets	Agency	Numerous small owner-managed firms	Personal direction and control
1850–	Limited, standardised product or service line. Regional/national markets	Functional	Carnegie Steel	Central plan and budgets
1900–	Diversified, changing product or service line. National/ international markets	Divisional	General Motors, Sears, Roebuck, Hewlett Packard	Corporate policies and division profit centres
1950–	Standard and innovative products or services. Stable and changing markets	Matrix	Several aerospace and electronic firms	Temporary teams and lateral resource allocation devices such as internal markets, joint planning systems, *etc.*
2000–	Product or service design. Global, changing markets	Dynamic network	International/ construction firms. Global consumer goods companies. Selected electronic and computer firms (*e.g.* IBM)	Broker-assembled temporary structures with shared information systems as basis for trust and co-ordination

hierarchy of corporate esteem. However, in contrast to Sturgeon's belief, it is argued here that this is also the case for European firms. For example, most companies still regard efficiency as the main objective of their production departments in a survey amongst Spanish companies [2.18]. Consequently, during the 1960s and 1970s the make-or-buy decision was at the heart of operations management research. Then, in the 1980s, the interest in Japanese manufacturing techniques, including partnerships with suppliers, sparked the next step towards models for collaboration and supply chain management using JIT principles, while in the early 1990s the concept of core competencies led to renewed interest in outsourcing models. Later the over-the-wall tactics of outsourcing made companies examine the networks they had created while managing these from a traditional cost perspective [2.19]. In the end, the increasing attention paid to networks has not challenged the proposition of Skinner that manufacturing is of paramount importance to industrial performance; and it has not altered that the most common view of manufacturing (including manufacturing networks) is the one taken from the traditional cost perspective.

2.1.2 The Impact of Globalisation

The awareness that has been created that manufacturing strategy comprises more than cost-driven objectives, *e.g.* also meeting customer demands, has created a wider array of perspectives for manufacturing; these perspectives on manufacturing strategy, complemented by the influence of advances in information and communication technology together with globalisation and specialisation, foster the specific characteristics of industrial networks, *i.e.* collaboration to deliver products and services, decentralisation of decision-making among the agents and inter-organisational integration across companies involved to meet imposed performance requirements in competitive markets (adapted from O'Neill and Sackett [2.20], see p. 42). In these three fields, each change in itself requires adaptations by companies and the influence of several of these shifts leverage the need for adequate responses. For example, collaboration not only requires solutions in advanced software, it should also account for the management of industrial networks in an international context whereby individual companies set their own course and develop over time (decentralisation). Conversely, efficient international collaboration depends on the appropriate deployment of information and communication technology. The intricate interdependencies of these characteristics transform industrial networks into dynamic, collaborative efforts that have a large number and wide variety of continuously evolving resources at their disposal especially to meet a greater range of customer demands.

This has caused a change in the prevailing attitude towards resource allocation due to the emergence of the industrial network paradigm. The need for proximity of supply, following the theories about co-production, has required a strong interaction between customers and suppliers. Consequently much research has focused on the need for economic clusters, *e.g.* [2.21]. Carter and Narasimhan [2.22] (pp. 17–20) note that already co-location of suppliers has become one of the least significant trends and there are examples from industry of these tendencies changing, like Daimler Chrysler's announcement in 2000 that suppliers need to deliver in six days (rather than 1–2 days previously, with close geographical proximity). It illustrates the different approaches towards supplier selection and purchasing management that are now emerging; these attitudes allow a greater independence of suppliers to some extent. These different views support the notion that the supplier base should be considered as a network rather than a set of individual actors linked to one firm (which also follows from Carter and Narasimhan's study).

Not only has the scene for suppliers to any industry changed but many more countries have also followed an active path towards developing relevant economic and industrial competencies, reinforcing the establishment of supply networks. For example, the Thai government has deliberately set out to strengthen its automotive sector by attracting foreign companies in that industry [2.23]. By contrast, during the

1990s, MIT undertook a study that led to a warning about the decline of manufacturing industry in the USA [2.24]. However, more recently the USA has adopted a more progressive approach with the study on visionary manufacturing challenges [2.25], the UK government has stimulated the creation of innovative manufacturing research centres [2.26], and for the first time the Dutch government set out a research strategy to support the manufacturing industry [2.27]. Consequently, a complex pattern has emerged with the industrial base undergoing shifts by moving to developing countries, emerging countries entering the manufacturing arena, and a revival of some traditional industrialised countries, thus making the situation more dynamic than ever before. In the end, these national policies have only encouraged more extended industrial networks.

At the same time, the make-up of industrial network has also undergone changes. The external drivers (such as the move from make-or-buy to co-production or alliances and the drive for flexibility of manufacturing), as well as the internally oriented concepts (such as the attempts to apply computer integrated manufacturing and the use of production cells), demonstrate a continuous move towards more loosely connected industrial entities for manufacturing. See also Brown *et al.* [2.28] for arguments and examples and Smith *et al.* [2.29] for geographically dispersed capacity and OEMs. The requirement for greater flexibility also impacts on the trend to increase the amount of customisation and production of goods on-demand [2.30]. Contemporary changes in industries point to a further repositioning along the dimension of loosely connected entities, with increasing pressure to respond to market opportunities and to increase flexibility.

2.1.3 Scope of Chapter

Following the moves made by companies that have been previously identified, this chapter explores the concept of industrial networks for manufacturing. It aims to visualise an approach for industrial networks of the future, *i.e.* for the next 15 years and beyond, based on ongoing research and additional considerations. Firms are operating increasingly as parts of industrial networks, *e.g.* [2.1, 2.31]. Although the situation is extremely fluid and the stage has not yet been reached where networks are configured optimally and network operations have reached a stage of maturity. Ritter *et al.* [2.32] (p. 118) even state that current understanding of networks is limited and consequently, the chapter also aims at contributing to the research agenda and making a contribution to foundations for generating grounded theory about industrial networks.

Initially, in Section 2.2, this chapter examines the types of traditional networks that have been identified, together with the reasons they have been formed and their advantages and weaknesses. This includes a critique of the traditional keiretsu and chaibol networks based on conglomerate structures that formed the basis of Japan's and Korea's economic success. Section 2.3 addresses how future networks will be shaped by discussing four contributory and related topics, *i.e.* network configuration, manufacturing as commodity, added value within networks and sustainability of supply chains. The chapter then moves to present the outlines of a research agenda in Section 2.4 and implications for practice in Section 2.5. This contribution to directing research into industrial networks uses a blend of illustrations (from the business literature) and findings of previous studies by others, together with results from research by the authors, to construct a picture of how future networks might look and behave.

2.2 Traditional Views about Networks

The study of networks as a key aspect of industrial organisation goes back to the 1980s with the seminal work of Håkansson at Uppsala University who defined networks as sets of more or less specialised, interdependent actors involved in exchange processes [2.8, 2.33]. Around the same time the study of urban, networked organisations in the industrialised regions of northern Italy recognised the importance of networks for improving logistical efficiency [2.34, 2.35]. Simultaneously, writings appeared on *strategic networks*, which are defined as long-term, purposeful arrangements among distinct, but related, for-profit organisations that allow members to gain or sustain competitive advantage over their competitors outside the arrangement [2.36] (see also p. 32 in [2.37,]). According to this view, strategic networks are merely a superior method of managing the process necessary for the generation and sale of a chosen set of products like in [2.38]; this applies also to innovation and new product development, *e.g.* [2.39]. It should be noted that some authors associate the term strategic networks with the concept of networked organisations in general, *e.g.* [2.40], and some with supply chains, *e.g.* [2.41]. The participation of companies in these networks depends on managing product development, both at the level of the network and the individual companies, and on managing manufacturing processes.

Within the overall primary process of most companies the connection between product development and manufacturing strategy has yet to result in conceptual approaches for establishing this vital link, with only Sharifi *et al.* [2.42] connecting a product strategy to conceptual design of the supply chain. Conducting a study into sequential and simultaneous approaches to engineering new products, Riedel and Pawar [2.43] highlight that the concepts of design and manufacturing are not linked in the literature and that the interaction of product design and manufacturing strategy is under-researched. Spring and Dalrymple [2.44] came to a similar conclusion when examining two cases of product customisation, where manufacturing issues received little attention during design and engineering. The only concept that addresses these issues so far is the one of order entry points (more commonly known as order decoupling points; see Figure 2.1). Order entry points and modular product architecture typically concern the optimisation of make-to-order production concepts and might include product development and engineering activities [2.45]. Introducing a different perspective, Smulder *et al.* [2.46] proposed a typology of intra-firm and inter-firm interfaces, therewith also connecting product development and production; yet this typology has still to be adopted in practice. Henceforth, the emerging paradigm of industrial networks, if it is to be successful, should address this matter of creating a link between manufacturing strategy and product development.

But do we find this link included as part of the current concepts for industrial networks? Four mainstream operations management and logistic concepts in this area dominate thinking about the industrial network paradigm: core competencies, agile manufacturing, keiretsu and chaibol arrangements and supply chain management. Other concepts such as strategic networks and the resource-based view come about through strategic concepts and can be associated with the thinking about core competencies (see pp. 4–5 in [2.47]). As it appears in the next four subsections, these concepts focus mainly on issues of manufacturing and less on product development, except in general terms.

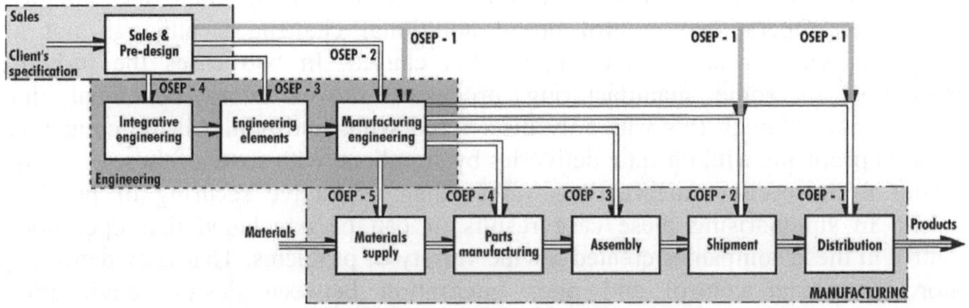

Figure 2.1. Position of the order entry points in the primary process of design, engineering, manufacturing and logistics. To simplify the figure, points of stock (inventory) have been omitted. OSEP-1 (order specification entry point) indicates that customer requirements are directly transferred into production instructions, while OSEP-4 points to engineering-to-order. Similarly in the material flow: COEP-1 (customer order entry point) tells that orders are delivered directly from stock, while COEP-5 marks make-to-order.

2.2.1 Core Competencies and Outsourcing

According to Friedrich [2.48], focusing on core competencies [2.49] and outsourcing [2.50] raises the key issue of which areas of production are needed to maintain the value chain and on which areas the company should concentrate for achieving optimal performance. Prahalad and Hamel [2.49] subtly expand the view of technology from a broadly described concept, the importance of which is determined by its support of the corporate mission, to a specific source of corporate uniqueness. In Prahalad and Hamel's view, core competencies represent the collective learning of the organisation, especially concerning how to coordinate diverse production skills and integrate multiple streams of technology. However, the application of this theory does not lead directly to a clearly defined strategy for global manufacturing or manufacturing networks. And often this thinking about core competencies leads to outsourcing mostly based on a cost perspective for manufacturing (as present in [2.51]). Only when core competencies are linked to decision-making will a manufacturing strategy be found that offers guidelines on decision-making for resource acquisition and capacity management [2.52].

Given the (often unquestioned) popularity of the concept of core competencies and its implications, how does industry manage the increasing scope of outsourcing? A study by Dekkers [2.53] based on six case studies (four in the Netherlands, one in China and one in Indonesia) points to poor control of outsourcing by industrial companies. Most of the case companies, with primary processes based mainly on engineering-to-order and make-to-order, experienced problems with implementing manufacturing strategies. Ideally, the manufacturing strategy of these companies should address their core competencies and opportunities for outsourcing. All the case companies, except one, had done so, implicitly or explicitly; but mostly this strategy had not been transferred to guidelines for implementation, which is why decision-making occurred at random or opportunistically. There was no feedback to the stages of design and engineering about suppliers' performance, so sometimes problems would recur regularly. None of the companies followed an active approach towards supplier networks for the purpose of expanding their technological capabilities. Operational control posed additional challenges, although not all companies were aware of the impact this caused. In two cases the in-house production of some manufacturing processes proved more beneficial than outsourcing, although this was only discovered with hindsight. All the companies reported problems with on-time deliveries by suppliers, with some of these problems arising from reactive interventions rather than pro-active securing of purchase orders. In summarising these case results, it can be concluded that operational control in these companies created a wide variety of problems. That is evidenced by poor operational control and poor integration between design, engineering, purchasing and manufacturing; additionally, it indicates that the simplified view of core competencies and outsourcing might have strong limitations.

Still today, even though insight into effective manufacturing strategies has progressed, many approaches for outsourcing rely on the deployment of criteria derived from traditional make-or-buy decisions. However, the rise of industrial networks creates the need for frameworks that take account of early supplier involvement, collaboration, and inter-organisational integration. Also, decision-making concerning the allocation of resources has shifted from making one-time decisions to continuous evaluation and reallocation. Current outsourcing approaches rarely account for this, and hence there is a need for expansion of criteria to include those suitable for networks. Practices for management and control of outsourcing still focus largely on minimising costs and meeting delivery schedules, while research into outsourcing has not yet investigated the specific impact of industrial networks [2.19].

2.2.2 Keiretsu and Chaibol Networks

Unlike the networks of Western companies that resulted from the make-or-buy decision and later outsourcing, the keiretsu and chaibol networks that formed the basis of Japan's and Korea's economic success were based on conglomerate structures. However, more recently these structures have proved less capable of meeting the need for speed of change, flexibility, and cost reduction that have been the key aspects of industrial management following the Asian economic crisis of the late 1990s [2.54]. At the same time, organisations that attempted to replicate the keiretsu concept outside Japan have encountered severe problems, making them rethink their plans to create similar supply networks [2.55].

A major weakness of the traditional keiretsu and chaibol networks has been their domestic focus and cross-ownership between companies in the network. This has hindered how they can respond effectively to the globalisation of manufacturing

[2.56]. It has also created difficulties as end-product manufacturers have moved offshore and taken them beyond the reach of domestically based network members. Also, the burden of debt resulting from borrowing to support cross-ownership has restricted their ability to develop and fully support international operations. As a consequence of this situation, Renault, on taking a controlling interest in Nissan, sought to dismantle its keiretsu supplier network by selling off most of its financial stakes in almost 1,400 companies [2.57]. This indicates that companies deploying traditional networks are searching for different concepts to manage their suppliers.

However, despite these concerns, a study by McGuire and Dow [2.58] still shows that throughout the first half of the 1990s the keiretsu system remained strongly in place. At the same time, they conclude that the continued move towards globalisation of capital markets in Japan and ongoing regulatory change may potentially impact networking and performance implications. Apart from the problems that can arise when there is cross-ownership between companies, the main criticism of the keiretsu relates to its lack of flexibility and responsiveness. The answer to this criticism has therefore been to propose the creation of agile networks [2.59].

2.2.3 Agile Manufacturing Networks

In contrast to the concept of outsourcing and keiretsu and chaibol networks, the approach of agile manufacturing relies more strongly on the exploitation of loosely connected networks than earlier concepts such as lean production [2.60–2.62]. Co-makership (and subsequently lean production) had already introduced a higher degree of outsourcing and improved control through supply chain management, although here the networks used were more closely connected keiretsu or chaibol types involving cross-ownership. In contrast to the internal focus of lean production, the paradigm of agile manufacturing has an external focus and is concerned primarily with the ability of enterprises to cope with unexpected changes, to survive against unprecedented threats from the business environment, and to take advantage of changes as opportunities [2.63]. Similarly, Kidd [2.64] recognises two main factors within the concept of agility, *i.e.* responding to changes in appropriate ways, and in due time, taking advantage of the opportunities resulting from change. This means that an agile manufacturing enterprise marshals the best possible resources to provide innovative (and often customised) products, with the flexibility to adjust the product and offer rapid delivery, and with the high level of efficiency required to be competitive and profitable (see p. 19 in [2.65]. The concept of agile manufacturing stresses two interconnected main processes:

1. the development of innovative products;
2. the manufacturing and distribution of these products.

These two processes should meet lead-time requirements (time-to-market, time-to-volume and delivery time) and flexibility requirements (to meet market opportunities and respond to market demands) [2.66]. A reconfigurable structure becomes a prerequisite for optimising the capabilities of an organisation for each business opportunity [2.67], which itself requires more loosely connected entities.

However, even the new types of agile manufacturing networks often are not designed within an international context and may still be suboptimal where acquisitions have taken place resulting in an inherited supplier base. Therefore, the notion of building international manufacturing networks is now a prevalent concern where competitiveness derives from an ability to garner and integrate resources from a number of different geographical sources. The basic principles for building a manufacturing network have been described by Mraz [2.68] who identifies four categories of resources (*i.e.* players) that can be used within the network: industrial design consultants, product development consultants, contract manufacturers, and original equipment manufacturers (OEMs). These last two players also demonstrate the options available for the production of complex products and their relative advantages and disadvantages, with the contract manufacturing approach typically involving external industrial design and product development, and the OEM approach typically retaining these activities in-house. A hybrid of these two forms can be found in the case of the Brazilian aircraft manufacturer Embraer (Empresa Brasiliera de Aeronáutica SA), which, with its network of risk sharing partners, was able to greatly accelerate the development and launch of the ERJ-170/190 series of regional jets. Hence, adequate suppliers' bases, with possibly an international dimension, reinforce performance during product development (reduced time-to-market) and manufacturing (improved performance to deliver) to the advantage of OEMs and their supplier networks.

The international dimension to designing agile manufacturing networks is also considered by Lee and Lau [2.30], who use the example of firms in Hong Kong and the Pearl River Delta to provide a factory-on-demand concept within the context of manufacturing networks. Shi and Gregory [2.69] have contributed by proposing the mapping of configurations for international manufacturing networks as a means of providing support for decision-making. Presentations by companies at the 9th Annual Cambridge International Manufacturing Symposium in 2004, organised by the University of Cambridge, have shown that there are two strategic directions for international manufacturing networks: rationalisation (with manufacturing units, sometimes including product development, specialising on product ranges) and globalisation (taking the opportunity to outsource operations or establish alliances). As frequently evidenced in the literature, *e.g.* [2.70], the current drive for globalisation by companies places its emphasis more on optimisation within existing conditions and less on capturing new market opportunities, even for the opportunities these international manufacturing networks offer.

2.2.4 Supply Chain Management

Likewise, within the concepts for supply chain management, agility has become a major issue. For example, Helo *et al.* [2.71] (p. 1059) see agility as the key for customisation within supply chains. In addition, Towill and Christopher [2.72] (p. 308) contend that agile and lean contribute to meeting performance demands imposed by the market. Gunasekaran [2.73] states that key enablers of agile manufacturing include: (i) tools and metrics for virtual enterprise formation; (ii) physically distributed manufacturing architecture and teams; (iii) tools and metrics for rapid partnership formation; (iv) concurrent engineering; (v) integrated information systems for products, manufacturing and business; (vi) rapid prototyping tools; and (vii) electronic commerce. Sanchez and Nagi [2.74] in their review of 73 papers reiterate these points, albeit in a different way. These works are all building on the concepts for agility introduced by Goldman and Nagel [2.65] and Goldman *et al.* [2.63]. Within the context of this chapter about networks, it is worth mentioning that the relation to engineering has a central role (see order entry points in the introduction to this section) and that collaboration in relation to information and communication technology seems pivotal.

So far, concepts for supply chain management rely heavily on applications of information and communication technology. For example, Akkermans *et al.* [2.75] present results from an exploratory study on the impact of enterprise resource planning (ERP) systems on supply chain management. They report the following key limitations of current ERP systems:

1. their insufficient extended enterprise functionality in crossing organisational boundaries;
2. their inflexibility to ever-changing supply chain needs;
3. their lack of functionality beyond managing transactions;
4. their closed and non-modular system architecture.

As they state, these limitations stem from the fact that the first generation of ERP products has been designed to integrate the various operations of an individual firm. However, since the unit of analysis, in their words, has become a network of organisations, these limitations render ERP products inadequate for the challenges that are posed; in this respect, Stadtler [2.76] (p. 586) draws a similar conclusion for inter-organisational integration in supply chain management. The open source solution from Helo *et al.* [2.71] is a step in this direction, given its flexibility to operate in conjunction with ERP, WMS (warehouse management system) and EDI (electronic data interchange). But that is only one step in the direction of decentralised decision-making and inter-organisational integration as key characteristics of industrial networks.

2.2.5 Traditional Views on the Wane

Despite the theoretical ability of agile manufacturing to provide greater flexibility and responsiveness than traditional network concepts (supply chain management, keiretsu and chaibol arrangements and networks born out of outsourcing), there are still questions about whether it can address the characteristics of networks, *i.e.* collaboration to deliver products and services, decentralisation of decision-making amongst the agents and inter-organisational integration across companies involved to meet imposed performance requirements in competitive markets. The special issue on dispersed manufacturing networks underlines the fact that progress is being made slowly [2.77]. The questions around the paradigm for networks that consist of loosely connected entities only demonstrate that we still know little about their behaviour. Nevertheless, many developments in information technology and data-communication allow interfacing in networked manufacturing; for example, as Boeing has done for the 787 Dreamliner. The current problems with production in this case can be traced back to selection processes of suppliers (even supported by sophisticated software applications that failed to solve the process of interaction). Generally speaking, the lack of synchronisation between the possibilities of information technology and the limited understanding of the actual behaviour of entities (or agents for that matter) have only increased instability in relationships, giving greater cause for instabilities in relationships. At the same time, inter-relationships have become more demanding and limited the capabilities of parties to operate within each other's constraints. Industrial companies demand partnerships, but these sometimes appear to be forcibly driven by strategy rather than being based on a true bilateral relationship. With the reduced capability to maintain long-term relationships, partners in industrial networks need different ways of interacting, sometimes facilitated by applications in information technology and data-communication (extending to both the domain of manufacturing and the domain of product development and engineering).

2.3 Future Networks

Contemporary manufacturing networks with more loosely connected entities have come about through two mechanisms. First, the manufacturing networks have emerged as a result of collaboration between loosely connected entities, or so-called *collaborative networks* (see p. 439 in [2.78]). This mostly concerns SMEs that coordinate either globally or regionally [2.79] but with the explicit aim to have a wider reach; the latter resembles the regional networks labelled Third Italy by Biggiero [2.80] and Robertson and Langlois [2.81] (p. 549). The second mechanism of manufacturing is the global production networks that come about through OEMs (see Kuhn [2.82] and Doner *et al.* [2.83] for the automotive industry and Smith *et al.* [2.29] for a survey), as similarly described by Ernst [2.84] with his focus on the electronics industry, Riis *et al.* [2.85] for six Danish companies and Sturgeon [2.17] for the American industry. The characteristics of global production networks correspond to those of strategic networks, as discussed in Dekkers [2.47] (pp. 4–5). These networks are often associated with power and trust that dominate these types of network relationships [2.86–2.89]. Hence, these strategic networks came into existence through strategic objectives of one or more of the partners, which makes it necessary to collaborate and which create tensions in inter-organisational relationships. On the dimensions of Robertson and Langlois [2.81], strategic networks and networks evolving from the resource-based view score high on ownership integration (*e.g.* holding companies and the Chandlerian firm). However, contemporary industrial networks rely less on ownership but require some degree of collaboration and coordination. At the heart of this chapter are the challenges these two forms of more loosely connected organisations face as they evolve towards collaborative networks and global production networks.

There are now many emerging possibilities offered by information technology and data-communication methods. Some of these include planning methodologies [2.90], smart supply chains [2.91], globalisation of markets [2.1] and the ongoing specialisation of firms. They drive companies to concentrate on core competencies, even given the flaws in this theory, and, consequently, enable them to move from centralised, vertically integrated and single-site manufacturing facilities to geographically dispersed networks of resources [2.66]. These simultaneous developments foster the specific characteristics of (international) networks, which require adaptations by companies to fit these characteristics.

2.3.1 Network Configuration

The dilemma with these networks extends to the problem of achieving a balance between having independent agents and controlling processes to meet performance, which requires a strong interaction between these agents. Virtual organisations, which can be considered as a further manifestation of networks, might display instability between the model of pure outsourcing and the establishment of more traditional alliances [2.92]. Even alliances, which are perceived as more stable relationships between firms, usually dissolve over time or result in mergers [2.93]. The network is optimised locally and creates power shifts if the balance moves towards independence of agents, depending on the uniqueness of their resources, [2.94]. Also, flexibility might be lost in the short and medium term through the creation of alliances or mergers [2.95]. Therefore, research needs to be undertaken to reveal whether this dilemma of balance between control and change in networks can be resolved.

The principal characteristic of industrial networks is their ability to capture market opportunities and to adapt to changes in the environment. Collaboration with other companies has a significant impact on the capabilities of a network. Hitherto, the dynamic capability has equated to changeability, which Milberg and Dürrschmidt [2.96] define as the sum of (i) flexibility, defined as the capability to operate in a wider space on certain dimensions of business management, and (ii) responsiveness, defined as the ability to handle emerging changes in the environment. Thus changeability is a measure of the total changes the environment demands of an organisation or network [2.9]. That changeability resembles the concept of dynamic capabilities introduced by Teece *et al.* [2.97]. In their paper, Möller and Svahn [2.98] expand on this, although their thinking seems much more directed at strategic networks. Sometimes, the sacrifices in a given production system to obtain flexibility (*i.e.* capturing market opportunities and adapting) exceed the derived benefits.

Each market opportunity requires an adequate response from an industrial network. The flexibility of a network relies on the deployment of resources to capture these market opportunities and thereby needs a control structure and organisational structure that fits the actual demand. Theory about organisational design distinguishes the process structure, the control structure, the organelle structure, and the hierarchy [2.99]; the organelle structure is based on the grouping of (business) process or activities to address performance requirements. The methodology for the design of organisations assumes a linear process when designing each of these structures consecutively (see Figure 2.2), even though this process should be considered iterative. In this approach, the design of the organelle structure is the key to meeting performance demands by customers; that leads Dekkers and van Luttervelt [2.100] (p. 13) to propose a model for reconfiguration of networks (see Figure 2.3). Industrial networks provide the opportunity to optimise each of the four structures independently and that through the connections between

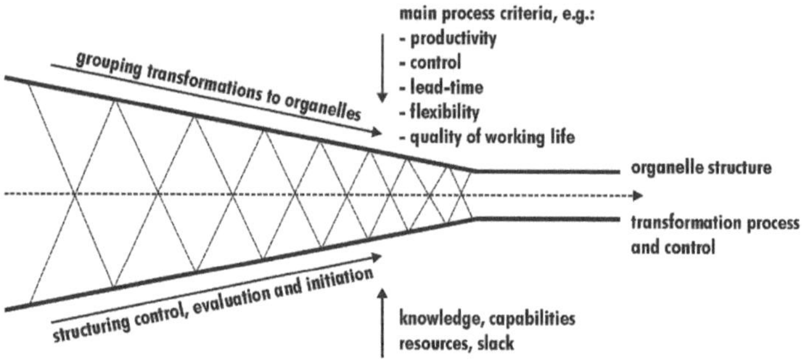

Figure 2.2. Design process for the organelle structure (see pp. 183–188 in [2.101]). The organelle structure affects both the grouping of tasks in the primary process as well as the control processes. By subsequent integration and iteration, the design of the organelle structure meets performance requirements.

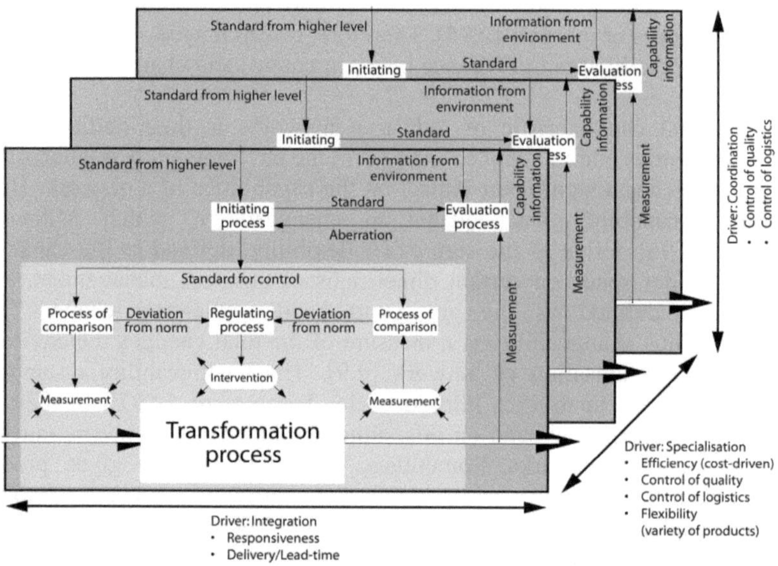

Figure 2.3. Model for reconfiguration within networks. Based on different drivers, market opportunities call for either integration, specialisation or coordination to meet performance requirements. Through predefined organelles for both the primary process and the control processes, reconfiguration becomes a preset decision-making process allowing quick responses to changing conditions.

these structures, as present in the value chain and as individual agents, network optimisation will occur over time.

Another phenomenon is the increasing participation of SMEs in international manufacturing networks [2.102], which has been enabled through the factors identified by Lall [2.103] as contributing to the increase in SME competitiveness. Bennett and Ozdenli [2.104] have studied the role of several SMEs in international manufacturing networks. The SMEs were based in industrialised countries, developing countries and transition economies. The analysis of the cases shows that they are motivated largely by the desire to extend their reach and a wish to begin establishing a global presence. It also shows that control and commitment are two major determinants for SMEs and international manufacturing networks, so managers must think carefully about how much control they want to have (or should have) within the network. This concerns the electronic and virtual integration of companies, so calling on totally new models for dealing with networks [2.105]. These include matchmaking and brokerage through web services [2.106, 2.107] and electronic contracts; these will enable companies to move away from the control paradigm for the monolithic company towards management approaches that fit the emergent properties of networks [2.108, 2.109]. The concept of complex networks with emerging properties strongly relates to the proposed idea of *open innovation systems* [2.110, 2.111]; the increased interaction between actors in networks requires a rethinking how it happens at all [2.112], whether it concerns manufacturing or product development.

2.3.2 Manufacturing as a Commodity

An important development influencing the shift in power within manufacturing networks has been the increasing importance of OEMs and, more recently, brand owners [2.113]. Sturgeon [2.17] argues that the revival of the American industry during the 1990s can be attributed to what he calls turnkey production networks. Essentially, these incorporate the trend towards outsourced manufacturing and an emphasis on branding. To demonstrate this concept, Sturgeon uses the example of the electronics industry, particularly the case of Apple Computer Inc. that contracted SCI Systems for a large part of its manufacturing operations in 1996. A system like a turnkey production network is highly adaptive because it uses turnkey relationships to weave various key production clusters into a global-scale production network based on external economics for OEMs and brand-owners.

With the rise in OEMs, especially in the electronics and automotive industries, the concept of outsourcing the production of complete systems and subsystems started to become a common phenomenon. In this way the idea of tiering in the supply network was created [2.114], with power generally reducing towards the lower tiers (with possible exceptions where suppliers are part of much larger companies involved with leading edge technologies). Along with this trend has also materialised the idea of manufacturing capacity as a commodity rather than a unique capability for "pushing" products onto growing markets. At the same time, the focus of technology has also moved upstream with suppliers increasingly turning to advanced manufacturing technologies as a means of competing for orders, while OEMs, especially those based offshore, have tended largely to rely on low-technology assembly techniques for enabling greater agility.

This trend has been taken further under the more recent, and increasingly dominant, regime of brand ownership. A characteristic is the separation of brand from origin of production and the virtually complete transition of manufacturing to a commodity with power residing almost totally with the brand owner; that often causes the brand to be more dominant than the actual product [2.115]. In turn, this has led to manufacturing becoming increasingly footloose with international mobility being an important aspect of network design. In particular, this has resulted in the transfer of production capital away from the traditional industrial economies to the low factor cost economies of the Far East and the transition economies of Eastern Europe [2.116].

2.3.3 Added Value of Industrial Networks

Collaborative efforts, whether or not they are crossing borders, are not only seen as an approach to decrease manufacturing cost; cooperation between network companies is increasingly seen as a means for lowering development costs, accelerating product and process development, and maximising commercialisation opportunities in innovation projects. The capability of building and maintaining inter-organisational networks, such as joint ventures, license agreements, co-development (between suppliers and customers) and strategic alliances has led to more product and process innovations [2.117]; see Figure 2.4. This also covers the extension of capabilities, with manufacturing services as a newly emerging trend and the capabilities embedded in manufacturing services partly answering the demand for customisation.

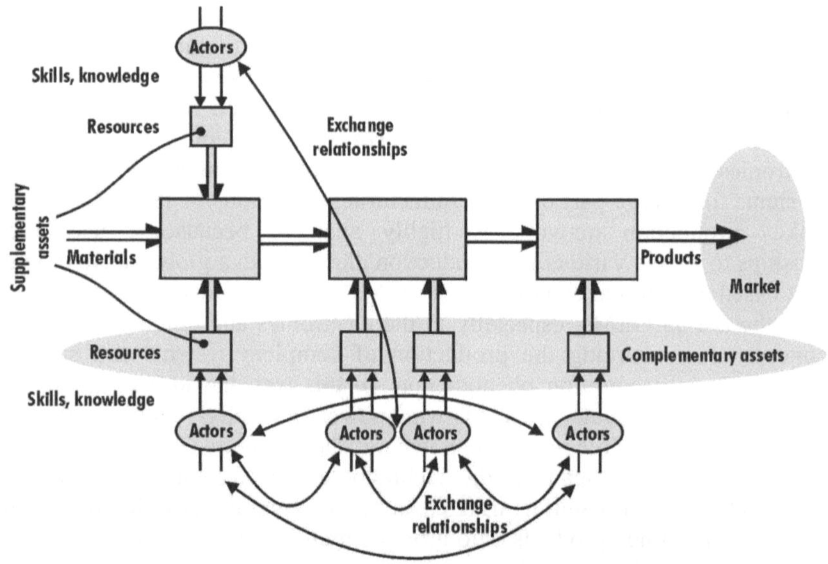

Figure 2.4. Collaboration model for the value chain (see p. 330 in [2.118]). Vertical collaboration indicates the capability of actors to manage the supply chain. Horizontal collaboration contributes to the dynamic capability of the network by reallocating resources or creating substitution.

Both horizontal and vertical collaboration require managing the relationships between actors in the network. Burt [2.119] and Uzzi [2.120] have demonstrated the general mechanisms by which relationships between firms in supply chains and networks can be explained. As starting point, they use two different aspects of networks, namely the positioning of firms in the structure of the network and the nature of the mutual relationships. Burt's reasoning implies that the chance of achieving completely radical innovations may decrease if companies establish strong mutual contractual links, such as in supply chains. Links with other companies in the supply chain might be so strong that they prevent a company from successfully implementing an innovation, even if it is in a strategic position to do so. Typically, a successful cooperation strategy consists of three basic elements, *i.e.* selection of a suitable partner, formulation of clear-cut agreements (getting the project underway) and management of the ongoing relationship. Carefully selecting future cooperation partners can prevent many problems and, according to Hagedoorn [2.121], the aim should be similarity balanced by complementarity, with similarity referring to the firm's size, resources and performance. However, of more importance are the required complementarities offered by the cooperation partner, *i.e.* the combination of complementary activities, knowledge and skills to realise the desired synergy. The literature on strategic partnerships offers many models to evaluate potential cooperation partners, *e.g.* [2.122]. Based on a study of 70 UK-based firms in different industry sectors, Bailey *et al.* [2.123] even concluded that selecting partners based on their track record in previous collaborations turns out to be a poor basis for future collaboration. These signals indicate that how collaborations can be exploited effectively has not yet been settled.

2.3.4 Sustainability of Supply Chains

For the more loosely connected networks that are even emerging in supply chains, but nevertheless call on collaboration, the key to managing the business processes is the monitoring of the capability of individual participating entities (see pp. 45–47 in [2.79]); this is called self-criticality by Kühnle [2.124] (pp. 62–66). It comes back in the central role of hubs that enabled by information and communication technologies exert that capability; the distributed plant automation, PADABIS, is an example [2.124], based on the notion of spaces-of-activity. Montreuil *et al.* [2.125] also propose a framework that they call NetMan but they are less explicit about the central role of monitoring. That capability of monitoring facilitates learning of the network and adaptation to changing circumstances; it strongly resembles the concept of process capability in the steady-state model that is mentioned by Dekkers [2.118] (p. 431). That then calls for reconfiguration, either by self-similarity based on fractals (see p. 67 in [2.124]) or by optimisation of the organelle structure (Section 2.3.1); note that the base for those reconfiguration approaches – the integration of business processes: physical flows and information flows – is the same. Therefore, the self-criticality in relation to reconfiguration constitutes a core capability of industrial networks and might be even the dynamic capability. However, these concepts of hubs and spaces-of-activity could become the cornerstone of future information and communication technologies for managing the supply chain; such a development will enhance collaboration and coordination across these chains as more loosely connected networks.

In the context of supply chains, Barratt [2.126] (p. 39) makes a similar remark: 'many of the problems related to ... collaboration are due to a lack of understanding'. Seuring [2.127] (p. 1069) places this notion in the context of environmental issues for supply chains: integrated supply chain management and understanding of interaction between actors in that chain are a prerequisite for achieving sustainability. Zhu and Cote [2.128] (p. 1033) report a similar finding for their case study of the Guitang Group in China. In addition, Srivastava [2.129] (p. 70) remarks that a paradigm shift is needed for *green supply chain management*. Even though sustainability might be linked to performance improvement, according to Rao and Holt [2.130], many have viewed green supply chain management as a constraint rather than an opportunity or a different *modus operandi* (see p. 70 in [2.129]). The calls for a more integrative framework for supply chain management seem to coincide with the rethinking necessary for concepts that address collaborations in the supply chains as networks. Collaboration might constitute that paradigm shift that is needed for sustainable and green supply chains based on integrative supply chain management and interaction between agents in the networks.

2.4 Research Agenda for Industrial Networks

The four themes described in the previous section – network configuration, manufacturing as commodity, added value of networks, and sustainability of supply chains – appear not to be congruent with most of the ongoing research into industrial networks. Nassimbeni [2.131] (p. 539) remarks that the bulk of available research on networks is devoted to the contractual aspects and social dynamics of inter-organisational relationships, while the dynamic forms of communication and coordination have been neglected, so requiring more attention from researchers. Most likely this originates in the conversion from the hierarchical firm, with direct control of resources and a cross-ownership strategy towards suppliers, to networks with more loosely connected entities, which is a view also found in Smulder *et al.* [2.46].

However, the shift towards more loosely connected entities requires additional theory, models and tools to cope with issues of collaboration, inter-organisational integration and decentralisation of decision making. It is probably more than a decade since the beginnings of academic research into the networked organisation (which initially looked at the extended enterprise, *etc.*). This research mainly has used models from the monolithic company – decision-making on make-or-buy and social dynamics – to further research. Reported findings of research argue that studies should pay more attention to modelling the interaction between agents [2.81], meaning that a more integrated approach becomes necessary. Therefore, research should consider taking different routes:

- The recent insights in natural sciences and the application of principles of complex systems theory to collaborative enterprise networks as socio-technical systems might yield these complementary approaches. Six themes emerge from this point of view (see pp. 71–73 in [2.105]):

  i. the dynamic description of networks (to respond to market opportunities and shifting demands and to capture the stability of networks themselves);
  ii. coordination possibilities (the networks consist of loosely connected

entities, each with their own strategy, and dependent on each other for delivery of products and services);

iii. radical and integrative innovation (the capturing of new market opportunities and technological prospects, and at the same time taking advantage of individual agent's knowledge and skills);

iv. path dependency in the evolution of networks (the concepts of evolutionary approaches and concepts like co-evolution and symbiosis applied to industrial networks);

v. sharing of information across agents (the network as a community of entities that evolve together);

vi. modelling and representation of industrial networks (to stretch beyond taxonomies and static approaches).

This might serve as a base for an interdisciplinary research approach, answering the call of Camarinha-Matos and Afsarmanesh [2.78] (pp. 443–444) for new approaches.

- Networks operate in dynamic environments and require dynamic approaches, so reflecting Ashby's law of requisite variety [2.132]. Perhaps even instability rather than stability is a rule, which requires that optimisation models should rely on insight from other sciences. Although neural networks incorporate some of these ideas, the explicit criteria of optimisation, dispersal, and bifurcation describe the evolution of networks [2.47]. In that perspective, industrial networks could be viewed as complex adaptive systems, similar to Biggiero [2.80] and Andriani [2.133] do for regional networks in Italy. Kühnle [2.124] builds on the proposal for the behaviour of complex systems by adding self-criticality and self-similarity as essential ingredients; *e.g.* Song *et al.* [2.134] consider self-similarity as a keystone for scale-free networks. Dekkers [2.112] offers an outlook on how to combine this complex systems view with evolutionary models, co-evolution, game-theoretical approaches and network theories. During the years to come, we might expect that further elaboration of the complex systems view in its widest sense will add to the understanding of agents' behaviour in industrial networks (*e.g.* Iansiti and Levien [2.135] (pp. 55–58) and Surana *et al.* [2.136] follow similar reasoning) and to the improvement of coordination mechanisms between loosely connected entities.

- The efficacy of industrial networks relies on the use of information and communication technology for collaborative engineering, computer-aided production planning, supply or value chain management and communication [2.137, 2.91], so exceeding the need for logistics integration, which is the main argument of Stock *et al.* [2.66]. Also, the optimisation of structures can be supported by information technology. Helo *et al.* [2.71] propose an integrated web-based logistics management system for agile supply demand network design, allowing interfacing different scheduling agents from different actors. The concept of hubs and spaces-of-activity might even lead to new generations of ERP or new information technologies that fit with the characteristics and coordination possibilities of industrial networks. Nevertheless, a lot of development work needs to be done to obtain methodologies, methods and tools to sustain industrial networks as loosely connected entities [2.47].

- Reconfiguration, for which a method still should be developed, allows a more appropriate approach for capturing market opportunities and optimising performance of networks (see Dekkers and van Luttervelt [2.100] (p. 19) and Section 2.3.1).
- The link between product development and manufacturing needs to be investigated more closely. So far research has concentrated on Order Entry Points, product families, *etc.*; but these concepts have limited reach, although they are addressing an important capability of networks: (mass) customisation. Particularly, the impact of the interface between product development and manufacturing on networks has not been well-researched.

Although the specific research into approaches for networks has progressed, further advances should create insight into optimisation and tools to support industrial networks; this is congruent with the remark of Camarinha-Matos and Afsarmanesh [2.78] (pp. 443–444) that research into collaborative networks constitutes a new interdisciplinary domain.

2.5 Implications for Practice

For managerial practice it follows that industrial networks requires a change in mind-set from three perspectives. First, the concepts embedded in the thinking about networks as an extension of the monolithic company will yield only marginal benefits. Besides it carries the danger that this management approach will result in issues of power and trust for industrial networks (see, *e.g.* [2.87]), much like the thoughts of the strategic network perspective and resource-based view (Section 2.2). Otherwise, the management of networks might suffer from fragmentation and its impact on decreasing the effectiveness of networks, as is so characteristic for the construction industry [2.138]; even though others take a contrasting position [2.139]. Second, the distribution of private and common benefits needs attention, where traditionally pricing and costs are focus of managerial attention. Although for part, it resembles the embeddedness in networks, *e.g.* see pp. 54–61 in [2.120], it does not imply that companies need to sacrifice. Rather they might benefit from the increased reach and responsiveness the networks offer on the long-term, albeit again through different mechanisms than traditional methods applicable for the Chandlerian or monolithic firm. Third, collaboration in networks has in some sense put smaller and bigger companies at equal footing. That implies that both smaller and bigger companies compete at a global scale with a greater flexibility and changeability. That in itself has accelerated the necessity to operate within networks: the emergence of networks going hand in hand with the necessity. Henceforth, networks have become a reality for many companies. Despite the changes that these three perspectives bring about, methods and tools have not fully been settled, that way calling also on managers to contribute to further insight and to collaborate with academics to advance both practice and theory.

2.6 Conclusions

There is little doubt that the issue of industrial networks has been an important concern to companies needing to compete in the dynamic competitive climate that has demanded greater flexibility, responsiveness and variety as well as responding to pressure on costs. The traditional networks of the past, especially those based on keiretsu or chaibol principles, have less use in today's business conditions and, as a consequence, more loosely connected agile networks have emerged. However, there has been very little research aimed at establishing the patterns that underlie their emergence and there remains the question of what support such research should provide for practitioners.

This chapter has identified a number of key issues concerning the future of networks, which have been based on a review of the relevant literature and additional considerations. First, network configurations require a control structure and organisational structure that fits actual demand, so companies have started to move away from the control paradigm of the monolithic company towards managing the emergent properties of networks. Second, with the move towards OEMs as network players there has been a greater tendency for manufacturing to become a commodity, which has accelerated under the regime of brand ownership. Third, the added value of industrial networks includes more product and process innovations and the extension of capabilities with manufacturing services. Fourth, the emergence of industrial networks has a strong impact on underlying theory, methods and tools, including applications of information and communication technology; researchers and practitioners should direct their efforts to develop more adequate approaches that fit the characteristics of industrial networks. Finally, a number of different routes have been identified that research should take if it is to properly reflect and support the real needs of industrial networks in the coming decade and beyond.

References

[2.1] Karlsson, C., 2003, "The development of industrial networks: challenges to operations management in an extraprise," *International Journal of Operations & Production Management*, **23**(1), pp. 44–61.

[2.2] Cánez, L.E., Platts, K.W. and Probert, D.R. (2000), "Developing a framework for make-or-buy decisions," *International Journal of Operations & Production Management*, **20**(11), pp. 1313–1330.

[2.3] Humphreys, P., McIvor, R. and Huang, G., 2002, "An expert system for evaluating the make or buy decision," *Computers & Industrial Engineering*, **42**(2–4), pp. 567–585.

[2.4] Probert, D., 1997, *Developing a Make or Buy Strategy for Manufacturing Business*, Institution of Electrical Engineers, London.

[2.5] Fischer, D. and Hafen, U., 1997, "Immer wieder die gleichen Fehler," *io Management Zeitschrift Industrielle Organization*, **66**(6), pp. 38–45.

[2.6] Micklethwait, J. and Wooldridge, A., 1996, *The Witch Doctors – What the Management Gurus are Saying, Why it Matters and How to Make Sense of it*, Heinemann, London.

[2.7] Capello, R., 1996, “Industrial enterprises and economic space: the network paradigm,” *European Planning Studies*, **4**(4), pp. 485–498.

[2.8] Håkansson, H. and Johanson, J., 1988, “Formal and informal co-operation strategies in international industrial networks,” In *Cooperative Strategies in International Business*, Contractor, F.J. and Lorange, P. (eds.), Lexington Books, Lexington, MA, pp. 369–379.

[2.9] Wiendahl, H.-P. and Lutz, S., 2002, “Production in networks,” *Annals of the CIRP*, **51**(2), pp. 1–14.

[2.10] Mayntz, R., 1993, “Modernisation and the logic of interorganisational networks,” *Knowledge and Policy: The International Journal of Knowledge Transfer and Utilization*, **6**(1), pp. 3–16.

[2.11] Miles, R.E. and Snow, C.C., 1984, “Fit, failure and the Hall of Fame,” *California Management Review*, **26**(3), pp. 11–28.

[2.12] Skinner, W., 1969, “Manufacturing – missing link in corporate strategy,” *Harvard Business Review*, **47**(3), pp. 136–145.

[2.13] Penrose, E.T., 1963, *The Theory of the Growth of the Firm*, Basil Blackwell, Oxford.

[2.14] Chandler, A.D., Jr., 1977, *The Visible Hand: The Managerial Revolution in American Business*, Harvard University Press, Cambridge, MA.

[2.15] Skinner, W., 1974, “The focused factory,” *Harvard Business Review*, **52**(3), pp. 113–121.

[2.16] Skinner, W., 1986, “The productivity paradox,” *Harvard Business Review*, **64**(4), pp. 55–59.

[2.17] Sturgeon, T.J., 2002, *Modular Production Networks: A New American Model for Industrial Organization*, Industrial Performance Center, Massachusetts Institute of Technology, Cambridge, MA.

[2.18] Avella, L., 1999, “Focal points in manufacturing strategic planning in Spain,” *International Journal of Operations & Production Management*, **19**(12), pp. 1292–1317.

[2.19] Dekkers, R., Wang, L. and Zhang, S. (2002), “Management and control of outsourcing in dispersed manufacturing networks,” In *Proceedings of the 10th International Manufacturing Conference in China*, Xiamen, China, 11–13 October.

[2.20] O’Neill, H. and Sackett, P., 1994, “The extended manufacturing enterprise paradigm,” *Management Decision*, **32**(8), pp. 42–49.

[2.21] Porter, M.E., 1990, *The Competitive Advantage of Nations*, Free Press, New York.

[2.22] Carter, J.R. and Narasimhan, R., 1996, “A comparison of North American and European future purchasing trends,” *International Journal of Purchasing and Materials Management*, **32**(2), pp. 12–22.

[2.23] Katayama, H., 1999, “Design of a global car production-logistics system for a future Asian-China region,” *International Journal of Operations and Production Management*, **19**(5–6), pp. 582–603.

[2.24] Lester, R.K., 2003, *China, America, and the Global Competition for Industry*, MIT Working Paper, IPC-03-005, Massachusetts Institute of Technology, Cambridge, MA.

[2.25] Bollinger, J.G., 1998, *Visionary Manufacturing Challenges for 2020*, National Academy Press, Washington DC.

[2.26] Dekkers, R. and Wood, K.L., 2007, *Innovative Manufacturing Programme – Overall Mid-Term Review of the Innovative Manufacturing Research Centres*, University of Paisley, Paisley.

[2.27] de Vaan, M.J.M., Verweij, M.J., van der Laan, J. and Wigman, J., 2002, *Voorstudie Innovatief Onderzoeksprogramma* 'Integratie in Ontwikkeling and Productie in de Maakindustrie', Berenschot, Utrecht.

[2.28] Brown, J.S., Durchslag, S. and Hagel III, J., 2002, "Loosening up: how process networks unlock the power of specialisation," *McKinsey Quarterly*, No. 2 (Special Edition), pp. 59–69.

[2.29] Smith, S.A., Petty, D.J., Trustum, D.G. and Labib, A.W., 2009, "Planning in companies with dispersed capacity," In *Dispersed Manufacturing Networks: Challenges for Research and Practice*, Dekkers, R. (ed.), Springer, Heidelberg, pp. 179–194.

[2.30] Lee, W.B. and Lau, H.C.W., 1999, "Factory on demand: the shaping of an agile production network," *International Journal of Agile Management Systems*, **1**(2), pp. 83–87.

[2.31] Kothandaraman, P. and Wilson, D.T., 2001, "The future of competition: value-creating networks," *Industrial Marketing Management*, **30**(4), pp. 379–389.

[2.32] Ritter, T., Wilkinson, I.F. and Johnston, W.J., 2004, "Managing in complex business networks," *Industrial Marketing Management*, **33**(3), pp. 175–183.

[2.33] Håkansson, H. and Johanson, J., 1992, "A model of industrial networks," In *Industrial Networks: A New View of Reality*, Axelsson, B. and Easton, G. (eds.), Routledge & Kegan Paul, London, pp. 28–38.

[2.34] Camagni, R., 1988, "Functional integration and locational shifts in the new technology industry," In *High-technology Industry and Innovative Environments: The European Experience*, Aydalot, P. and Keeble, D. (eds.), Routledge, London, pp. 48–64.

[2.35] Camagni, R., 1993, "Inter-firm industrial networks: the costs and benefits of cooperative behaviour," *Industry & Innovation*, **1**(1), pp. 1–15.

[2.36] Ireland, R.D., Hitt, M.A. and Vaidyanath, D., 2002, "Alliance management as a source of competitive advantage," *Journal of Management*, **28**(3), pp. 413–446.

[2.37] Jarillo, J.C., 1988, "On strategic networks," *Strategic Management Journal*, **9**(1), pp. 31–41.

[2.38] Freiling, J., 1998, "Kompetenzorientierte Strategische Allianzen," *io Management*, **67**(6), pp. 23–29.

[2.39] Deeds, D.L. and Hill, C.W.L., 1996, "Strategic alliances and the rate of new product development: an empirical study of entrepreneurial biotechnology firms," *Journal of Business Venturing*, **11**(1), pp. 41–55.

[2.40] Gulati, R., Nohria, N. and Zaheer, A., 2000, "Strategic networks," *Strategic Management Journal*, **21**(3), pp. 203–215.

[2.41] New, S. and Mitropoulos, I., 1995, "Strategic networks: morphology, epistomology and praxis," *International Journal of Operations & Production Management*, **15**(11), pp. 53–61.

[2.42] Sharifi, H., Ismail, H. and Reid, I., 2009, "Framework for developing an agile future-proof supply chain," In *Dispersed Manufacturing Networks: Challenges for Research and Practice*, Dekkers, R. (ed.), Springer, Heidelberg, pp. 131–153.

[2.43] Riedel, J. and Pawar, K.S., 1998, "The strategic choice of simultaneous versus sequential engineering for the introduction of new products," *International Journal of Technology Management*, **6**(3–4), pp. 321–334.

[2.44] Spring, M. and Dalrymple, J.F., 2000, "Product customisation and manufacturing strategy," *International Journal of Operations & Production Management*, **20**(4), pp. 441–467.

[2.45] Dekkers, R., 2006, "Engineering management and the order entry point," *International Journal of Production Research*, **44**(18–19), pp. 4011–4025.

[2.46] Smulder, F.E., Boer, H., Hansen, P.H.K., Gubi, E. and Dorst, K., 2002, "Configurations of NPD – production interfaces and interface integration mechanisms," *Creativity and Innovation Management*, **11**(1), pp. 62–73.

[2.47] Dekkers, R., 2009, *Dispersed Manufacturing Networks: Challenges for Research and Practice*, Springer, Heidelberg.

[2.48] Friedrich, S.A., 2000, "Was ist "core" und was ist "non-core"?," *io Management*, **69**(4), pp. 18–23.

[2.49] Prahalad, C.K. and Hamel, G., 1990, "The core competence of the corporation," *Harvard Business Review*, **68**(3), pp. 79–91.

[2.50] Gilley, K.M. and Rasheed, A., 2000, "Making more by doing less: an analysis of outsourcing and its effects on firm performance," *Journal of Management*, **26**(4), pp. 763–790.

[2.51] Arnold, U., 2000, "New dimensions of outsourcing: a combination of transaction cost economics and the core competencies concept," *European Journal of Purchasing & Supply Management*, **6**(1), pp. 23–29.

[2.52] Hayes, R.H. and Pisano, G.P., 1994, "Beyond world-class: the new manufacturing strategy," *Harvard Business Review*, **72**(1), pp. 77–85.

[2.53] Dekkers, R., 2005, "Outsourcing revisited," In *Proceedings of the 18th International Conference on Production Research*, Salerno, 31 July–4 August.

[2.54] Editorial, 1999, "Time to overhaul Japan's Keiretsu model," *Business Week*, No. 3620, pp. 114.

[2.55] Stein, T., 2002, "Keiretsu comedown," *Investment Dealers' Digest*, **68**(6), pp. 28–36.

[2.56] Bennett, D., 2002, "Current and future challenges for international manufacturing networks," In *Proceedings of the VIII International Conference on Industrial Engineering and Operations Management*, UFRGS/ABEPRO, Porto Alegre, Brazil, October.

[2.57] Zachary, K., 2001, "Dismantling the Keiretsu," *China-Britain Trade Review*, **37**(5), pp. 14–16.

[2.58] McGuire, J. and Dow, S., 2003, "The persistence and implications of Japanese Keiretsu organisation," *Journal of International Business Studies*, **34**(4), pp. 374–388.

[2.59] Tian, G.Y., Yin, G. and Taylor, D., 2002, "Internet-based manufacturing: a review and a new infrastructure for distributed intelligent manufacturing," *Journal of Intelligent Manufacturing*, **13**(5), pp. 323–338.

[2.60] Nagel, R.N., Dove, R., Goldman, S. and Preiss, K., 1991, *21st Century Manufacturing Enterprise Strategy: An Industry Led View*, Iacocca Institute/Lehigh University, Bethlehem, PA.

[2.61] Burgess, T.F., 1994, "Making the leap to agility: defining and achieving agile manufacturing through business process redesign and business network redesign," *International Journal of Operations & Production Management*, **14**(11), pp. 23–34.

[2.62] Katayama, H. and Bennett, D.J., 1999, "Agility, adaptability and leanness: a comparison of concepts and a study of practice," *International Journal of Production Economics*, **60–61**(1), pp. 43–51.

[2.63] Goldman, S.L., Nagel, R.N. and Preiss, K., 1995, *Agile Competitors and Virtual Organisation; Strategy for Enriching the Customer*, Van Nostrand Reinhold, Princeton.

[2.64] Kidd, P.T., 1994, *Agile Manufacturing, Forging New Frontiers*, Addison-Wesley, Wokingham.

[2.65] Goldman, S.L. and Nagel, R.N., 1993, "Management, technology and agility: the emergence of a new era in manufacturing," *International Journal of Technology Management*, **8**(1–2), pp. 18–38.

[2.66] Stock, G.N., Greis, N.P. and Kasarda, J.D., 1999, "Logistics, strategy and structure: a conceptual framework," *International Journal of Physical Distribution & Logistics*, **29**(4), pp. 224–239.

[2.67] Ross, E.M., 1994, "The twenty-first century enterprise, agile manufacturing and something called CALS," *World Class Design to Manufacture*, **1**(3), pp. 5–10.

[2.68] Mraz, R., 1997, "Building a manufacturing network," *Machine Design*, **69**(16), pp. 85–89.

[2.69] Shi, Y. and Gregory, M., 1998, "International manufacturing networks – to develop global competitive capabilities," *Journal of Management*, **16**(2–3), pp. 195–214.

[2.70] Shi, Y., 2003, "Internationalisation and evolution of manufacturing systems: classic process models, new industrial issues, and academic challenges," *Integrated Manufacturing Systems*, **14**(4), pp. 357–368.

[2.71] Helo, P., Xiao, Y. and Jiao, J.R., 2006, "A web-based logistics management system for agile supply demand network design," *Journal of Manufacturing Technology Management*, **17**(8), pp. 1058–1077.

[2.72] Towill, D. and Christopher, M., 2002, "The supply chain strategy conundrum: to be lean or agile or to be lean and agile," *International Journal of Logistics: Research and Applications*, **5**(3), pp. 299–309.

[2.73] Gunasekaran, A., 1998, "Agile manufacturing: enablers and an implementation framework," *International Journal of Production Research*, **36**(5), pp. 1223–1247.

[2.74] Sanchez, L.M. and Nagi, R., 2001, "A review of agile manufacturing systems," *International Journal of Production Research*, **39**(16), pp. 3561–3600.

[2.75] Akkermans, H.A., Bogerd, P., Yücesan, E. and van Wassenhove, L.N., 2003, "The impact of ERP on supply chain management: exploratory findings from a European Delphi study," *European Journal of Operational Research*, **146**(2), pp. 284–301.

[2.76] Stadtler, H., 2005, "Supply chain management and advanced planning – basics, overview and challenges," *European Journal of Operational Research*, **163**(3), pp. 575–588.

[2.77] Dekkers, R., 2006, "Guest editorial," *Journal of Manufacturing Technology Management*, **17**(8), pp. 1017–1021.

[2.78] Camarinha-Matos, L.M. and Afsarmanesh, H., 2005, "Collaborative networks: a new scientific discipline," *Journal of Intelligent Manufacturing*, **16**(4–5), pp. 439–452.

[2.79] Noori, H. and Lee, W.B., 2009, "Dispersed network manufacturing: an emerging form of collaborative networks," In *Dispersed Manufacturing Networks: Challenges for Research and Practice*, Dekkers, R. (ed.), Springer, Heidelberg, pp. 39–58.

[2.80] Biggiero, L., 1999, "Market, hierarchies, networks, districts: a cybernetic approach," *Human Systems Management*, **18**(2), pp. 71–86.

[2.81] Robertson, P.L. and Langlois, R.N., 1995, "Innovation, networks, and virtual integration," *Research Policy*, **24**(4), pp. 543–562.

[2.82] Kuhn, J., 2009, "Developing a worldwide production network," In *Dispersed Manufacturing Networks: Challenges for Research and Practice*, Dekkers, R. (ed.), Springer, Heidelberg, pp. 159–177.

[2.83] Doner, R.F., Noble, G.W. and Ravenhill, J., 2004, "Production networks in East Asia's auto part industry," In *Global Production Networking and Technological Change in East Asia*, Yusuf, S., Altaf, M.A. and Nabeshina, K. (eds.), World Bank Publications, Washington, pp. 159–208.

[2.84] Ernst, D., 2002, "Global production networks and the changing geography of innovation systems – implications for developing countries," *Economics of Innovation and New Technology*, **11**(6), pp. 497–523.

[2.85] Riis, J.O., Johansen, J. and Waehrens, B.V., 2007, "Strategic roles of manufacturing," *Journal of Manufacturing Technology Management*, **18**(8), pp. 933–948.

[2.86] Das, T.K. and Teng, B.-S., 2001, "Trust, control, and risk in strategic alliances: an integrated framework," *Organization Studies*, **22**(2), pp. 251–283.

[2.87] Huemer, L., 2004, "Balancing between stability and variety: identity and trust trade-offs in networks," *Industrial Marketing Management*, **33**(3), pp. 251–259.

[2.88] Thorelli, H.B., 1986, "Networks: between markets and hierarchies," *Strategic Management Journal*, **7**(1), pp. 37–51.

[2.89] Vangen, S. and Huxham, C., 2000, "Building trust in inter-organisational collaboration," In *Proceedings of the Academy of Management Research*, Toronto, Canada, 4–9 August.

[2.90] Frederix, F., 2001, "An extended enterprise planning methodology for the discrete manufacturing industry," *European Journal of Operational Research*, **129**(2), pp. 317–325.

[2.91] Noori, H. and Lee, W.B., 2002, "Factory-on-demand and smart supply chains: the next challenge," *International Journal of Manufacturing Technology and Management*, **4**(5), pp. 372–383.

[2.92] Roosendaal, C., 2000, *The Concept of the Virtual Organization*, Delft University of Technology, Delft.

[2.93] Kogut, B., 1989, "The stability of joint ventures: reciprocity and competitive rivalry," *The Journal of Industrial Economics*, **38**(2), pp. 183–198.

[2.94] Medcof, J.W., 2001, "Resource-based strategy and managerial power in networks of internationally dispersed technology units," *Strategic Management Journal*, **22**(11), pp. 999–1012.

[2.95] Mody, A., 1993, "Learning through alliances," *Journal of Economic Behaviour and Organization*, **20**(2), pp. 151–170.

[2.96] Milberg, J. and Dürrschmidt, S., 2002, "Planning methodology for changeable logistic systems," *Production Research*, **9**(1), pp. 63–68.

[2.97] Teece, D.J., Pisano, G. and Shuen, A., 1997, "Dynamic capabilities and strategic management," *Strategic Management Journal*, **18**(7), pp. 509–533.

[2.98] Möller, K. and Svahn, S., 2003, "Managing strategic nets: a capability perspective," *Marketing Theory*, **3**(2), pp. 209–324.

[2.99] Dekkers, R., 2000, "Organelle structures: strategic choice of performance management of manufacturing," In *Proceedings of the 1st World Conference on Production and Operations Management*, Sevilla, Spain, 27 August–1 September.

[2.100] Dekkers, R. and van Luttervelt, C.A., 2006, "Industrial networks: capturing changeability?" *International Journal of Networking and Virtual Organizations*, **3**(1), pp. 1–24.

[2.101] Bikker, H., 1993, *Lecture Notes wb5413*, Delft University of Technology, Delft, The Netherlands.

[2.102] Tesar, G., Boter, H. and Bohman, H., 2003, "Network participation and internationalisation of smaller manufacturing enterprises in Central Europe: a case for research," In *Proceedings of the 19th Annual IMP Conference*, Lugano, Switzerland, 4–6 September.

[2.103] Lall, S., 2000, "Strengthening SMEs for international competitiveness," *Workshop "What Makes Your Firm Internationally Competitive"*, ECES Working Paper 44, In *Third Meeting of the Mediterranean Development Forum,* Cairo, Egypt, 5–8 March.

[2.104] Bennett, D.J. and Ozdenli, O., 2004, "International manufacturing networks and the role of small and medium enterprises," In *Proceedings of the ENEGEP X International Conference on Industrial Engineering and Operations Management and XXIV National Conference on Production Engineering*, Florianopolis, Brazil, 3–5 November.

[2.105] Dekkers, R., Sauer, A., Schönung, M. and Schuh, G., 2004, "Collaborations as complex systems," In *Proceedings of the 9th Annual Cambridge International Manufacturing Symposium*, IMNet/CIM, Cambridge, UK, 25–26 September.

[2.106] Field, S. and Hoffner, Y., 2003, “Web services and matchmaking,” *International Journal of Networking and Virtual Organizations*, **2**(1), pp. 16–32.

[2.107] Molina, A., Mejía, R. and Velandia, M., 2003, “Core processes, methods and e-services to support virtual enterprise brokerage,” *International Journal of Networking and Virtual Organizations*, **2**(1), pp. 33–49.

[2.108] Angelov, S. and Grefen, P., 2003, “The 4W framework for B2B e-contracting,” *International Journal of Networking and Virtual Organizations*, **2**(1), pp. 78–97.

[2.109] Barata, J. and Camarinha-Matos, L.M., 2003, “Coalitions of manufacturing components for shop floor agility – the CoBASA architecture,” *International Journal of Networking and Virtual Organizations*, **2**(1), pp. 50–77.

[2.110] Chesbrough, H.W., 2003, *Open Innovation: The New Imperative for Creating and Profiting from Technology*, Harvard Business School Press, Boston, MA.

[2.111] Tidd, J., 1995, “Development of novel products through intraorganizational and interorganizational networks,” *Journal of Product Innovation Management*, **12**(4), pp. 307–322.

[2.112] Dekkers, R., 2009, “Distributed manufacturing as co-evolutionary system,” *International Journal of Production Research*, **47**(8), pp. 2031–2054.

[2.113] Kotabe, M. and Murray, J.Y., 2004, “Global sourcing strategy and sustainable competitive advantage,” *Industrial Marketing Management*, **33**(1), pp. 7–14.

[2.114] Sadd, R.J. and Bennett, D.J., 1999, “Automotive supply logistics: myth and reality,” In *Proceedings of the 4th International Symposium on Logistics*, Servizi Grafici Editoriali, Padova, Italy, 11–14 July.

[2.115] Joo, M.-S., Nakamoto, K. and Nelson, J.E., 2003, “The shielding effects of brand image against lower quality countries-of-origin in global manufacturing,” *Journal of Business Research*, **56**(8), pp. 637–646.

[2.116] Bennett, D.J., Liu, X., Parker, D., Steward, H.F. and Vaidya, K.G., 2001, “Technology transfer to China: a study of strategy in 20 EU industrial companies,” *International Journal of Technology Management*, **21**(1–2), pp. 151–182.

[2.117] Ritter, T. and Gemünden, H.G., 2003, “Network competence: its impact on innovation success and its antecedents,” *Journal of Business Research*, **56**(9), pp. 745–755.

[2.118] Dekkers, R., 2005, *(R)Evolution, Organizations and the Dynamics of the Environment*, Springer, New York.

[2.119] Burt, R.S., 1992, *Structural Holes: The Social Structure of Competition*, Harvard University Press, Cambridge, MA.

[2.120] Uzzi, B., 1997, “Social structure and competition in interfirm networks: the paradox of embeddedness,” *Administrative Science Quarterly*, **42**(1), pp. 35–67.

[2.121] Hagedoorn, J., 1990, “Organizational modes of inter-firm co-operation and technology transfer,” *Technovation*, **10**(1), pp. 17–30.

[2.122] Souder, W.E. and Nassar, S., 1990, “Choosing an R&D consortium,” *Research Technology Management*, **33**(2), pp. 35–41.

[2.123] Bailey, W.J., Masson, R. and Raeside, R., 1996, “Choosing successful technology development partners: a best practice model,” In *Proceedings of the 6th International Forum on Technology Management*, Amsterdam, The Netherlands, 15–18 October.

[2.124] Kühnle, H., 2009, “Self-similarity and criticality in dispersed manufacturing: a contribution to production networks control,” In *Dispersed Manufacturing Networks: Challenges for Research and Practice*, Dekkers, R. (ed.), Springer, Heidelberg, pp. 59–76.

[2.125] Montreuil, B., Frayret, J.-M. and D'Amours, S., 2000, “A strategic framework for networked manufacturing,” *Computers in Industry*, **42**(2–3), pp. 299–317.

[2.126] Barratt, M., 2004, "Understanding the meaning of collaboration in the supply chain," *Supply Chain Management: An International Journal*, **9**(1), pp. 30–42.
[2.127] Seuring, S., 2004, "Integrated chain management and supply chain management comparative analysis and illustrative cases," *Journal of Cleaner Production*, **12**(8–10), pp. 1059–1071.
[2.128] Zhu, Q. and Cote, R.P., 2004, "Integrating green supply chain management into an embryonic eco-industrial development: a case study of the Guitang Group," *Journal of Cleaner Production*, **12**(8–10), pp. 1025–1035.
[2.129] Srivastava, S.K., 2007, "Green supply-chain management: a state-of-the-art literature review," *International Journal of Management Reviews*, **9**(1), pp. 53–80.
[2.130] Rao, P. and Holt, D., 2005, "Do green supply chains lead to competitiveness and economic performance?" *International Journal of Operations & Production Management*, **25**(9), pp. 898–916.
[2.131] Nassimbeni, G., 1998, "Network structures and co-ordination mechanisms: a taxonomy," *International Journal of Operations & Production Management*, **18**(6), pp. 538–554.
[2.132] Ashby, W.R., 1956, *An Introduction to Cybernetics*, Wiley, New York.
[2.133] Andriani, P., 2003, "Evolutionary dynamics of industrial clusters," In *Complex Systems and Evolutionary Perspectives on Organizations*, Mitleton-Kelly, E. (ed.), Pergamon, Oxford, pp. 127–148.
[2.134] Song, C., Havlin, S. and Makse, H.A., 2005, "Self-similarity of complex networks," *Nature*, **433**(7024), pp. 392–395.
[2.135] Iansiti, M. and Levien, R., 2002, *The New Operational Dynamics of Business Ecosystems: Implications for Policy, Operations and Technology Strategy*, Working Paper 03-30, Harvard Business School, Boston.
[2.136] Surana, A., Kumara, S., Greaves, M. and Raghavan, U.N., 2005, "Supply-chain networks: a complex adaptive systems perspective," *International Journal of Production Research*, **43**(20), pp. 4235–4265.
[2.137] Maropoulos, P.G., 2003, "Digital enterprise technology – defining perspectives and research priorities," *International Journal of Computer Integrated Manufacturing*, **16**(7–8), pp. 467–478.
[2.138] Dubois, A. and Gadde, L.-E., 2002, "The construction industry as a loosely coupled system: implications for productivity and innovation," *Construction Management and Economics*, **20**(7), pp. 621–631.
[2.139] Dorée, A.G. and Holmen, E., 2004, "Achieving the unlikely: innovating in the loosely coupled construction system," *Construction Management and Economics*, **22**(8), pp. 827–838.

3

Agile Manufacturing in Complex Supply Networks

Henry Xu

UQ Business School, The University of Queensland
St Lucia, Queensland, 4072, Australia
Email: h.xu@business.uq.edu.au

Abstract
Today's manufacturing industries are characterised by complex distributed supply networks, which require agile manufacturing strategy to compete as a whole in the volatile global market. Key to an agile supply chain is fast material, information and decision flows across the supply network, which can be achieved through the close integration and coordination of both internal and external supply chains of a firm in the network. This chapter presents the important aspects of an IST project: Co-OPERATE, which aims to develop a Web-based system for improved coordination of manufacturing planning and control activities across the supply network. These include review of the currently available commercial solutions for supply network coordination (SNC), analysis of the challenges and the business and systems requirements of SNC in complex supply networks (*e.g.* in the automotive industry), a framework for SNC, design of the overall Co-OPERATE system, and implementation and evaluation of the system. The key contributions of the presented work include a research framework for SNC, a decentralised system architecture for SNC, the high-level design of four focused coordination processes, and the implementation of the Co-OPERATE system.

3.1 Introduction

The rapid growth of global outsourcing and manufacturing requires an agile supply chain as a strategy to compete in the volatile global market. Key to an agile supply chain is fast material, information and decision flows across the supply network, which can be achieved through the close integration/coordination of both internal and external supply chains of a firm in the network. In the supply chain execution process, the response time is basically the time it takes to close the gap from when an event occurs until actions are taken, which will rely not so much on equipment, but on information and knowledge workforce [3.1–3.6] as shown in Figure 3.1.

It is thus recognised that the effectiveness and efficiency of any supply chain relies on the visibility of demand, inventories and material flows throughout the pipeline. Time lapses in information flows are directly translated into inventory [3.7]. These problems tend to be larger at the interfaces between companies, where more disparities and uncertainties exist. Therefore, it is paramount to facilitate fast,

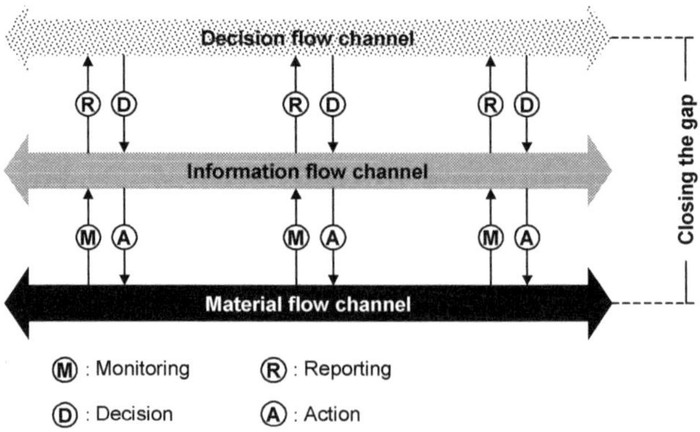

Figure 3.1. Closing the gap – the MRDA cycle in an agile supply chain

accurate information exchange and efficient coordination between companies to achieve the synchronisation of material and information flows.

This chapter is organised as follows: in the next section currently available commercial solutions for SNC are reviewed. Section 3.3 outlines the challenges and business and systems requirements of manufacturing coordination in complex supply networks such as in the automotive and semiconductor industries. Based on an analysis of business and systems requirements and a review of existing commercial solutions, a framework for the improved coordination of manufacturing planning and control activities in the supply network is presented in Section 3.4. Section 3.5 describes the design of the overall Co-OPERATE system, which is followed by the implementation and evaluation of the system in Section 3.6. Finally, the presented research work is concluded and future work suggested.

3.2 An Overview of Commercial Solutions for SNC

If the manufacturing resources planning (MRP II) system is recognised as an early attempt for internal integration, electronic data interchange (EDI) systems can be regarded as a pioneering effort for external integration [3.8]. Enterprise resources planning (ERP), advanced planning and scheduling (APS) and supply chain management (SCM) can generally be regarded as enterprise information systems. Business-to-business (B2B) systems (*e.g.* B2B marketplaces, B2B portals) go a step further beyond the traditional enterprise information systems to support multi-enterprise, multi-tier connection. However, these systems do not provide full support for the integration and coordination of production planning and control activities in complex supply networks [3.9–3.12].

ERP was mainly focused internally and thus did not contribute much to the elimination of barriers between business partners [310, 3.13, 3.14]. Unrealistic assumptions of MRP II/ERP planning logic (*e.g.* infinite capacity and fixed lead-times) led to the development of APS engines, which lie at the heart of the so-called SCM systems [3.15]. However, most of the current APS/SCM systems do not manage the supply chain outside an organisation due to the centralistic view of hierarchical planning that underlies these systems [3.16]. The recently emerged concept ERP II claims to support the idea of an extended enterprise [3.14]. However, it is still not clear whether ERP will become a module within some broader system, or evolve into the all-encompassing ERP II system or something else [3.17]. Moreover, the costs of such systems are normally beyond what most small and medium-sized enterprises (SMEs) can afford [3.18].

Industry-wide standards are necessary for reducing the costs in relation to B2B connectivity. RosettaNet [3.19] is a pioneer in the field of open B2B electronic commerce standards, which are complementary to many SCM applications such as those offered by i2 [3.20]. Its partner interface processes (PIPs) are specialised system-to-system extensible markup language (XML)-based dialogs that define business processes between trading partners, and apply to eight core processes, including administration, order management, *etc.* As RosettaNet is currently focused only on the computer and electronics industries and does not define very complex business processes [3.21], its process standards are still evolving and leave some gaps between defined standards and comprehensive industry requirements [3.22].

Web-based B2B systems such as B2B e-marketplaces and B2B portals can be used to promote cooperation between business partners. However, most of these tools are still in their early stages in terms of supply chain coordination [3.9, 3.11, 3.23, 3.24]. For example, it is still not clear how e-marketplaces can be used to achieve the goal of full network coordination in terms of business process changes and technical arrangements. Recent developments of process portals are mainly focused on the development of an integration architecture for inter-organisational information systems [3.11]. Therefore, it is highly desirable to develop an innovative framework that captures key business processes that promote close collaboration and coordination in the supply network.

3.3 Challenges and Requirements of SNC

Although integration and coordination are important issues for almost all types of supply chain, few industries offer as many challenges and opportunities for supply chain integration and coordination as the automotive and the semiconductor industries [3.25, 3.26]. These two industries are characterised by complex distributed manufacturing networks, high demand and supply uncertainties, and heterogeneous local planning systems [3.26].

Despite their differences in structure and technology, the fundamental operational problem of today's supply networks is the lack of synchronisation between demand and supply [3.27]. On the demand side, the most detrimental factor is the 'bullwhip effect' that can create excesses or shortages of inventories. In extreme cases the 'bullwhip effect' may even cause small suppliers to go out of business, or to be acquired [3.28]. On the supply side, unreliable execution processes, which often incur late deliveries [3.29], can be the primary cause of inventory shortages and can also be another reason for holding excessive inventories. However, demand and supply variations are inevitable in today's ever-changing industrial environment, which requires that the supply network must be flexible and responsive.

The analysis of business and system requirements was based on an extensive literature review and the results of in-company interviews with managers and operational staff of six companies in the automotive and semiconductor industries, covering first, second and third-tier suppliers and SMEs in the Co-OPERATE project [3.30]. The major business requirements for a supply network coordination system are in three areas: (1) reducing the bullwhip effect, (2) enhancing supply chain reliability, and (3) improving supply chain responsiveness.

The major common features for most industries' supply networks have structural and organisational requirements, which must be met by SNC systems. Particularly, (1) the system architecture must be compatible with the heterogeneous and distributed industrial environment in terms of different local legacy systems and geographically separated manufacturing locations, (2) the system needs to be scalable and quickly re-configurable to cater for growth and changes in the network structure, and (3) the system should provide decision-support information, especially in exceptional cases.

3.4 A Research Framework for SNC

3.4.1 Seven Coordination Processes

On the basis of the above requirements analysis, a framework for manufacturing coordination in the supply network was proposed, comprising seven interrelated business processes, and this has formed the potential scope of the Co-OPERATE project. The focus of each coordination process can be summarised as follows:

1. Long-term production planning
    - Communication of long-term forecast of demand and associated risks.
    - Communication of new product introduction, product variants, product engineering changes and other events that significantly influence the future demand or the trend of demand of products.
    - Capacity checking and related decision-making for the affected nodes in the supply network (*e.g.* purchasing of new machinery or hiring of new employee).
2. Operational ordering and planning
    - Visibility of order schedules, consumption schedules and single orders to the upstream companies in the supply network.
    - Visibility of the current buffer stock levels and delivery schedules of certain products to the downstream companies in the supply network.
    - Detection and resolution of planning problems early through feedback loop in the regular operational planning process.
3. Request and feasibility study
    - Fast check for capacity and material availability across the supply network when new orders or large order changes are expected.

4. Visibility of the order progress
   - Visibility of orders in supply fulfilment of the upstream companies to their immediate customers.
   - Production monitoring and early detection of production disturbance in the upstream companies.
   - Early warning of delivery problems from the upstream companies to their immediate customers.

5. Exception handling
   - Rush order handling through the fast request–quotation–review–order confirmation process.
   - Resolution of delivery problems (*e.g.* partial delivery, delayed delivery) through fast information exchange and negotiation process to solve the problems quickly.
   - Establishment of containment rules and escalation routes for disturbances to reduce their negative effects to a minimum and to enhance schedule stability in the supply network.

6. Multi-sourcing coordination
   - Medium- or long-term coordination of new orders or large order changes to enhance the capability of the whole supply network in filling the gap.
   - Short-term coordination of urgent demand to increase the success rate of order fulfilment and to enhance the delivery reliability of suppliers.

7. Network performance management
   - Agreement on a common set of network performance indicators (*e.g.* forecast accuracy, delivery reliability).
   - Evaluation and analysis of network performance indicators and consequent action programs.

Generally, long-term production planning (LTP), operational ordering and planning (OOP), and request and feasibility study (ReFS) work on the planning level, addressing issues before actual orders are placed. At this level, LTP is a strategic production planning process in the long term, OOP is mainly focused on regular ordering and production planning processes in the short to medium term, and ReFS is actually an efficient planning tool for LTP and OOP when new orders or large order changes are expected in the medium or long term. As OOP covers most of inter-company production planning activities and its outputs serve as important inputs for control of the supply chain execution process, it is regarded as the core of the coordination framework.

On the execution level, visibility of order progress (VOP) and exception handling (eXH) address issues after actual orders are placed. Basically, VOP is focused on monitoring the supply chain execution process, providing the up-to-date information on order fulfilment and giving early warnings to the affected customers in case of anticipated problems in delivery. Accordingly, eXH mainly deals with various delivery problems (*e.g.* partial delivery, delayed delivery) through standard exception handling processes, which try to resolve conflicts between the customer's demand and the supplier's delivery capability in a timely, cost-effective way.

Therefore, eXH complements VOP for enhancing the reliability and responsiveness in the supply chain execution process.

In contrast to LTP, OOP, ReFS, VOP and eXH, multi-sourcing coordination (MSC) and network performance management (NPM) work both on the planning level and on the execution level, and cover the entire time horizon. Actually, LTP, OOP and ReFS can employ MSC for medium- or long-term coordination of new orders or large order changes, while eXH can utilise MSC for short-term coordination of urgent demand. NPM is based on a predefined set of network performance indicators (*e.g.* forecast accuracy, delivery reliability).

It should be noted that the proposed framework described above captures only the most important business processes for improved coordination of production planning and control activities in the supply network. Actually, the functionality of each coordination process can be modified and/or extended to meet the specific business requirements of network members. On the other hand, both at the planning and execution levels and along the entire time horizon, new business processes may be identified and integrated into the existing framework, if necessary, to cover the special business requirements of a given industry.

To achieve maximum possible results with available resources, the seven coordination processes were prioritised by all project partners based on a set of criteria, such as information visibility, delivery performance, *etc.* Among the seven business processes, OOP, VOP, eXH and ReFS were chosen for further development and implementation, which comprised the main modules of the Co-OPERATE system.

3.4.2 Functional Relationship Between the Focused Processes

Figure 3.2 presents the functional relationship between OOP, VOP, eXH and ReFS [3.31]. Conceptually, the Co-OPERATE processes start with OOP, which extracts short- to medium-term planned consumption and forecasts in the form of order schedules, consumption schedules and single orders from the ERP system or local planning system (LPS) in an individual company. The planned consumption and forecast information is communicated to the upstream companies. Based on the consumption information from the customer, detailed delivery schedule for the immediate coming weeks with time buckets per week, or per day or per shift for JIT cases is generated by the supplier and transmitted to the customer. The delivery schedule has to be checked and approved by the customer before being accepted.

The planning process described above assumes that customer demand evolves over time and fluctuates only to certain predefined limits. In reality, however, there are two major issues complicating the planning process in the medium to long term: (1) new orders unforeseen in the previous demand plan, and (2) large order changes in a planned order according to predefined business rules, *e.g.* over 30% quantity change in month 2. In such cases, it is desirable to have a fast feasibility check across the relevant part of the network within a short time period (*e.g.* 1–2 days) before actually placing new orders or initiating large order changes. Therefore, ReFS complements OOP in functionality for supply chain planning. As the process entails the propagation of request on demand and quotation on delivery capabilities across the relevant part of the network, an agent-based approach has been adopted to facilitate the desired fast system response.

Once the plans are established, the focus of manufacturing coordination in the network shifts from the demand side to the supply side. It is the suppliers' responsibility to provide the customer with meaningful, formalised information on the status of their orders. VOP achieves this purpose by continuously monitoring the production process in the upstream company and alert the customer to exceptions when certain delivery cannot be fulfilled as promised in the delivery schedule. In this case, it triggers the exception handling process, which focuses on resolving unexpected demand and supply problems in the short term, *i.e.* rush orders and missed deliveries. Essentially, a rush order is a special single order that requires less than the standard delivery lead-time.

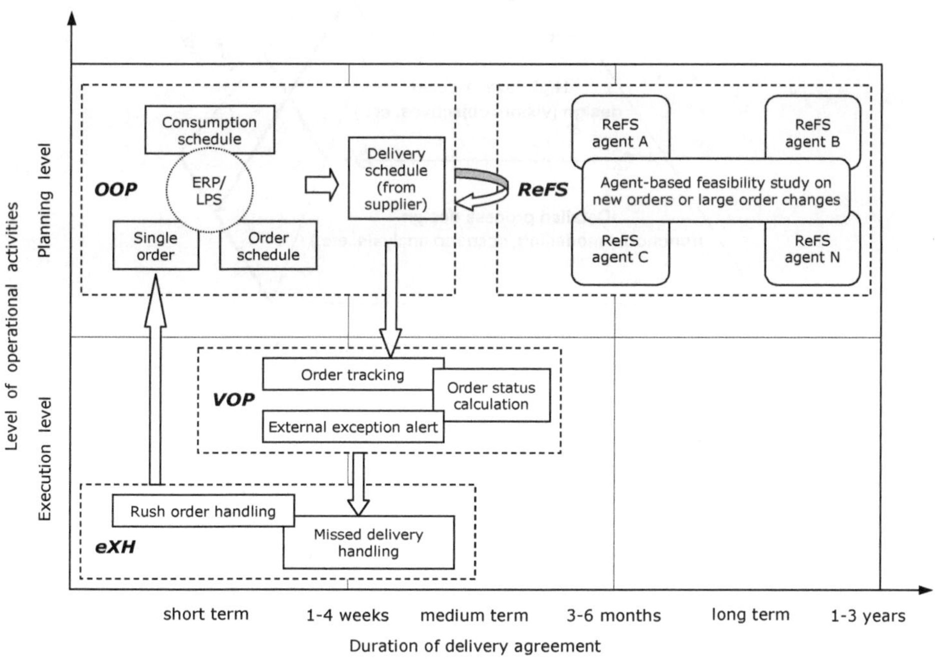

Figure 3.2. Functional relationship between the focused processes

3.5 The Overall Co-OPERATE System

3.5.1 System Design Approach

As with engineering design, a top-down approach was adopted for the design of the Co-OPERATE system as illustrated in Figure 3.3. At the top level, a research framework for SNC is outlined in Section 3.4. In Section 3.5.2, three basic system architectures for SNC are discussed. Consequently, the completely decentralised system architecture is chosen after the evaluation of the three alternatives. High-level design of the four focused coordination processes (*i.e.* OOP, VOP, eXH and ReFS) is presented in Sections 3.5.3–3.5.6, which include vision, objectives, performance indicators and process outline for each process. Detailed business process design is accomplished through functional modelling and scenario analysis, *etc.*, which are summarised in Section 3.6.

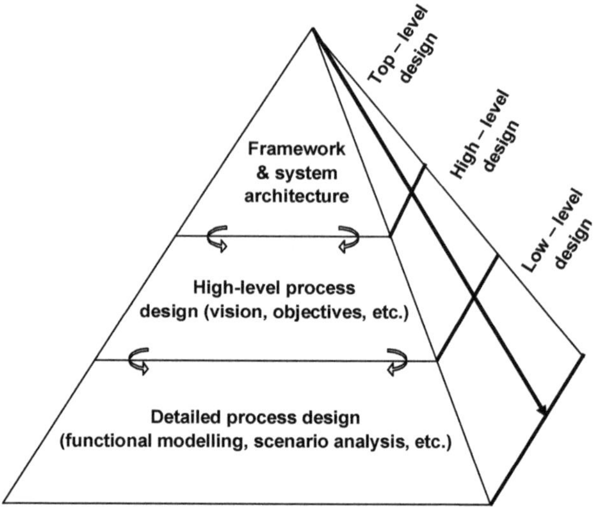

Figure 3.3. The top-down system design approach

3.5.2 Network Coordination Architecture

According to current research (*e.g.* European research projects X-CITTIC [3.32], and PRODNET II [3.33]) and existing software systems, such as ERP and SCM systems, there are three reference models for network coordination [3.34]:

- centralised coordination system with local data access;
- hybrid coordination system with some distributed local functionality and a central coordination module;
- completely decentralised coordination system without central coordination.

As each architecture has specific characteristics, strategic choice can be made by evaluating the strengths and weaknesses of each reference model and matching them to the characteristics of the complex supply networks.

Centralised Coordination Architecture

This traditional architecture has been used extensively in all kinds of enterprise systems (*e.g.* MRPII and ERP systems). As illustrated in Figure 3.4, the central coordination module runs a global coordination algorithm for the whole supply network. Users access to the central module through thin clients that have very limited functionality. Generally, the central module is usually accompanied by a central database, and is managed by a responsible department or an appropriate partner. This architecture is typically adopted by ERP systems such as SAP R/3. Besides, most SCM systems have a similar architecture, although they may extract data from more than one source.

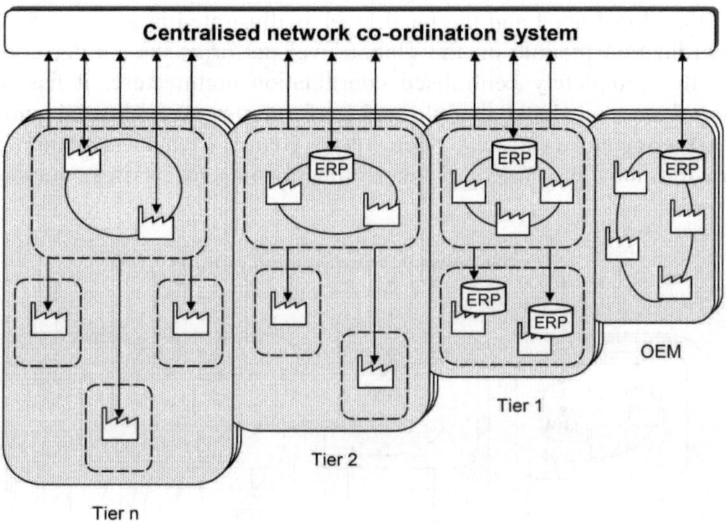

Figure 3.4. Centralised system architecture

Advantages:

- Simple system structure provides fast information processing.
- Easy to achieve a wide range of integration to cover major business processes.
- Central database renders data readily accessible and easy to maintain.

Disadvantages:

- Global control compromises local decision autonomy.
- Possible overlap in functionality with existing enterprise systems.
- Central database incurs data security issues.
- Central coordination prevents partners from participating in more than one supply network.
- Central coordination algorithms make it difficult to incorporate the differential specifics of individual business partners.

Conclusions:

Though the centralised architecture has been widely employed by ERP and SCM systems due to its attractive benefits and available information technologies, it is inherently unsuitable for the development of a network coordination system simply because no partner would agree to give up all their decision-making autonomy.

Moreover, it is technically extremely difficult, if not impossible, to achieve information synchronisation especially in a complex supply network.

Hybrid Coordination Architecture

The hybrid coordination architecture divides the whole task of coordination into two levels, *i.e.* the global level and the local level as illustrated in Figure 3.5. While the central coordination module on the global level performs the similar coordination task as in the completely centralised coordination architecture, it has less input information from each local module and performs less complicated optimisation, leaving each company primarily independent. Hence, each local planning module (LPM) provides local planning and control functionality for its own company.

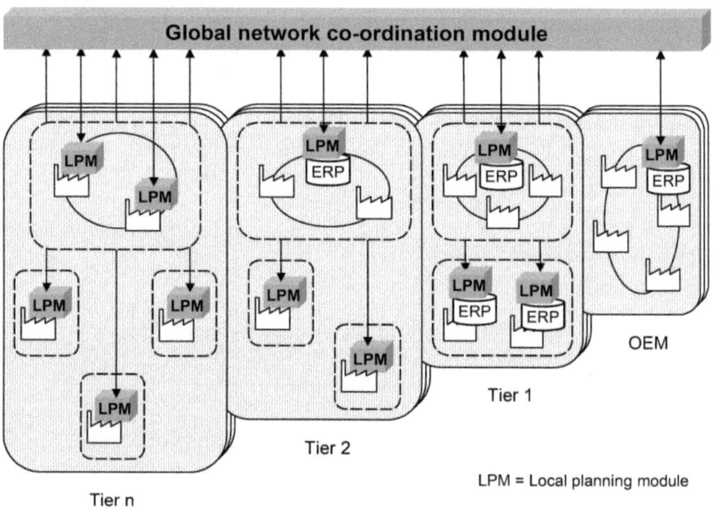

Figure 3.5. Hybrid system architecture

In a sense, this hybrid architecture is an evolution of the completely centralised architecture and combines local decision making with global optimisation. It seems that this hybrid architecture has been adopted in the current SCM systems such as SAP's APO and i2's SCM application. For example, i2's SCM software package offers multi-enterprise, multi-tier visibility, planning and execution capability, which could use this hybrid system architecture. However, it is not clear whether it is actually hybrid or completely centralised.

Advantages:

- Local planning modules allow local decision making autonomy within individual companies.
- Local modules can in principle be plugged into several central coordination modules for different supply networks if they use compatible communication standards.

- Any overlap in functionality can be avoided through an interface between the local module and existing local enterprise systems.

Disadvantages:

- Limited suitability for industries (*e.g.* the automotive industry) that have different manufacturing networks for different products.
- Need for complete and exact BOMs (bill of materials) and routing information for complex products.
- Heavy network loading for data exchange during run times.

Conclusions:

By virtue of local planning modules tailored to the need of each node, the hybrid architecture overcomes some problems of the centralised coordination architecture, such as data privacy and system flexibility. However, the centralised coordination still poses some difficult problems, such as the ownership of the central module. Additionally, heavy network loading may cause the delay of information flow. Therefore, the hybrid architecture is well suitable for industries with close, long-term relationships, such as the semiconductor industry, but much less for the automotive, machinery and telecommunications equipment industries.

Decentralised Coordination Architecture

Unlike the centralised and hybrid architectures, the decentralised architecture has no central coordination module. As illustrated in Figure 3.6, this architecture consists of a self-coordinating unit, called a unit coordination module (UCM), for each local unit, which can be a company or a plant within a company. If a company has its own local planning system, it can perform local planning for the company while connecting via a coordination module to other companies in the supply network. If there is no local planning system in a company, the network coordination can also work on the plant level. This flexibility makes network coordination suitable for companies with local planning systems, especially for SMEs without large-scale integrated ERP systems.

The UCM coordinates both the company for which it performs coordination and its immediate suppliers and customers. This architecture seems to have been employed in the latest development of commercial e-business marketplaces. For example, the recent development of i2's Tradematrix Open Commerce Network (OCN) implies a similar decentralised architecture. However, most B2B marketplaces provide very limited functionalities for SNC.

Advantages:

- Each partner has complete control of its local planning function and thus can maintain its local decision-making autonomy intact.
- Each partner can easily participate in several networks with compatible communication standards.
- Highest data protection and process integrity.

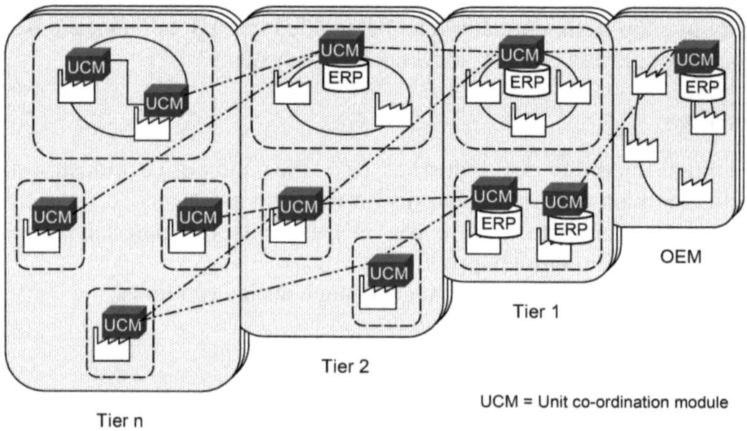

Figure 3.6. Decentralised system architecture

Disadvantages:

- New and potentially complex coordination algorithms need to be developed.
- Possible heavy network loading for data exchange between network members.

Conclusions:

From the network coordination perspective, the decentralised architecture with self-coordinating units has some unique characteristics from the two previous reference models as it corresponds directly with the many-to-many structure of the network. This makes the network coordination pattern compatible with the current coordination practices well established and proven in industries and hence can be easily accepted by all partners in the network.

Chosen Coordination Architecture

From the above description and evaluation, it can be seen that the completely centralised architecture is most suitable for a single company as it provides a simple and a relatively quick way for the synchronisation of information and processes. The hybrid architecture is best suited for large and diverse companies or for networks with long-lasting relationship between partners. For both types of architecture, their similar central coordination modules make them unsuitable for varying products with differing network configurations, which features the automotive supply industry.

Consequently, the fully decentralised architecture was chosen as the network coordination architecture for the Co-OPERATE system. Its highest flexibility and versatility mirror the nature of the real supply network. The architecture also complies with modern management principles such as delegation of decision making to the same level as operative responsibility. Nevertheless, as this coordination architecture has been much less explored in comparison with the other two, it still needs to prove its suitability for manufacturing coordination in complex distributed supply networks.

3.5.3 Operational Ordering and Planning

Vision

To create feasible and synchronised manufacturing plans across the supply network while resolving planning-related problems through fast feedback loops (see Figure 3.7).

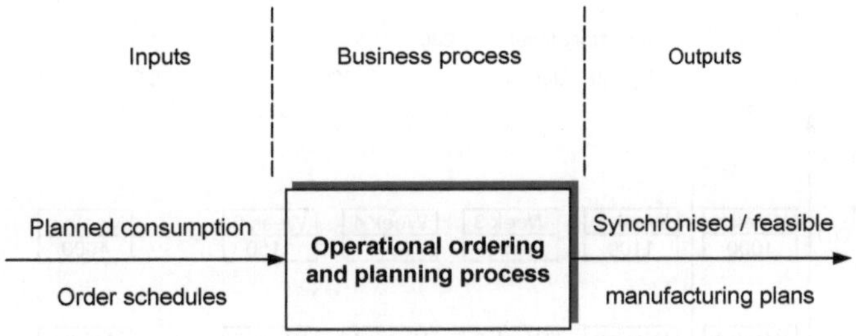

Figure 3.7. Basic diagram of the OOP process

Objectives

- Reduce the 'bullwhip effect' in the supply network through the provision of planned consumption to the upstream companies and feedback of delivery schedules to the downstream companies.
- Detect and resolve order fulfilment issues early at the collaborative planning stage.
- Improve customer service through the creation of mutually agreed delivery schedules.

Performance Indicators

- Percentage of parts managed through consumption driven supply management.
- Average buffer stock levels at different locations in the network.
- Percentage of planning-related problems undetected and/or unresolved.

Process Outline

The process aims to reduce the 'bullwhip effect' in the network through the transition from the traditional push model to the pull model based on consumption driven supply management. Figure 3.8 presents the basic operational ordering and planning process in a three-tier business case. To avoid the tendency for high buffer levels and over-ordering, the downstream company communicates to the upstream company their planned consumption in the coming week(s) with a time bucket per day or even per shift in JIT cases, together with the on-hand inventory levels. The planned consumption can be updated daily. Meanwhile, forecasts in the form of order schedules with time bucket per week or per month reaching up to six months into the future coupled with some qualitative information (*e.g.* prediction of market changes and capacity variations) are also transferred to the upstream company for strategic or tactical decision making.

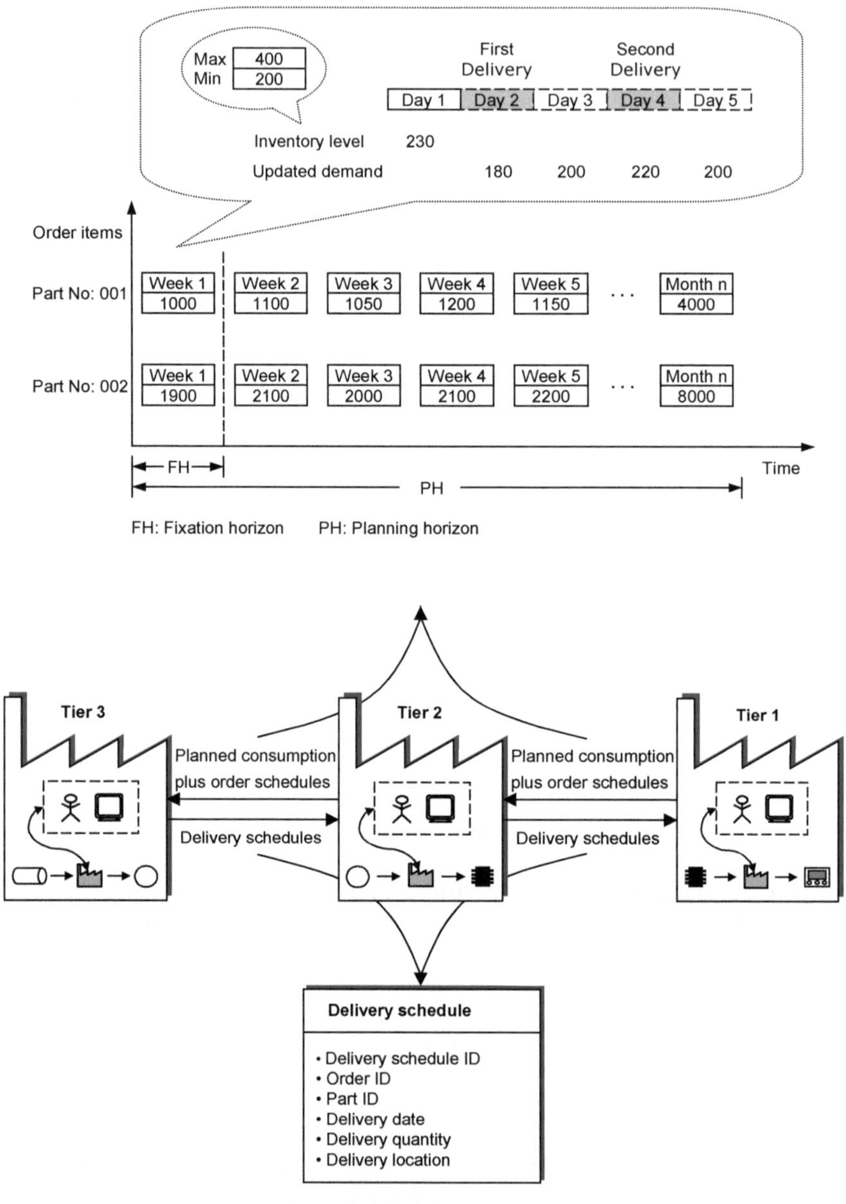

Figure 3.8. OOP process outline

Based on the accurate demand and inventory information, the upstream company can create the delivery schedule for a specific part taking into account their current production and inventory status and manufacturing capabilities. This delivery schedule is fed back as a response to the downstream company. In case of difficulties, the upstream company should indicate how much they can fulfil and solutions for the remaining quantities and reasons for difficulties. In such cases, it triggers a negotiation process between the upstream and the downstream companies. Predefined business rules such as the fixation horizon, standard workflow structure and formalised information exchanges will improve the negotiation process.

It should be noted that the collaborative planning process does not intend to replace the internal production planning systems (*e.g.* ERP systems) within individual network companies. Instead, it takes advantage of local planning systems and communicates with them by extracting and writing back relevant data, which accords with the decentralised coordination architecture for the Co-OPERATE system.

3.5.4 Visibility of Order Progress

Vision

To enhance the visibility of order progress by continuously monitoring the production process and immediately providing meaningful, formalised information on order status, while alerting customers in case of exceptions (Figure 3.9).

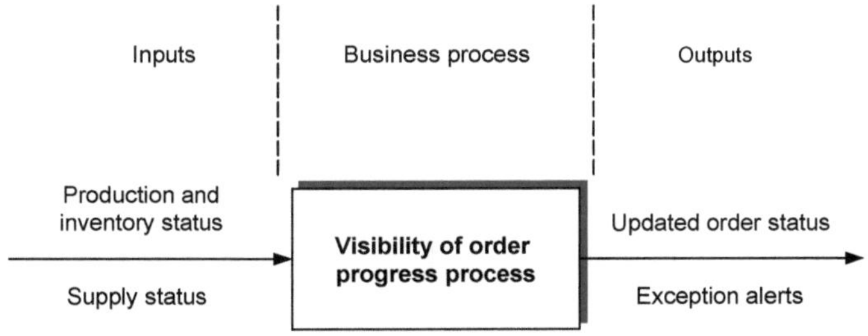

Figure 3.9. Basic diagram of the VOP process

Objectives

- Improve customer service (fulfilled deliveries) by monitoring and providing order status in near real time across the network.
- Increase the length of time for response through early detection and alerting of delivery problems.
- Reduce manual work by providing a formalised way of reporting order status.

Performance Indicators

- Percentage of early warnings sent to customers for delivery problems.

- Average length of advance time of early warnings for delivery problems.
- Percentage of fulfilled deliveries.

Process Outline

If the *operational ordering and planning* process is to make customers' demand clearer to their suppliers, the main objective of the *visibility of order progress* process is for suppliers to provide their customers with near-real-time, formalised supply information. The combination of these two processes makes demand and supply information visible across the supply network. Figure 3.10 outlines the VOP process, which is mainly for the supplier to provide the customer with up-to-date order status information and to alert the customer in case of delivery problems.

To make supply information visible in near real time, the supplier needs to continuously track work orders and inventory levels to provide relatively accurate and up-to-date production and inventory information. The production information is

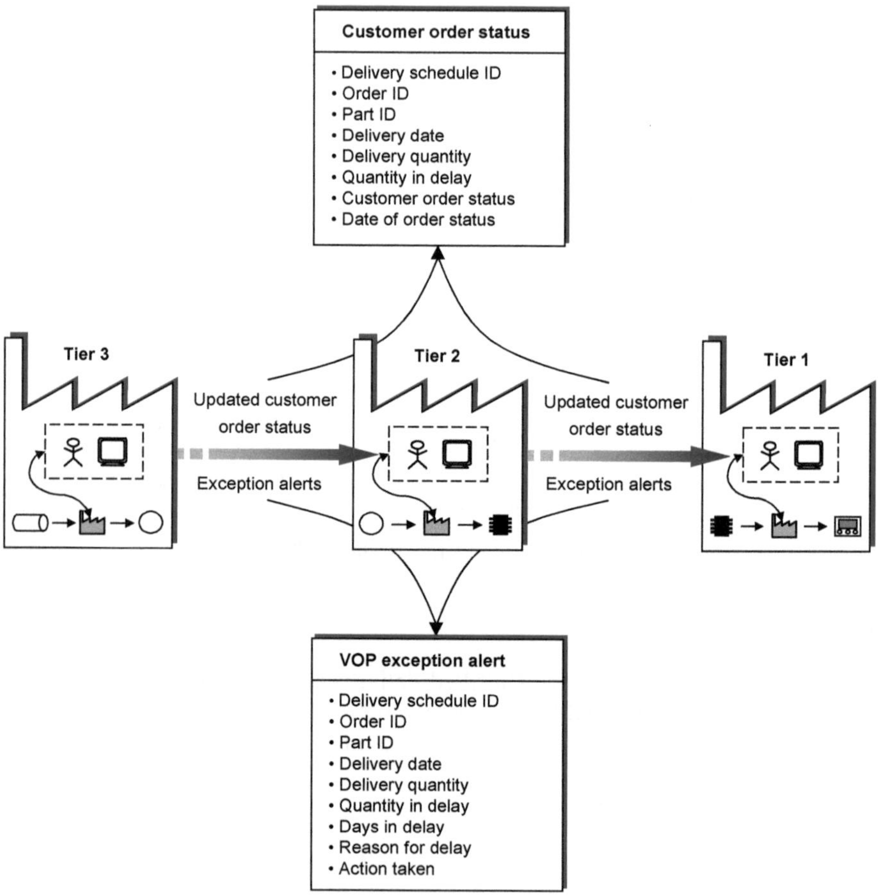

Figure 3.10. VOP process outline

usually extracted from local shop floor management systems, such as manufacturing execution systems or other internal tracking systems, or even a manual system in the case of a few production stages, while inventory information is pulled out of local ERP systems or warehouse management systems.

However, tracking work orders and inventory levels is not sufficient for the supplier to provide meaningful information on the progress of their customer orders. To achieve this, there are two major issues to be addressed by the supplier. The first one is how to dynamically match external customers' orders with internal manufacturing batches, which is especially critical in a consumption-driven supply network. Generally, an external customer order does not correspond directly to a manufacturing batch since the supplier may have several customers ordering the same or a functionally compatible product with similar requirements on delivery time. In such cases, a matching table between external customer orders and internal inventory and manufacturing batches coupled with corresponding rules will automate or at least significantly improve the allocation process. Nevertheless, for highly customised products (one product for only one customer), the automation of the matching process is quite straightforward.

Once this dynamic matching process has been established, the second issue is how to detect the delay of order progress early enough and give an 'early' warning (*e.g.* at least one week before the due date) to the affected customer whenever warranted. Thus, a milestone approach has been employed to track the internal progress against standard lead-times of the milestones, which are determined according to the local production process for a specific product and the criticality of the product. If the actual order progress is detected to be lagging behind the standard process, an internal exception alert is first generated so that the production manager can analyse the consequence of the delay. If the results of analysis indicate that the delay cannot be recovered at the later stages of production and/or be complemented by available buffers, an external exception alert is to be sent to the customer, which triggers the eXH process.

Compared with current order tracking practices in industries, which mainly focus on the logistics process, such as Dell Computers and Amazon.com, this process

Table 3.1. Traffic light approach to express customer order status

Colour	Meaning	Actions to be taken
Green	The customer order is progressing normally	No actions need to be taken by the supplier or the customer
Yellow	The customer order is slightly delayed or the quantity falls short, but can be recovered at the later stages of production or can be accommodated by buffer stock	The supplier needs to take some internal actions to recover the delayed customer order (*e.g.* by expediting the order), but the customer does not need to take actions
Red	The customer order is severely delayed, and can neither be recovered at the later stages of production nor accommodated by buffer stock	In this case, the customer order status needs to be reviewed by the supplier expeditor before an external exception alert is sent to the customer, which triggers the collaborative exception handling process

emphasises the holistic view of a manufacturer's delivery capability, including their own suppliers' delivery capabilities. On the other hand, instead of requiring the customer to understand the supplier's internal processes in detail, it provides simplified, meaningful information on order progress. Hence, the traffic light approach is employed to signal the general status of order progress, as shown in Table 3.1.

3.5.5 Exception Handling

Vision

To provide both customer and supplier expeditors an efficient, robust and formalised tool when dealing with rush orders and missed deliveries (Figure 3.11).

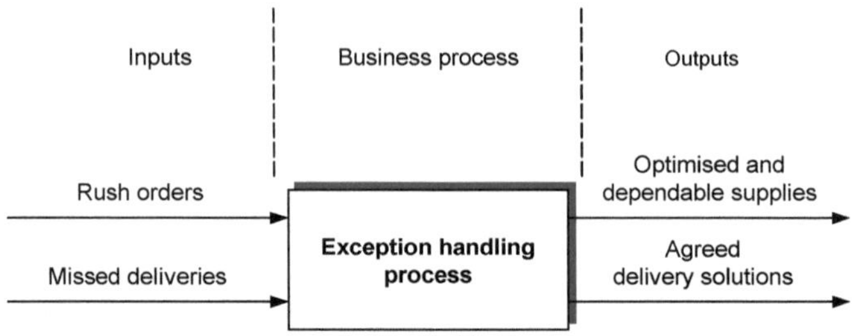

Figure 3.11. Basic diagram of the eXH process

Objectives

- Reduce the 'fire fighting' pressure and the handling time by providing formalised ways of information exchange, including information transferring and processing and negotiation.
- Optimise the delivery lead-time, cost and success rate of rush order fulfilment by considering current stock levels, costs and locations of available suppliers.
- Reduce the frequency of material shortages and the associated costs with them by resolving missed deliveries as early as possible.

Performance Indicators

- Frequency of material shortages.
- Percentage of rush orders that are successfully fulfilled.
- Percentage of missed deliveries that are successfully resolved.

Process Outline

The visibility of order progress process performs only the production monitoring functionality for production control. To achieve the full production control of the network, it needs a clear route to resolve delivery problems detected and reported by

VOP. Therefore, the *exception handling* process is logically the next step of the VOP process to close the control loop. Figure 3.12 illustrates two basic exceptional situations in the eXH process (*i.e.* missed deliveries and rush orders).

Usually, unforeseen events or 'exceptions' in the supply chain (*e.g.* late deliveries) have heavier impact on the supply chain than do normal activities (regular deliveries). Therefore, supply chain planning and execution processes need to be managed by exceptions, allowing more management attention focused on handling exceptions. This is also in line with the idea of supply chain event management (SCEM). Basically, SCEM is an application that monitors the supply chain based on business rules and automatically alerts individuals when important events occur. In general, exceptions in the supply planning and execution process can be in the short term or medium to long term, and on the demand or the supply side. To minimise their negative effects, exceptions should be detected and resolved at the planning stage. In other words, demand and supply exceptions in the medium to long term should be identified and resolved by OOP and ReFS during the collaborative planning process. Therefore, the eXH process should focus on handling short-term demand and supply exceptions, *i.e.* rush orders and missed deliveries.

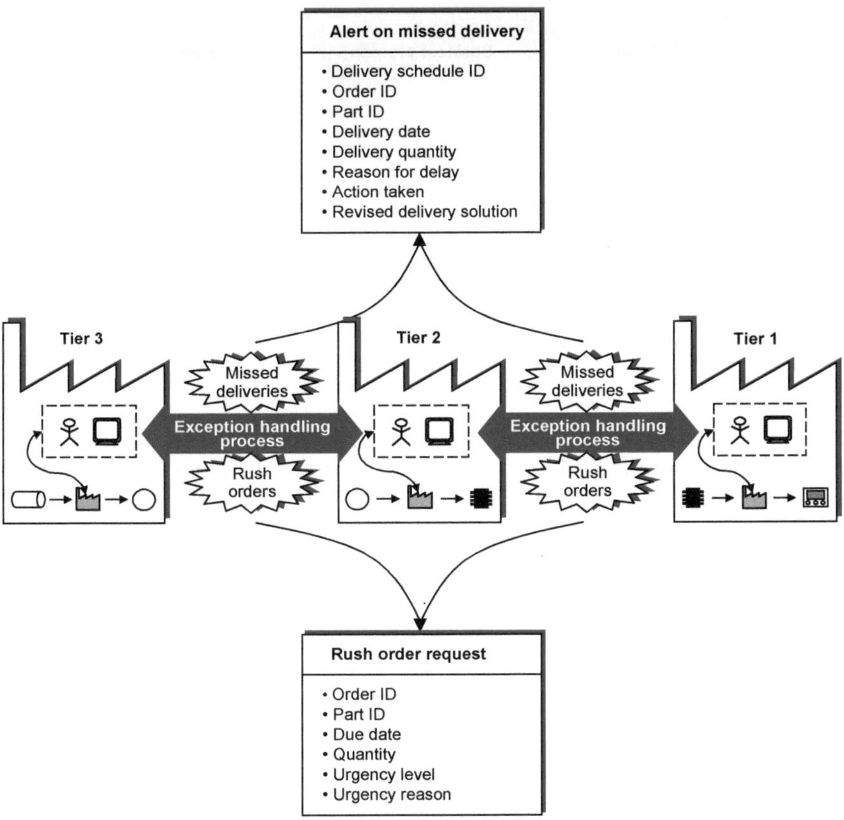

Figure 3.12. eXH process outline

There are two major issues in handling exceptions in the supply network. One is that there is a need to establish standard exception handling processes via fast and effective communications, which facilitate quick and correct decision making. The other one is that clear escalation routes for alerting the affected business partners should be established to allow fast propagation of exception alerts across the relevant part of the network. Nevertheless, exceptions should be contained at the lowest level to minimise their negative effects on the network level. For example, if an internal exception alert signals that certain demand fulfilment has slipped behind schedule, the node should take internal measures (*e.g.* expediting late batches) to contain the problems. Only when internal efforts are not sufficient to recover the problem can the node alert the affected customer.

3.5.6 Request and Feasibility Studies

Vision

To provide a fast response to request on new orders or large order changes across the relevant part of the supply network (Figure 3.13).

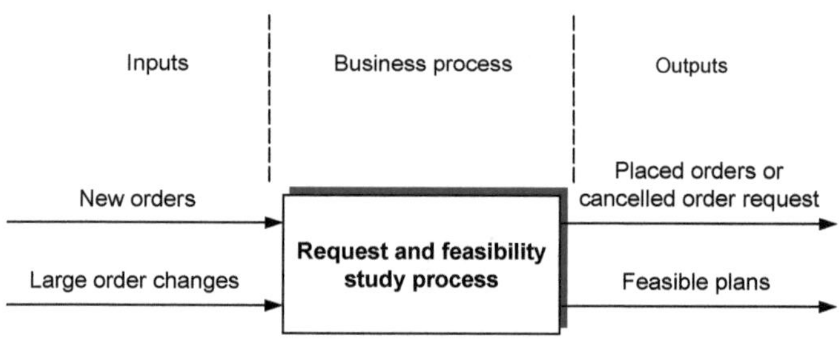

Figure 3.13. Basic diagram of the ReFS process

Objectives

- Reduce the response time to request on new orders or large order changes.
- Facilitate more feasible and dependable plans in the operational ordering and planning process.

Performance Indicators

- The average response time for an incoming inquiry from a customer.
- The average response time for the network answering an inquiry from the OEM.

Process Outline

The *request and feasibility studies* process is complementary in functionality to the *long-term production planning* and the *operational ordering and planning* processes. It is a planning tool that helps to make more feasible plans especially in the medium or long term. Within a node, this process is in correspondence with available-to-promise (ATP), which determines if a customer's order request can be met from existing inventory and production orders, and capable-to-promise (CTP), which checks if a customer's request date can be met from available plant capacity. However, ReFS focuses more on a fast feasibility check on the network level than on the node level. Figure 3.14 illustrates the ReFS process.

The process starts with a node in the network that is often an OEM or a first-tier supplier. It sends a request to its immediate suppliers for a new order or large order changes on due dates and/or qualities that override business rules stipulated in contractual agreements. Each node that has received a request breaks down the request into internal and external requirements using the BOMs and checks its internal capacity availability and other capabilities. The external supply requirements are transferred to their own suppliers who in turn perform the same process of internal material requirements planning and external supply requests. This way, each node queries its internal capacity model and production plans and then gives a quotation quickly to its direct customer, who consolidates the quotations from their suppliers to create a quotation for their own customer until a final answer is achieved. If the request is probably to be followed by the placement of an order, ReFS can reserve the capacity for the relevant part of the supply network.

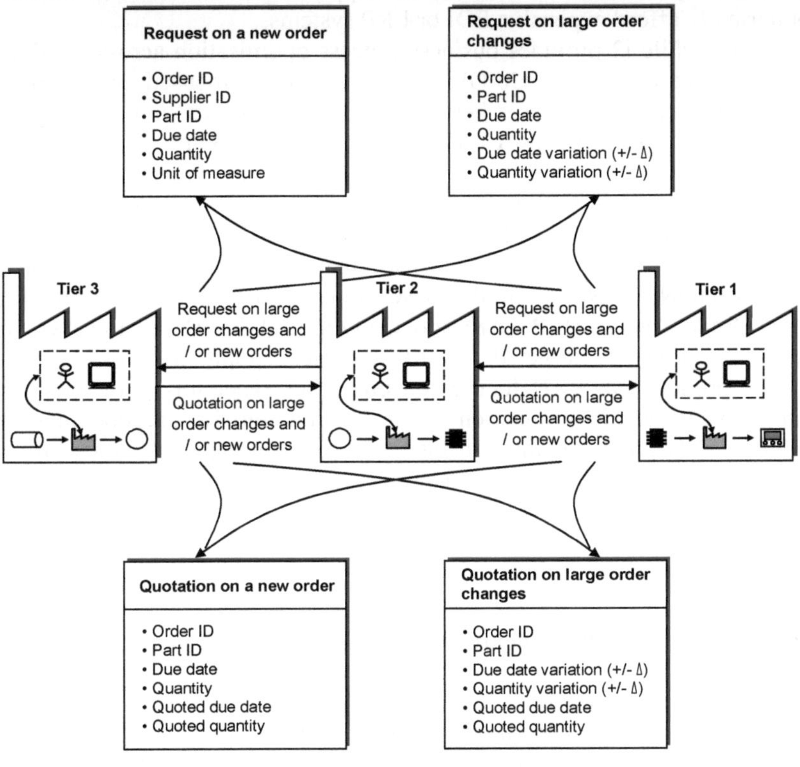

Figure 3.14. ReFS process outline

As the ReFS process often entails the propagation of request on demand and quotation on delivery capability across the relevant part of the network, an agent-based approach has been adopted so that the result of the ReFS process can be obtained in a reasonably short period. Among others, one key issue is how to construct a capacity model for a specific node, which should be configurable to capture the specificity of each node in the network. A practical, efficient way to do this is through an algorithm repository that contains flexible and user-configurable algorithms for capacity modelling. These configurable algorithms should consider specific business scenarios, level of detail of capacity models and the extent to which users' knowledge can be incorporated in the capacity models [3.35].

3.5.7 Comparison of Co-OPERATE with Other Solutions

The decentralised coordination architecture for the Co-OPERATE system has two important aspects: (1) there is no central optimisation for the total supply network in view of incommensurate decision variables (*e.g.* cost, delivery lead-time) and loss of local decision autonomy for each node in the supply network, and (2) Co-OPERATE supports existing local planning systems (*e.g.* ERP systems) as local business processes are extremely heterogeneous and so are local planning systems. By focusing on the inter-company planning and execution coordination processes, it is expected to get SMEs involved in the supply chain coordination process that cannot normally afford expensive EDI or ERP systems.

Therefore, while i2 provides business process optimisation across multiple tiers of suppliers and customers, Co-OPERATE is mainly focused on planning and execution coordination processes between a company and its immediate customers and suppliers. Meanwhile, as Co-OPERATE targets complex supply networks where companies have partnerships based on long-term contracts, it does not take into account pricing optimisation (a key feature of Manugistics EPO) or promotional activities that are covered by CPFR.

From the business process perspective, there is some overlap, for instance, between RosettaNet's Notify of Purchase Order Update (PIP 3A7) and Request Purchase Order Change (PIP 3A8) and Co-OPERATE's eXH. However, as Co-OPERATE is primarily targeted towards complex supply networks with highly customised products, its eXH focuses on providing expeditors with relevant local planning, inventory and production status information for decision making and streamlining eXH processes through fast, effective negotiations, which differs Co-OPERATE significantly from RosettaNet.

3.6 Implementation and Evaluation

3.6.1 Process Design and Implementation

The four focused coordination processes of the Co-OPERATE system (*i.e.* OOP, VOP, eXH and ReFS) were modelled with integration definition for function modelling (IDEF0). In alignment with the top-down design approach, the IDEF0 modelling technique uses a hierarchical approach for functional decomposition

[3.36]. While function modelling provides a detailed process view of the system through functional decomposition, scenario analysis tries to describe what the system should achieve from the user's point of view. It bridges the gap between business process design and software development. The result of scenario analysis is sequence diagrams that show interactions between systems (*e.g.* ERP systems) and company users (*e.g.* expeditors) in a time sequence.

The overall implementation process of the Co-OPERATE system consists of two stages and in the form of two prototypes: concept and final. The concept prototype were designed to demonstrate the basic business processes to the project partners and helped to get some feedback from partners' experts involved in the evaluation of the concept prototype. After the implementation of the concept prototype, the most popular implementation tools for Web application development were re-evaluated. Microsoft's active server pages (ASP) was finally chosen for implementing the final prototype of OOP, VOP and eXH. This was mainly because these three processes focus on improving short- to medium-term supply chain planning and execution activities. Meanwhile, the implementation of ReFS was based on an agent framework – FIPA-OS (Foundation for Intelligent Physical Agents – Open Source). The reasons for this choice were twofold: (1) ReFS necessitates fast responses from individual network companies, and (2) it can also be used as an independent application for decision support in network collaboration.

With reference to the ASP approach, a three-tier client/server architecture was adopted for the implementation of the coordination processes. The Web browser and the Web server comprise the first tier. Microsoft's Internet Explorer was the preferred Web browser. Microsoft's Internet Information Server (IIS) was chosen to implement the Web server. The application server represents the second tier, executing server-side business logic and accesses the relational database management system (RDBMS), which is the third tier. Microsoft's IIS and ASP engine implement this functionality and the integration between the Web server and the application server. Data exchanges between the RDBMS and ERP systems or local legacy systems can be conducted via XML files.

3.6.2 Pilot System Evaluation

To facilitate the elicitation and incorporation of feedback from the industrial partners in the project, a progressive, multi-phase evaluation strategy was employed for the Co-OPERATE system. Before the actual implementation process, the proposed coordination processes had been validated from a conceptual perspective through multiple communications with the industrial partners and verified by the evaluation of the concept prototype. On the other hand, a simulation study was conducted to provide a quantitative assessment of the potential benefits of the proposed consumption-based planning approach in comparison with the conventional transactional-order-based planning approach.

The evaluation of the Co-OPERATE system involves experts from industrial partners who are highly experienced with the existing methods and procedures. Company users were trained with user manuals and scenario scripts. After the user trial and verification phase, project members participated in an evaluation workshop to make sample runs of real-world business scenarios and to judge the system's performance under these conditions. The results of the scenario-based testing and evaluation were captured by using detailed questionnaires. The main strengths of this system evaluation approach include analysing workflow in relation to existing working practices and IT tools.

As the evaluation work was carried out by the company users in the project, it may be argued that objectivity of their evaluation was compromised. However, the project team was fully aware of such potential criticisms. They were counteracted by basing the evaluation on real, historical data, such that the proposed processes can be analysed in comparison with the existing ones. Meanwhile, attention was paid to engage staff who are highly experienced with the existing methods and procedures, and who can therefore be regarded as qualified experts.

The feedback from the business experts verified that the pilot system fulfilled all business and system requirements proposed in Section 3.3. From the business perspective, the evaluation results showed that the pilot system could improve demand and supply information visibility, provide early warning of disturbances, and enhance the capability of collaborative problem-solving. Particularly on the demand side, accurate demand information can be communicated in near real time through streamlined business processes across the supply network, reducing the bullwhip effect. On the supply side, delivery performance in network companies is expected to be significantly improved through early identification and handling of delivery problems, reducing inventory levels and material shortages.

From the system perspective, first, the pilot system could be scaled to cater for the growth of the supply network, or easily re-configured to reflect the changes of the network structure due to its open and flexible system architecture. Second, the integrated database of the pilot system ensures data consistency and minimises data redundancies, making it a very stable and reliable system. Third, the pilot system features user-friendly interfaces and secure access by virtue of data ownership and user authorisation on different levels (*e.g.* system administrators, expeditors and production planners). Finally, though it may be argued that the local connection to the Internet influences the speed in accessing and navigating the system, by focusing on the key network coordination processes and minimising client–server transactions, the speed of the pilot system was enhanced and was satisfactory from the users' perspective.

3.7 Conclusions and Future Work

There are many issues to be addressed to achieve agile manufacturing in complex supply networks in terms of organisations, people and technology. From the whole supply chain perspective, however, key to the success of an agile supply chain is the fast flow of material and information both upstream and downstream along the supply chain. Essentially, the problems in any supply chain process can be classified into: any time delay of material and information. If these two problems can be resolved, the entire supply chain and its constituent companies will stand to gain a significant competitive advantage. For this purpose, it is paramount to achieve an effective integration and coordination of all the participating companies at different levels of cooperation.

The presented research work represents an important step towards improved coordination of production planning and control activities in complex distributed supply networks, which is supported by collaborative business processes and corresponding Web-based systems. Specifically, the key contributions of the presented work include a research framework for SNC, a decentralised system architecture for SNC, the high-level design of four focused coordination processes, and the implementation of the Co-OPERATE system.

The implemented system has some important advantages over expensive commercial SCM systems. For example, by focusing on the key inter-company planning and execution coordination processes, it is expected to significantly reduce the cost of ownership of the Web-based SNC system. This way SMEs, who usually do not have expensive EDI, ERP or SCM systems, can be involved in the network coordination process. On the other hand, the developed framework for SNC, the decentralised system architecture and the system implementation approach are generally applicable to the coordination of production planning and control activities in a manufacturing network environment.

The implemented SNC system can be extended as a demonstrator to show how best practices in SCM can be achieved through enhanced information sharing and integration with business partners and what are the expected benefits of improved coordination to all the parties involved. For example, the missed delivery handling process will get more and more responsive to delivery problems when information sharing migrates from the modest level, *e.g.* only the supplier has the visibility of inventories on the customer side, to the high level, *e.g.* the supplier has full control of inventories and clear visibility of the customer's short-term production schedule.

It can be argued that the pilot system could be implemented in a different way. However, the presented research work has formed a fertile ground for future research in the area of network coordination. First, to get the system up and running in a real industrial environment with the maximum possible cost–benefits in a reasonable period, it is essential to address the issue of the integration of the SNC system with local legacy systems. Second, the system could be extended to include other important coordination processes, *e.g.* long-term business planning and network performance management.

Acknowledgement

The author would like to acknowledge all the comments and suggestions made by his colleagues in the Co-OPERATE project.

References

[3.1] Sheridan, J.H., 1993, "Agile manufacturing: stepping beyond lean production," *Industry Week*, **242**(8), pp. 30–37.

[3.2] Goldman, S.L. and Nagel, R.N., 1993, "Management, technology, and agility: the emergence of a new era in manufacturing," *International Journal of Technology Management*, **8**(1–2), pp. 18–38.

[3.3] Gunasekaran, A., 1998, "Agile manufacturing: enablers and an implementation framework," *International Journal of Production Research*, **36**(5), pp. 1223–1247.

[3.4] Gunasekaran, A., 1999, "Agile manufacturing: a framework for research and development," *International Journal of Production Economics*, **62**, pp. 87–105.

[3.5] Yusuf, Y.Y., Sarhadi, M. and Gunasekaran, A., 1999, "Agile manufacturing: the drivers, concepts and attributes," *International Journal of Production Economics*, **62**, pp. 33–43.

[3.6] Schönsleben, P., 2000, "With agility and adequate partnership strategies towards effective logistics networks," *Computers in Industry*, **42**, pp. 33–42.

[3.7] Korhonen, P., Huttunen, K. and Eloranta, E., 1998, "Demand chain management in a global enterprise – information management view," *Production Planning and Control*, **9**(6), pp. 526–531.

[3.8] Stevens, G.C., 1989, "Integrating the supply chain," *International Journal of Physical Distribution & Materials Management*, **19**(8), pp. 3–8.

[3.9] Murtaza, M.B., Gupta, V. and Caroll, R.C., 2004, "E-marketplaces and the future of supply chain management: opportunities and challenges," *Business Process Management Journal*, **10**(3), pp. 325–335.

[3.10] Møller, C., 2005, "ERP II: a conceptual framework for next-generation enterprise systems?," *Journal of Enterprise Information Management*, **18**(4), pp. 483–497.

[3.11] Puschmann, T. and Alt, R., 2005, "Developing an integration architecture for process portals," *European Journal of Information Systems*, **14**(1), pp. 121–134.

[3.12] Karkkainen, M., Laukkanen, S., Sarpola, S. and Kemppainen, K., 2007, "Roles of interfirm information systems in supply chain management," *International Journal of Physical Distribution & Logistics Management*, **37**(4), pp. 264–286.

[3.13] Huang, A., Yen, D.C., Chou, D.C. and Xu, Y., 2003, "Corporate applications integration: challenges, opportunities, and implementation strategies," *Journal of Business and Management*, **9**(2), pp. 137–150.

[3.14] Loh, T.C., Koh, S.C.L. and Simpson, M., 2006, "An investigation of the value of becoming an extended enterprise," *International Journal of Integrated Manufacturing*, **19** (1), pp. 49–58.

[3.15] Gould, L.S., 1998, "Introducing APS: getting production in lock step with customer demand," *Automotive Manufacturing and Production*, May, 54–58.

[3.16] Stadtler, H., 2005, "Supply chain management and advanced planning – basics, overview and challenges," *European Journal of Operational Research*, **163**, pp. 575–588.

[3.17] McGaughey, R.E. and Gunasekaran, A., 2007, "Enterprise resources planning (ERP): past, present and future," *International Journal of Enterprise Information Systems*, **3**(3), pp. 23–35.

[3.18] Zheng, S., Yen, D.C. and Tarn, J.M., 2000, "The new spectrum of the cross-enterprise solution: the integration of supply chain management and enterprise resources planning systems," *Journal of Computer Information Systems*, **41**(1), pp. 84–93.

[3.19] http://www.rosettanet.org/

[3.20] http://www.i2.com/

[3.21] Booker, E., 1999, "XML greases supply chain," *Internet Week*, **778**, pp. 1–2.

[3.22] Olson, E.G. and Williams, P.R., 2001, "E-commerce with web-based standards: a case study in implementing RosettaNet process standards at Avnet, Inc." *Technical White Papers*, http://www.rosettanet.org/.

[3.23] Sodhi, M.S., 2001, "Applications and opportunities for operations research in Internet-enabled supply chains and electronic marketplaces," *Interfaces*, **31**(2), pp. 56–69.

[3.24] Skjøtt-Larsen, T., Kotzab, H. and Grieger, M., 2003, "Electronic marketplaces and supply chain relationships," *Industrial Marketing Management*, **32**, pp. 199–210.

[3.25] Hahn, C.K., Duplaga, E.A. and Hartley, J.L., 2000, "Supply-chain synchronisation: lessons from Hyundai Motor Company," *Interfaces*, **30**(4), pp. 32–45.

[3.26] Xu, H.Q., Besant, C.B. and Ristic, M., 2003, "System for enhancing supply chain agility through exception handling," *International Journal of Production Research,* **41**(6), pp. 1099–1114.

[3.27] Vokurka, R.J. and Lummus, R.R., 2000, "The role of just-in-time in supply chain management," *The International Journal of Logistics Management*, **11**(1), pp. 89–98.

[3.28] Baliga, J., 2001, "Supply chain collaboration will determine future success," *Semiconductor International*, January, pp. 81–86.

[3.29] Maltz, A.B., Grenoble, W.L., Rogers, D.S., Baseman, R.M., Grey, W. and Katircioglu, K.K., 2000, "Lessons from the semiconductor industry," *Supply Chain Management Review*, November/December, pp. 42–52.

[3.30] Co-OPERATE deliverable D3, 2000, *Solution Scope*.

[3.31] Xu, H.Q., Ristic, M., Besant, C.B. and Pradoux, C., 2005, "A Web-based system for manufacturing co-ordination in complex supply networks", *International Journal of Production Research,* **43**(10), pp. 2049–2070.

[3.32] Rupp, T.M. and Ristic, M., 2000, "Fine planning for supply chains in semiconductor manufacture," *Journal of Materials Processing Technology*, **107**(1–3), pp. 390–397.

[3.33] Camarinha-matos, L.M., Afsarmanesh, H., Garita, C. and Lima, C., 1998, "Towards an architecture for virtual enterprises," *Journal of Intelligent Manufacturing*, **9**, pp. 189–199.

[3.34] Co-OPERATE deliverable D9, 2000, *System Architecture*.

[3.35] Co-OPERATE deliverable D10, 2000, *High-level System Design*.

[3.36] Rensberg, A.V. and Zwemstra, N., 1995, "Implementing IDEF techniques as simulation modelling specifications," *Computers and Industrial Engineering*, **29**(1–4), pp. 467–471.

4

Enterprise Network and Supply Chain Structure: the Role of Fit

Federica Cucchiella* and Massimo Gastaldi

Department of Electrical and Information Engineering
Faculty of Engineering, University of L'Aquila
Monteluco di Roio, 67040 L'Aquila, Italy
Emails: federica.cucchiella@univaq.it; massimo.gastaldi@univaq.it

Abstract
Nowadays, it is necessary to organise an enterprise network according to the integration of all enterprise operations and to develop a structure where knowledge is organised in order to identify the need for changes in the enterprise. This chapter mainly focuses on an enterprise architecture with the scope to define the structure and operation of an organisation. The intent of the enterprise architecture is to determine how an organisation can most effectively achieve its current and future objectives and, more specifically, in this chapter the enterprise architecture development is viewed as largely the process of decision making under uncertainty and incomplete knowledge. Taking value maximisation as the primary objective of the enterprise architecture decision-making process, this chapter attempts to develop guidelines for value enhancement. It is assumed that part of the value of the enterprise architecture initiative is in the form of embedded options (real options), which provide architects with the flexibility to change operation plans when uncertainties are resolved over time.

4.1 Introduction

Nowadays, the concept of centralised business systems planning has become less popular; on the contrary, the rapid change connected in the e-business environments, along with the more decentralised nature of organisational resources, requires not only an increasing flexibility and adaptability but also a more cohesive and value-creating role of information systems infrastructure and its management [4.1–4.6]. More specifically, it is critical to synchronise business goals and strategies, governance principles, organisational structures, processes and data, business applications, their systems and databases, and network infrastructure (internal and external to the enterprise).

* Dedicated to my brilliant and handsome husband without whom I would be nothing.

Only in such a way it is possible to proceed with the correct management of the decision-making process related to the investment in manufacturing capacity that has a relevant rule for the successful management of a firm. A firm is often called upon to make difficult decisions related to its manufacturing organisation, and, to the firm's operations executives, these decisions have, indeed, a significant impact on the firm's ability to compete positively in the actual business environment, which is characterised by a high level of complexity and uncertainty. One of the biggest challenge that a firm – also when organised in a network – has to face is related to the manufacturing capacity required for products that are under development. In this situation, it needs to optimise the problem from among multiple alternatives, which include building internal capabilities, outsourcing capabilities to a contract manufacturing organisation, or a combination of the two.

These decisions also involve significant capital investment and the opportunity cost of allocating funds away from other important initiatives, which can have relevant consequences for all the network actors. For these reasons, this chapter aims to present a new methodology that, using a real option approach, explores whether the management can optimise the strategic structure investment and choose the best strategy for the manufacturing management [4.7–4.10].

Traditional project evaluation based on discounted cash-flow analysis ignores the upside potentials to an investment from managerial flexibility and innovations. A real option approach that borrows ideas from financial options offers a fresh perspective. It views investment strategy as a series of options that are continually being exercised to achieve both short- and long-term returns on investment. Advocates of real options suggest that the thinking behind financial options may be extended to opportunities in real markets that offer, for a fixed cost, the right to realise future payoffs in return for further fixed (that is, independent of the asset value) investments, but without imposing any obligation to invest. Real options are important in strategic and financial analysis because traditional valuation tools such as net present value (NPV) ignore the value of flexibility. Viewing a corporation as a set of businesses, each with an NPV, creates a static picture of the existing investments and opportunities.

To facilitate understanding, after considerations of the enterprise architectures in Section 4.2, the main historical developments of these architectures are discussed in Section 4.3 with some generic enterprise architectures analysed. Sections 4.4 and 4.5 present an overview of the available approaches and recommendations, respectively, regarding what academic, industrial and standards communities should do in order to overcome the complexity arising from integrating the information and material flow throughout the enterprises. Through a generic model, each major component of an enterprise architecture is defined, and their purpose and use are introduced in Section 4.6.

Nowadays, in the highly competitive environment, more and more enterprises are organised into a network type structure. It is necessary to optimise not only the internal enterprise processes but also the relationships that link each network enterprise. As a consequence, a firm's enterprise architecture must be developed in an extended way useful for meeting the needs of an extended network structure. In this case, the management has to face new difficulties (see Section 4.7) that may be solved through the use of the real option theory (given in Section 4.8). Traditional approaches in investment analysis fail to capture the flexibility, risks and contingencies that have the potential to impact on business decisions. Indeed, it is possible to optimise an organisation's decision to investment in an enterprise architecture if the investment includes actions embedded with options. This chapter presents in detail a real option framework developed to manage investments in an extended enterprise architecture.

4.2 Relevance of Enterprise Architecture

One of the most important characteristics of today's enterprises is that they are facing a rapidly changing environment with no long-term provisions. In this situation, the only way to modify the enterprise organisation to continuously fit the market trend is to create a reactive firm where the changes and adaptations are dynamic than something that is occasionally forced to happen inside the enterprise. In order to meet such a requirement, it is necessary to remove all organisational barriers and increase the interoperability for creating a synergic situation among all the firms operating inside the network. Only in this way can the firm operate in a more efficient and adaptive manner. It is also necessary to organise today's network by integrating all the enterprise operations and develop a structure where knowledge is organised in order to identify the need for changes in the enterprise.

An enterprise architecture is a mechanism that allows the increasing complexity that nowadays typifies a firm to be addressed. It is a challenging and confusing concept based on various heterogeneous architecture proposals. Moreover, there is no agreed terminology, and probably for this reason it is difficult to find an efficient application [4.11, 4.12]. On the basis of the contextual usage, there are several meanings of *architecture*:

- a formal description of the component belonging to a system;
- the structure of the components, their interrelationship, their guidelines for design and evolution in the future;
- organisational structure of a system or its component.

An enterprise architecture must be organised to successfully support the structure in its complexity, giving some indications on the actions to undertake on the whole system. It defines the components that can form the whole system and furnishes a program of actions; starting from these actions, it is possible to develop the whole system. The concept of architecture is understood in a purely engineering meaning and it is finalised to support the management of the complexity and risk of the whole system. Its utility is therefore even more evident in the actual competitive context where the sources of uncertainty are numerous, for example, technology, dimensions, interface, reference context, *etc*. For the organisation of large systems, it is necessary to proceed with a study conducted to an elevated level of abstraction for the guaranteed homogeneity to the whole structure. As a consequence, it is necessary to be able to describe the enterprise as a modular system, composed of subsystems that can be sub-divided. Starting with such an organisation, it is possible to analyse the actual state of the system (AS-IS model) and to understand (TO-BE model), as such a system must be modified in order to reach the desired state. To build an enterprise architecture with an elevated level of abstraction allows the fundamental principles for the future design of the structure to be planned, so that the points of strength and weakness of the whole system can be easily analysed. The software engineering community considers that architecture is the fundamental organisation of a system embodied in its components, their relationships to one another and to their environment, and the principles guiding its design and evolution. Enterprise architecture is seen as complementary to software architecture, to document the system-wide organisational and business context in which the software operates.

4.3 The IFIP–IFAC Task Force

In 1992, IFIP and IFAC established a joint task force to review the existing approaches to enterprise integration and to make recommendations to the industrial and research community. The task force was chaired by Professor Emeritus T.J. Williams of Purdue University (1992–1996), and by Professor Peter Bernus of Griffith University (1996–2002). Members comprised representatives from both the industrial and research communities, with several researchers coming from management or consultant positions within industry.

Enterprise integration has steadily evolved from the 1990s with the increasing need to integrate the information and material flows throughout an enterprise. There have been separate accomplishments in the area of manufacturing both in design and production, including numerical control (NC) systems, computer-aided design/manufacturing (CAD/CAM) systems, computer-integrated manufacturing (CIM) systems, manufacturing cells, material requirements planning (MRP) and production scheduling systems. In the area of business support, integrated systems were developed for accounting, financial planning, human resource management, decision support, *etc.*

By the mid-1980s, it had become evident that isolated efforts led to systems that could not easily communicate and thus elaborate islands of automation had to be maintained, which could not easily be integrated. Today, industry still feels the problems arising from these isolated efforts, with many isolated 'legacy' applications still in use.

At the same time, it was realised that considering only the automated parts of material and information processing was no longer tenable, because the human element in the enterprise was still the most important part, and thus an approach was needed that dealt with both the human and automated parts of the enterprise. Therefore, the complete enterprise, as any other human-made system, needed to be properly designed, which required methods to do this. Over time, two approaches have emerged.

The first approach was based on generic models or designs (called *architectures*) that could subsequently be implemented as information systems products (or families of products), by incorporating most or all information processing tasks in an enterprise (especially its management). The resulting systems are called enterprise resource planning (ERP) systems. Also, specifically for CIM systems, a number of CIM reference models were developed, which tried to systematise the functional building blocks of a CIM system. The problem, however, was that there were several dozen competing models, all of which failed to achieve any industry-wide acceptance, or standard status. The appeal of this approach, on the other hand, was that it could easily be turned into products (software systems).

The second approach was based on the recognition, similar to many engineering disciplines (such as chemical engineering, manufacturing engineering, software engineering, civil engineering, *etc.*), that enterprise engineering should be based on the so-called life-cycle approach. According to this approach, in order to create an integrated enterprise, the enterprise creation activities (and thus methodologies) must be extended to the whole life of the enterprise, from its inception until it was no longer operating (*i.e.* when it was decommissioned). Several such architectures were developed – some by groups with a manufacturing systems background, and some with an information systems background.

4.4 The First IFIP–IFAC Mandate

The first mandate of the IFIP–IFAC joint task force was an overview of the available approaches, which then made recommendations regarding what the academic, industrial and standards communities should do in order to overcome the complexity arising from its environmental context.

The results were published in a task force report entitled *Architectures for Enterprise Integration* [4.13] and presented to both IFIP and IFAC. This report contains contributions from several task force members, and summarises the findings and recommendations [4.14] of the task force. The major findings are as follows.

There are two types of 'architecture' available to support enterprise integration. Type I architectures are models of information systems, which integrate the information flow of an enterprise. Unfortunately, these models

- are of a very high level, and
- give rise to many incompatible solutions.

Type II architectures are life-cycle models of the enterprise, systematising the activities that are needed in order to create integrated enterprises. The Type II (life-cycle) architectures allow the enterprise to introduce all necessary methodological processes (including managerial and technical tasks) so as to evolve the enterprise in the desired direction.

It is also possible to categorise an enterprise architecture into two main groups:

- system architecture (also known as 'Type 1' architecture), which addresses the design aspects of the system; and
- enterprise-reference projects (also known as 'Type 2' architecture), which are related to the development and implementation of a project organised as an enterprise integration.

While the 'Type 1' architectures may describe a system from the point of view of its structure and behaviours, those of 'Type 2' are similar to some frameworks finalised to define all the activities and the necessary actions to draw and to build the system.

The elevated abstraction level of the enterprises architecture has consequences for the communications between stakeholders. It is possible, indeed, to represent their expectations in terms of the future path of the system, while abandoning the old setting based on detailed documentations. Only functions characteristics are described at this level, leaving data or resources to be specified in a later stage.

Architecture must be defined to serve the purposes of the network and show that such purposes can be reached and that the problems relating to the conflict among different actors who participate in the network can be solved. In order to meet this requirement, the architecture must have a structure that can be easily understood by all actors, easily checked and analysed, and easily described with a language that is comprehensible to every level of the business.

At the architecture elaboration phase, the principles used should be:

- generic when they must be applied to all the enterprises; and
- specific if instead they must be applied to a particular enterprise, and in such case they must be shared by the actors of the specific enterprise and represent the base on which future decisions can be found.

An additional aspect to be managed in the definition of an enterprise architecture is related to the architectural decisions, *i.e.* the decisions that have to be taken to organise the structure in a way rather than in another to adopt a general perspective.

According to [4.15], these decisions define the structural elements of the system and the relationships among the various actors.

4.4.1 The Historical 'Type 2' Architecture

The studies in the literature are mainly focused on the R&D of Type 2 architectures that define concepts, principles and assignments of integrated enterprises. Single enterprise is not represented in this kind of architecture, since it is already structured once it operates. Type 2 architectures do not define how operational processes, data or structure are organised. Their purpose is to define the base principles on which the structure is developed, but not to specify a system. Such models are essential to build a network from strongly integrated homogeneous elements. A brief description of the architectures available in the literature is presented here. Since such systems require strong economic commitment, the purpose of this chapter is also to show how such investments can be managed with the support of modern techniques developed for investment management. Within this context, we will examine some real cases of investments in new technologies and show how the investments can be positively managed by employing real options theory. To this end, although the enterprise architectures have been well studied in the literature, the optimal management of the required investments still constitute a field that remains to be explored.

The first study on enterprise architectures can be traced back some eighteen years when a group of European and American researchers began work on the first base frameworks for the development of enterprise architectures; among these structures, some are particularly noteworthy even today.

Three major Type 2 architectures were identified, *i.e.* Purdue Enterprise Reference Architecture (PERA), Computer Integrated Manufacturing Open System Architecture (CIMOSA), and GRAI Integrated Methodology (GIM) or Graphs with Results and Actions Inter-related Integrated Methodology. The conclusion of the joint task force was that these three architectures all contained components that were deemed necessary. Thus, there was a need for generalisation, which would allow each architecture to be further developed towards its completion. The first attempt at generalisation is the generalised enterprise reference architecture model (GERAM) [4.14]. The task force decided to base its further works on this proposal and bring the specification of GERAM to its completion.

Note that the aim was not to develop a fourth architecture to replace the existing ones, but to create a generalisation that allows users of the existing architectures to make their architectures more complete and also to demonstrate that they all satisfy the GERAM requirements.

GERAM was developed as the ontology of enterprise architecture, defining what enterprise architectures needed to contain. Although it created a vehicle for communications between different groups of practitioners as well as taking elements from one architecture and incorporating them into another, the definition of one architecture still holds more useful details than the other in some aspects.

CIMOSA was first referenced in [4.16, 4.17], followed by PERA [4.18], GIM and ARIS (Architecture of Integrated Information Systems). These models, all of Type 2, give insights on how to model, design and implement an integrated system.

Another well-known model is the Zachman Framework for Enterprise and Information Systems Architecture developed by John Zachman of IBM in the 1980s [4.19]. This framework borrows concepts from business design principles in manufacturing, and provides a means of classifying an organisation's architecture. It draws from Zachman's experience on how change is managed in complex products such as aircraft and buildings. In today's complex business environments, many large organisations have difficulty in responding to change. Part of this difficulty is due to the lack of internal understanding of other areas of the organisation, where legacy information about the business is locked away in the minds of specific employees or in business units. The framework provides a proactive business tool that can be used to model an organisation's existing functions, elements and processes and help to manage business change. It can also be used as a thinking tool, to help organisations understand complex issues and develop appropriate business strategies. It can be used for information systems architecture and is widely adopted by systems analysts and database designers. However, John Zachman stressed that it can be extended to the entire *enterprise* architecture, and is not restricted to *information* architecture. From Zachman's framework, other enterprise frameworks have been derived, such as the Federal Enterprise Architecture Framework (FEAF), The Open Group Architecture Framework (TOGAF) and the Department of Defence Architecture Framework (DoDAF).

The Zachman framework is now in the public domain and can be used by any organisation; it is a classification schema, represented visually as a table with columns and rows. Each cell within the schema provides a unique model or representation of the enterprise. The information in each row of the schema provides a unique perspective of the enterprise. Each cell in the schema must be aligned with the cells immediately above and below it. All the cells in each row must also be aligned with each other. Each cell is unique. Combining the cells in one row forms a complete description of the enterprise from that view.

The columns represent the *interrogatives*, or questions, that are asked of the enterprise:

- what (data) – what is the business data or business information?
- how (function) – how does the business work, *i.e.* what are the business' processes?
- where (network) – where are the businesses operations?
- who (people) – who are the people that run the business, what are the business units and their hierarchy?
- when (time) – when are the business processes performed, *i.e.* what are the business schedules?
- why (motivation) – why are the processes, people or locations important to the business, *i.e.* what are the business drivers or business objectives?

The Zachman framework enables complex subjects to be distilled into systematic categories, using these six basic questions. The answers to these questions may differ, depending on the audience perspective (represented in the rows).

Each row represents a distinct view of the organisation, from a unique audience perspective. A row is allocated to each of the following audiences:

- planner – understands the business scope and can offer a contextual view of the enterprise;
- owner – understands the business model and can provide a conceptual view of the enterprise;
- builder – develops the system model and can provide a logical view of the enterprise;
- designer – produces the technology model and can provide a physical view of the enterprise;
- integrator (sub-contractor) – understands detailed representations of specific items in the business, although they may have an out-of-context view of the enterprise;
- user – provides a view of the functioning enterprise, from the perspective of a user (*e.g.* an employee, a partner, or a customer).

As can be seen in Table 4.1, the Zachman framework consists of six functional voices, each being considered from a major player's point of view; as a result, there are 36 intersecting cells, each being a meeting point between a player's perspective and a descriptive focus. By moving horizontally in the grid while adopting the same player's perspective, it is possible to analyse different descriptions of the system; on the contrary, moving vertically in the grid, the focus is always constant but changes the perspectives from different players.

There are three suggestions that can be derived from the Zachman grid, which help the management to select the best architectural organisation.

First, every architectural artifact should live in one and only one cell. There should be no ambiguity about where a particular artifact lives. When there is the availability of more than one artifact, it is possible to use the Zachman grid to clarify the focus of each of these structures. Second, an architecture can be considered as a complete architecture only when every cell in the architecture is complete. A cell is complete when it contains sufficient artifacts to fully define the system for one specific player looking at one specific descriptive focus. When each cell is complete with the right artifacts, it is possible to assess whether a sufficient amount of details are available, which completely describe the system from the perspective of every player looking at the system from every possible point of view (descriptive focus). Third, cells in columns should be related to each other. For example, considering the data column (first column) of the Zachman grid, we understand that (1) from the business owner's perspective, data is the information about the business, and (2) from the database administrator's perspective, data is rows and columns in the database. Although the business owner thinks about data quite differently than the database administrator, there must be some relationships between their perspectives.

Table 4.1. The Zachman framework

	Data (what)	Function (how)	Network (where)	People (who)	Time (when)	Motivation (why)
Objective scope *Contextual* (planner)	List of things important	List of core business process	List of business locations	List of important organisations	List of events	List of business goals/strategies
Enterprise model *Conceptual* (owner)	Conceptual data/object model	Business process model	Business logistics system	Work-flow model	Master schedule	Business plan
System model *Logical* (designer)	Logical data model	System architecture model	Distributed systems architecture	Human interface architecture	Processing structure	Business role model
Technological model *Physical* (builder)	Physical data/class model	Technology design model	Technology architecture	Presentation architecture	Control structure	Rule design
Detailed representations *Out of context* (programmer)	Data definitions	Program	Network architecture	Security architecture	Timing definition	Rule specification
Functioning enterprise (user)	Usable data	Working function	Usable network	Functioning organisation	Implemented schedule	Working strategy

A comparison among the CIMOSA and ARIS architectures underlines an elevated level of similes among the two: both models are based on process-oriented approaches and are finalised to reach an integration level of the functions through the modelling and the monitoring of the actions developed in the network.

4.5 The Second IFIP–IFAC Mandate

The second mandate of the joint task force was the continuation of the first. Based on the recommendations made in the first mandate, GERAM was fully developed during this period.

The task was two-fold: (1) complete the definition of GERAM, and (2) develop an international standard specifying the requirements that an enterprise reference architecture must satisfy.

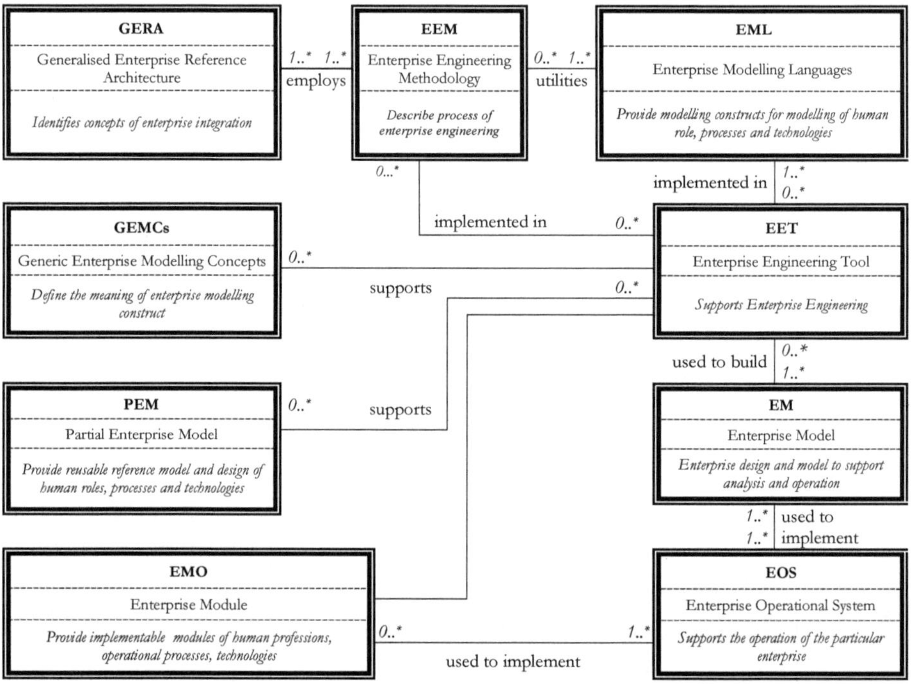

Figure 4.1. A possible GERAM framework

As mentioned earlier, GERAM is a generalised framework for enterprise integration and business process engineering. It identifies the set of components recommended for use in enterprise engineering [4.1]. This set of components is identified in Figure 4.1 and briefly described below. Starting from the defined concepts to be used in enterprise integration (GERA), GERAM distinguishes between the methodologies used for enterprise integration (EEM) and the languages used to describe the structure, contents and behaviour of the enterprise (EML).

As such, GERAM assists in the choice of tools and methodologies by providing criteria to be satisfied by them, rather than trying to enforce particular options. Used as a generic framework, GERAM may also assist in establishing the completeness and suitability of the solution to a particular change process (used as a checklist towards identifying potential gaps/uncovered areas). In other words, GERAM is 'a tool-kit of concepts for designing and maintaining enterprises for their entire life history': GERA(M) states the difference between life cycle (seen as 'the finite set of generic phases and steps [that] a system may go through over its entire life history') and life history ('the actual sequence of steps [that] a system has gone through during its lifetime') [4.20].

An essential component of GERAM is GERA (the reference architecture), which provides the critical architectural concepts needed in enterprise integration. GERA features a three-dimensional structure and contains several views in order to limit the complexity of the enterprise model.

The enterprise engineering methodologies (EEMs) are aimed at assisting the user in the enterprise modelling activity. EEMs may represent models of the engineering processes by making use of enterprise modelling languages (EMLs). Due to their limited scope, different EMLs (or several combinations) have to be used for various modelling viewpoints. GERAM provides guidelines for choosing a complete set of modelling languages. Generic enterprise modelling concepts (GEMCs) provide the necessary concepts and definitions for enterprise modelling (*e.g.* semantics for modelling languages).

GERAM describes three types of generic concepts: (1) glossaries, (2) meta-models, and (3) ontological theories. Partial enterprise models represent reusable and partially instantiated models of human roles (organisational), processes (describing only common functionality) or technology (resources, *e.g.* information technology (IT)).

Enterprise engineering tools (EETs) implement the modelling languages (and implicitly the enterprise engineering methodologies) in order to construct the desired enterprise model (EM).

GERAM defines essential requirements for EETs, such as support for analysis/design and enactment/simulation of the model, model upgrading capabilities, *etc.*, which should be particularly relevant to the EET developers.

Enterprise models (EMDs) are the ultimate purpose of the modelling activity. A complete model as described by GERAM should include enterprise operations and organisation but also its control and information systems.

GERAM sets out three main requirements for enterprise models: to enable decision support; to be a communication tool across various user groups; and to enable model-driven operation and control of the business processes. The implicit requirement is, however, that enterprise models should enable and support the change process that determined their creation in the first place, indeed the enterprise modelling represents the ontology of change [4.21].

The Enterprise operational system (EOS) represents a fully instantiated model, *i.e.* a model representing a particular enterprise. The modelling tools construct models by employing enterprise modules (EMOs). EMOs are implemented partial models, which may be used as plug-and-play components (*i.e.* typically no customisation is necessary).

4.6 The GERAM Model

The GERA defines the enterprise-related generic concepts recommended for use in enterprise integration projects. GERA is a life-cycle reference architecture, which is an architecture able to model activities involved in the implementation of a project spanning over part or whole of an entity life cycle (the opposite is in the system architecture that models the structure of a system [4.20]).

GERA has been developed along three main dimensions:

1. *Life cycle* models the enterprise entities according to the life-cycle activities. Seven activities are identified (inherited and extended from PERA [4.18] and CIMOSA [4.16]): identification; concept; requirements; preliminary/ detailed design; implementation; operation; decommissioning.
2. *Generality* accommodates various degrees of specialisation and possible instantiation of the models. For example, two partial models, ISO 15288 (Systems Life Cycles) and ISO 12207 (Software Life Cycles), are both in the *partial* (middle vertical) area of GERAM. However, ISO 12207 will be represented to the right of ISO 15288 because it is more specialised.
3. *View* provides visualisation of certain aspects of the whole (and complex) enterprise model in isolation. Views are also grouped based on:
   a. model content: function vs. information vs. resource vs. organisation (inherited from CIMOSA and GRAI-GIM [4.22]: models processes, data, technology and human);
   b. purpose: customer service vs. product management and control;
   c. implementation: mission support technology vs. human tasks vs. management and control technology (inherited from PERA: showing the role of human in the enterprise); and
   d. physical manifestation: software vs. hardware.

Business process-oriented modelling aims at describing the processes in the enterprise, capturing both their functionality (*i.e.* what has be done) and their behaviour (*i.e.* when things are done and in which sequence). In order to achieve a complete description of a process, a number of concepts have to be recognised; more specifically in GERA, these concepts include:

- life cycle;
- enterprise entity types (such as enterprise modelling with business process modelling);
- integrated model representation in different model views; and
- modelling languages for different users of the architecture (business users, system designers, IT modelling specialists, among others).

4.6.1 Life-cycle Concept

Life cycle provides for the identification of the life-cycle phases for any enterprise entity from entity conception to its final end. Figure 4.2 shows the GERA life-cycle phases of enterprise entities.

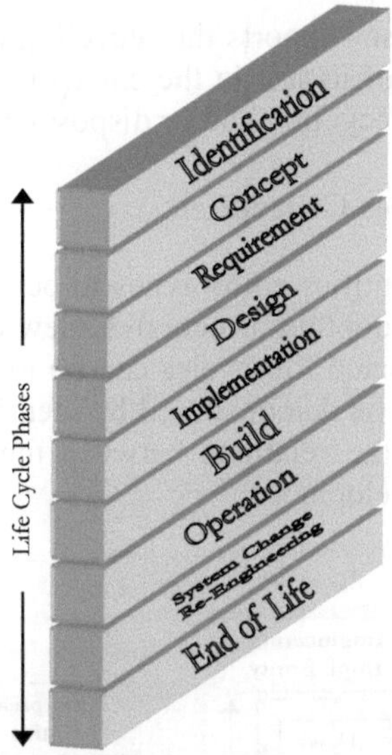

Figure 4.2. GERA life-cycle concept

A total of nine life-cycle phases have been defined:

- *Identification* – this phase allows the identification of the enterprise business or any part of it in terms of its relation to both its internal and external environment. This includes the definition of general commitments of the integration or engineering activities to be carried out in relevant projects.
- *Concept* – this phase provides for presentations of management visions, missions, values, operational concepts (build/buy, *etc.*), policies, and others.
- *Requirement* – this phase allows the description of operational processes and collection of all their functional, behavioural, informational and capability requirements.
- *Design* – this phase is the specification of an operational system, with all its components satisfying the above requirements. Processes and resources alternatives may be specified, which provide operational alternatives to be used during the operation.
- *Implementation* – this phase describes the real operational system, which may deviate from the designed system due to enterprise preferences or availability of components.
- *Build* – this phase supports the system manifestation, physical implementation of resources, testing and validation of the designed processes, and subsequent release for operation.
- *Operation* – this phase employs the released operational processes and the provided resources to support the life-cycle phases of the enterprise products.
- *System change/re-engineering* – this phase allows the modification or re-engineering of the operational processes according to the newly identified needs or capabilities provided by new technologies.

- *End of life* – this phase supports the recycling or disposal of the operational system at the ending of its use in the enterprise operation. This phase has to provide concepts for recycling and/or disposal of all or part of the system.

4.6.2 Enterprise Entity Types Concept

Enterprise entity concept identifies entity types to be used in enterprise engineering and enterprise integration. Adopting a recursive view of integration, altogether five entity types with their associated life-cycles can be identified. The recursiveness of the first four entity types can be demonstrated by identifying the role of the different entities, their products and the relations between them. Figure 4.3 illustrates the chain of enterprise entity developments.

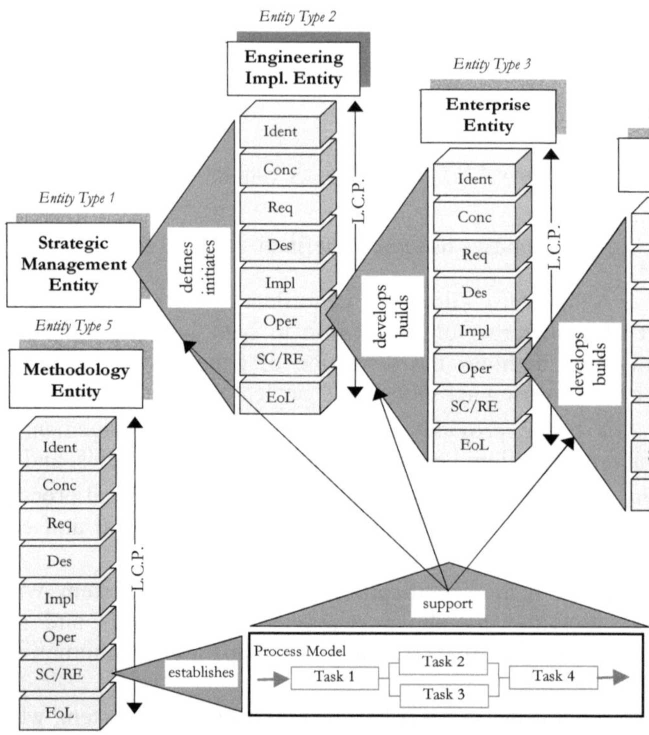

Figure 4.3. GERA enterprise entity types concept

A total of five enterprise entity types have been defined:

- *Strategic enterprise management entity* (*Type 1*) – defines the necessity and the starting of any enterprise engineering effort.
- *Enterprise engineering/integration entity* (*Type 2*) – provides the means to carry out the Type 1 enterprise entity. It employs methodologies (Type 5 entity) to define, design, implement and build the operation of the enterprise entity (Type 3 entity).

- *Enterprise entity* (*Type 3*) – the result of the operation of entity Type 2. It uses methodologies (entity Type 5) and the operational system provided by entity Type 2 to define, design, implement and build the products (services) of the enterprise (Type 4 entity).
- *Product entity* (*Type 4*) – the result of the operation of entity Type 3. It represents all products (services) of the enterprise.
- *Methodology entity* (*Type 5*) – represents the methodology to be employed in any enterprise entity type.

As shown in Figure 4.3, the Type 1 entity always starts first before the creation of any lower-level entities by identifying the goal, scope and objectives for the particular entity. Development and implementation of a new enterprise entity (or new business unit) will then be done by a Type 2 entity, whereas a Type 3 entity is responsible for developing and manufacturing a new product (Type 4 entity). With the possible exception of the Type 1 entity, all enterprise entities have an associated entity-life-cycle. However, it is always the operational phase of the entity-life-cycle in which the lower entity is defined, created, developed and built. The operation itself is supported by an associated methodology for enterprise engineering, enterprise operation, product development and production support. Figure 4.3 also shows the life cycle of the methodology (Type 5 entity) and the process model developed during the early life-cycle phases of the methodology. However, there must be a clear distinction between the life cycle of the methodology with its different phases and its process model. The latter is used to support the operational phase of a particular enterprise entity. The operational relations of the different entity types are also shown in Figure 4.4 (Type 3), which demonstrates the contributions of the different entities to the Type 3 entity life-cycle phases. The manufacturing entity itself produces the enterprise product in the course of its operation phase (Type 3 entity).

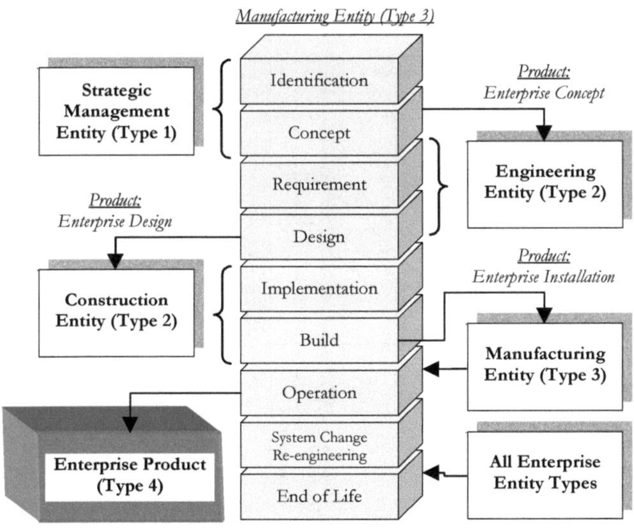

Figure 4.4. GERA enterprise entity concept (Type 3)

4.6.3 Enterprise Modelling Concept

Process-oriented modelling allows the operation of enterprise entities and entity types to be represented in all its aspects: functional, behaviour, information, resources and organisation. The resulting models can be used for decision support by evaluating operational alternatives or for model-driven operation control and monitoring. To hide complexity of the resulting models, it is necessary to present the models to users in different sub-sets (views). This view concept is shown in Figure 4.5. It is applicable during all phases of the life cycle. Note that the views are generated from the underlying integrated model via model manipulation. This means that any change done in one particular view will be reflected in all relevant aspects of the model. The GERA life-cycle model has defined four different views: function, information, decision/organisation, and resource/structure. Other views may be defined, if needed, and supported by the modelling tool. In addition, the life-cycle model of GERA provides for two different categories of modelling: operation control and customer service.

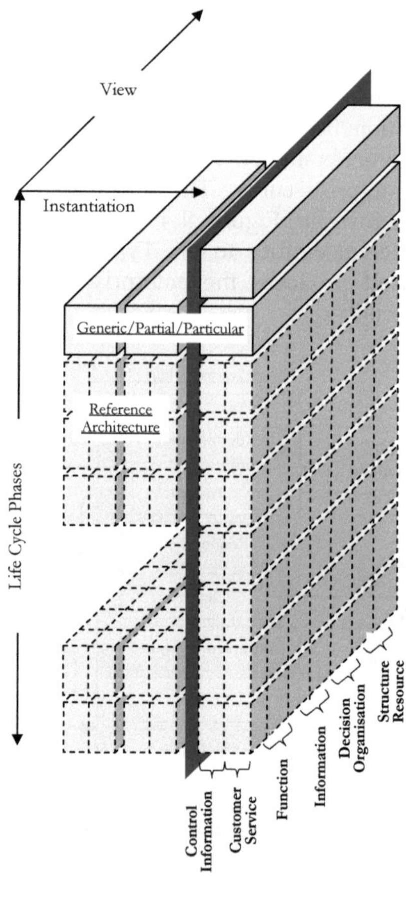

Figure 4.5. GERA generic reference architecture concept

4.6.4 Modelling Language Concept

Modelling languages increase the efficiency of enterprise modelling, and also allow a common representation of the enterprise operation. Modelling languages must accommodate different users of enterprise models; for example, business users, system designers and IT-modelling specialists. They must support the modelling of all entity types across all phases of their respective life cycles. Moreover, modelling languages must provide generic constructs as well as macro constructs (GEMs) built from generic ones. The latter can further enhance modelling productivity. Figure 4.5 highlights the reference architecture for those enterprise entity life-cycle phases that require generic constructs. The *partial* level shows the place of the GEMs in the reference architecture. The *particular* level indicates the life-cycle phases of the enterprise entity itself.

4.6.5 Generic Enterprise Engineering Methodologies

Enterprise engineering methodologies describe the process of enterprise integration, and according to the GERAM framework (Figure 4.1), they result in a model of an enterprise operation. The methodologies can guide users in the engineering task of enterprise modelling and integration. Different methodologies may co-exist, which guide the users through the different tasks required in the integration process.

Enterprise engineering methodologies should orient themselves on the life-cycle concept identified in GERA and support the different life-cycle phases shown in Figure 4.2. The enterprise integration process itself is usually directed towards the enterprise entity Type 3 operation and carried out as an enterprise engineering task by an enterprise entity Type 2 (Figures 4.3 and 4.4). The integration task may start at any relevant engineering phase (Figure 4.6) of the entity life cycle and may employ any of those phases.

Therefore, the processes relating to the different phases of enterprise engineering should be independent of each other to support different sequences of engineering tasks. Enterprise engineering methodologies may be described in terms of process models with detailed instruction for each step of the integration process. This not only allows a very good representation of the methodology for its understanding, but provides for identification of information to be used and produced, resources needed and relevant responsibilities to be assigned for the integration process.

4.6.6 Generic Enterprises Modelling Languages

Generic enterprise modelling languages define generic constructs (building blocks) for enterprise modelling. Generic constructs that represent the different elements of the operation improve both modelling efficiency and model understanding. These constructs must be adapted to the different needs of people creating and using enterprise models. Therefore, different languages may exist, which accommodate different users (*e.g.* business users, system designers, IT modelling specialists, *etc.*). Modelling the enterprise operation means to describe its processes and the necessary information, resources and organisational aspects. Therefore, modelling languages must provide constructs capable of capturing the semantics of enterprise operations.

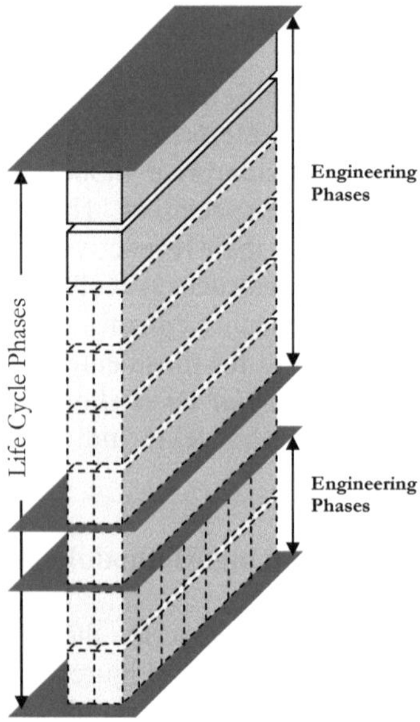

Figure 4.6. Enterprise engineering and life-cycle concept

This is especially important if enterprise models are to support the enterprise operation itself. Model-based decision support as well as model-driven operation control and monitoring require modelling constructs that support end users and represent the operational processes according to the users' perception.

Modelling languages increase the efficiency of enterprise modelling. They allow a common representation of an enterprise operation. Modelling languages must support the modelling of all entity types across all phases of their respective life cycles. In addition, modelling languages must provide generic constructs and macro constructs built from the generic ones. The latter will further enhance modelling productivity.

4.6.7 Generic Enterprise Modelling Tools

Generic enterprise modelling tools define the generic implementation of the enterprise integration methodologies, modelling languages, and other support for the creation and use of enterprise models. Modelling tools should provide user guidance for both the modelling process itself and for the operational use of the models. Therefore, enterprise modelling tools design must not only encompass the modelling methodology, but also provide model enactment capability for the simulation of operational processes. The latter should also include analysis and evaluation capabilities for the simulation results.

4.6.8 Enterprise Models

Enterprise models represent the enterprise operation mostly in the form of business processes. However, in certain cases, other representations may be suitable as well.

Business processes are represented using the generic modelling language constructs defined above for the relevant engineering methodology. Enterprise operations are usually rather complex and, therefore, difficult to understand if all relevant aspects of the operation are represented in a common model. In order to reduce the model complexity for users, different views should be provided which allow the users only to see the aspect of concern.

4.7 Architectural Structure and Life Cycle

From the above discussion, it is possible to conclude that the GERAM is a generic framework useful for the life-cycle management of an enterprise, from the starting point until the endpoint and through all the phases in its life. More specifically, inside the life cycle, it is possible to identify the following seven activities [4.19]: identification, conception, requirements, design, implementation, operation and decommission. These activities of a life-cycle process may be related to both business and ICT (information and communication technology) issues. Moreover, it is necessary to classify the activities according to who is making decisions; indeed, it is possible to find people that are involved in the execution of operations, or in the management of the operations. At the same time, three levels of management are identified, *i.e.* strategic, tactical and operational [4.23]. The framework for the architectural development of the enterprise (FADE) needs to take all these activities into account.

In today's highly competitive environment, more and more enterprises are organised according to the network type structure; as a consequence, it is necessary to optimise not only the internal enterprise processes but also the relationships that link each network enterprise [4.24]. Therefore, the FADE can also be adopted in an extended way, FADEE (framework for the architectural development of the extended enterprise), useful for meeting the needs of an extended network structure.

The main problem that the management has to face with the constitution of an extended enterprise is given by the way in which an extended enterprise is born; indeed, an extended enterprise is created by individual enterprise that are already operating on the market [4.19]. This has relevant consequences for the individual enterprises that have to come back at the requirement phase, the third phase of the seven that constitute the life-cycle history of an enterprise (Figure 4.7).

Normally, it is preferable to constitute the extended enterprise starting from the operative individual enterprises rather than redesigning the enterprise; this means that the individual enterprises are involved in two types of activities: requirement and operational phases. The framework for the constitution of a network enterprise must coordinate, simultaneously, not only the ICT side but also the business process that has to be modelled and redesigned [4.25].

The FADE can be extended in the form of FADEE (Figure 4.8), a building block useful for creating a roadmap to the IT-enabled (extended) enterprise [4.19]. The integration of the firms that constitute the extended enterprise can be analysed under several point of view: operational, tactical and strategic [4.26]. However, for a network firm, there are several issues connected at the definition of the enterprise architecture that restricts people and takes away their freedom. Sometimes, it just

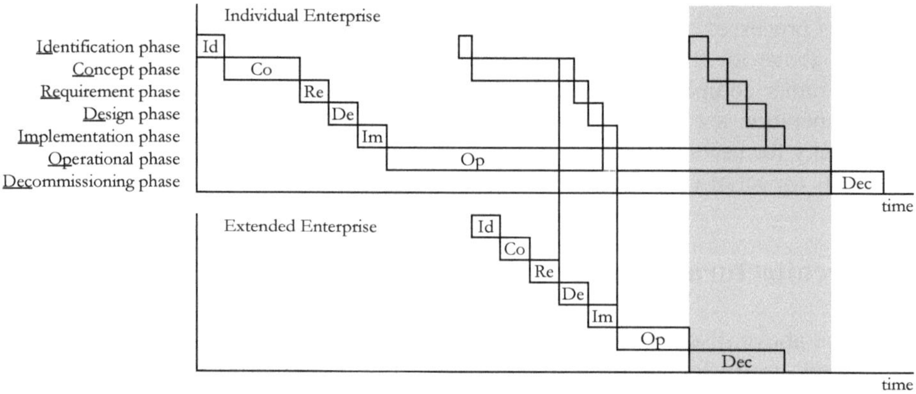

Figure 4.7. Life-cycle history of individual and extended enterprise

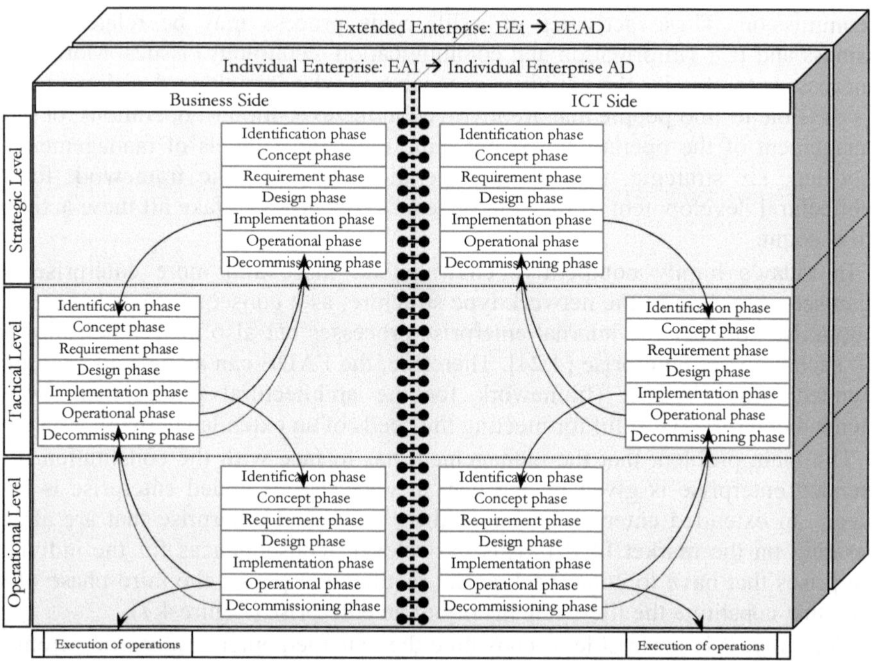

Figure 4.8. An FADEE framework

does not make any sense to give people too much freedom [4.19]. This is especially true in the case of an extended enterprise where companies are sharing a process and do not want surprise to happen. That is why it is a good practice to make all issues explicit in architecture description. For this reason, in Section 4.8.1, a model is developed for assisting the management to map out an extended enterprise architecture. Since the extension of the current architecture is not always possible, several strategies are delineated, and the most economic choice is individualised with the support of the real option theory.

4.8 Real Option and Enterprise Architecture

Architectural modelling efforts currently tend to focus overly on developing engineering and structural models where enterprise architecture is viewed as an engineering activity. Viewing enterprise architecture as an investment activity creates the need for building economic models. Traditional engineering-oriented enterprise architecture modelling focusing more on structure and technical perfection leads to lower total cost than value added (or asset productivity).

On the other hand, it is possible to adopt an economic view for the enterprise architecture development: in this case 'better return on existing information and reduced risk for future investments' [4.27]. The economic value of an enterprise is greatly influenced by its structure; it is, indeed, the structure that determines the behaviour of the firm (also under a flexibility point of view) and its ability in facing the changing and uncertain business environment. Under such conditions, flexibility in the architecture development process can provide great value, and potentially avoid risks and take benefits of new opportunities [4.28]. The maximisation of the organisational value can be reached by managing risks and uncertainties that are connected at enterprise architecture. In this way, an economic and financial model can be used so that by applying the options pricing theory, it allows building real enterprise architecture framework options. Table 4.2 provides a potential list of enterprise architecture option models that can be developed.

4.8.1 High-tech Manufacturing – Optimising Enterprise Network Architecture with Real Options

In this chapter, an option has a precise meaning; it represents a *right, but not an obligation, to do something under predefined arrangements*. The key feature of an option is that the cost of exercising the option, of using one's right to do an action, is somehow defined in advance. It is in this respect that an option has value. This is the feature that distinguishes an 'option' from a 'choice' or an 'alternative'.

In this section, the case of a high-tech manufacturing firm is examined. More specifically, the firm is interested to extend its boundary by changing its enterprise structure in a network structure. In doing so, the enterprise structure has also to be defined, changing from a 'traditional' enterprise structure type to an extended enterprise structure. Since this change can be economically expensive, it is needed to support the management with a real option framework.

Indeed, the definition of the right structure becomes more complex due to the risks that the firm faces, including market risks (related to product selling, number of product selling, price of selling) and private risks (for example, development of the right technology ahead of competitors). From an economic point of view, the consequences of these risks can be a serious disaster for the firm. For defining a new network enterprise architecture, the management can try the following options that are useful for minimising the potential damages of a wrong investment:

- Postpone the investment commitment on the enterprise architecture (EA) initiative in order to learn more about the potential investment outcomes, expected payoffs and costs. The organisation may defer the decision to

Table 4.2. Real option (RO) and enterprise architecture

Type of RO	Option elaboration and investment features	Conditions for options to be viable
Defer	An option to postpone investment commitment on EA initiative in order to learn more about the potential investment outcomes, expected payoffs and costs. In this scenario, an organisation may defer the decision to embark on an EA development journey if benefits out of EA are fuzzy and unclear. It is a feasible option when the linkage between EA development objectives and enterprise business objectives is not clear	• Investment opportunity is not a 'now or never' situation • Organisation is not exposed to overly competitive environment • Deferral is an explicit decision and not an implicit way to avoid decision • Deferral has the potential to resolve some uncertainties
Explore/ pilot	An option to realise EA implementation on a prototype/pilot scale, which has expected payoffs and associated costs. If the pilot is deemed successful, the investment can then be scaled up with a follow-up investment that has higher expected payoffs and associated costs	• Availability of investment avenues at a reduced scope and cost • Pilot can be performed using existing resources and avoiding full-scale investment • Some risks can be mitigated using an exploratory approach • Pilot findings are useful if a full-scale investment is the next step • Abandoning the pilot has no competitive, operational and regulatory consequences • Pilot should not be performed half-heartedly and failure in pilot is seen as learning in itself
Scale up/ down	An option to expand/contract the scope of EA initiative depending on observed conditions. Changes in operating scope could be achieved by: • Limiting the number of business units/entities where the EA is deployed; • Limiting the number of views to the architecture incorporated; • Limiting the role and importance of architecture governance	• Possible to enhance/lower investment without much negative consequences to the initiative • Full-scale implementation is decomposable into a series of stages that can be performed one at a time and fairly independently • Organisation can get benefits of the initiative, albeit reduced, even if full-scale implementation is not chosen • Expanding and contracting scale of initiative should have commensurate impact on the investment needs, payoffs and benefits
Compound (sequential)	An option involving two or more of the above options, where the value of an earlier option can be affected by the value of later options or *vice versa*. As both EA Maturity Framework and EA Management Maturity Framework are five-level systems, each of the above options is relevant to a particular level. This provides an option to realise EA implementation as a series of sequential implementation stages incrementally without initially committing to attain the highest maturity levels	• Possibility of combining any of the above two options • Phased investment is possible and investment lifecycle can be aligned with the architecture development lifecycle • Benefits of each phase can be clearly delineated and used as an input to decide on investment for the next phase • Not mandatory to commit to all phases upfront, as the investment is contingent upon perceived success of the preceding phase
Strategic (growth)	An option where EA investments provide the capability to create future investment opportunities as well as allow the organisation to respond quickly to regulatory and/or competitive threats	• Availability of growth options to take advantage of future opportunities • Capability to make pre-emptive moves to seize upcoming opportunities, by leveraging on strengths gained from original program

embark on an enterprise architecture development project if benefits out of enterprise architecture are fuzzy and unclear.

- Start with a test project and expand the scope of the EA initiative depending on observed conditions. The changes in operating scope could be achieved in the case under analysis, limiting the number of business units/entities where the EA is deployed.

- Buy a firm that represents a start-up project, which just adopts the enterprise structure that has to be extended to the entire network. This means creating an option where the enterprise architecture investments provide the capability to create future investment opportunities.

With respect to the third option (the start-up project), the investment seems interesting but a price of €75,000 has to be paid. The question is if the firm should acquire the firm and mitigate some development risk but still face the market risk and some residual development risk. Moreover, the start-up firm has the enterprise architecture only partially completed, because it is true that the start-up firm just adopts the enterprise type, but the enterprise architecture has to be extended to the entire network.

In order to resolve the problem, it is necessary to define how much the start-up firm is really worth compared to the price of €75,000 required for its acquisition. Moreover, it has to define if one or more options are available that can mitigate some of the market and development risks. At the same time, there can be additional opportunities, connected to these risks, in the market that the firm can take through the acquisition of the start-up firm. Indeed, the sources of uncertainties may lead to additional risks for the firm but, according the modern view of the real option theory, the same uncertainties can be the sources of opportunities [4.29–4.32].

Analysing the data available for the start-up firm, it is possible to define the discounted cash flow (DCF) of the firm. More specifically, the best estimated present value of the benefits is of €150,000. This means that, since a cost of €75,000 is required for the acquisition of the start-up firm, the NPV deriving from this buying is of €75,000. For the estimation of the DCF model, several binomial distributions are used to define the probability of technical success. Moreover, triangular distributions are used to simulate different market conditions and market positioning of the firm. The annualised volatility resulting from Monte Carlo simulation is found to be 25%, which represents a moderate level of risk.

In order to define the goodness of this estimated result, it is necessary to define another DCF that compares with the first one. Therefore, consider the case where the enterprise architecture is developed in-house without the acquisition of the start-up firm; here the cost required for converting the existing enterprise architecture into the new extended architecture is €60,000, which is much less than the acquisition cost of €75,000. According to this preliminary analysis, the in-house development of the new enterprise architecture seems more convenient, determining an NPV of €90,000. In reality, developing an enterprise structure in-house rather than buying a start-up firm with the technology already complete is more risky. For this reason, it is appropriate to set the volatility at a level of 30%, which is 5% higher than for the start-up case.

Now, the best choice must be made. To properly evaluate such a project, the use of options approach rather than the traditional DCF technique is more appropriate. The options approach considers all future investment opportunities along the value chain, allowing a more flexible assessment of strategic projects; whereas when traditional DCF methods are naively applied to evaluate strategic projects, future opportunities that create values are often ignored in the valuation process. This results in too little strategic investment. In contrast to naive DCF valuations, ROA (real option analysis) provides for better corporate strategic investment decisions in terms of value added. This means that it is necessary to define the real option types that can protect the firm against failures in the case of start-up acquisition, and against development risks in the case of in-house technology development.

A preliminary analysis is finalised to define what risks exist and how they can be reduced. Moreover, for developing a real option framework, a preliminary strategy tree analysis is performed (Figure 4.9).

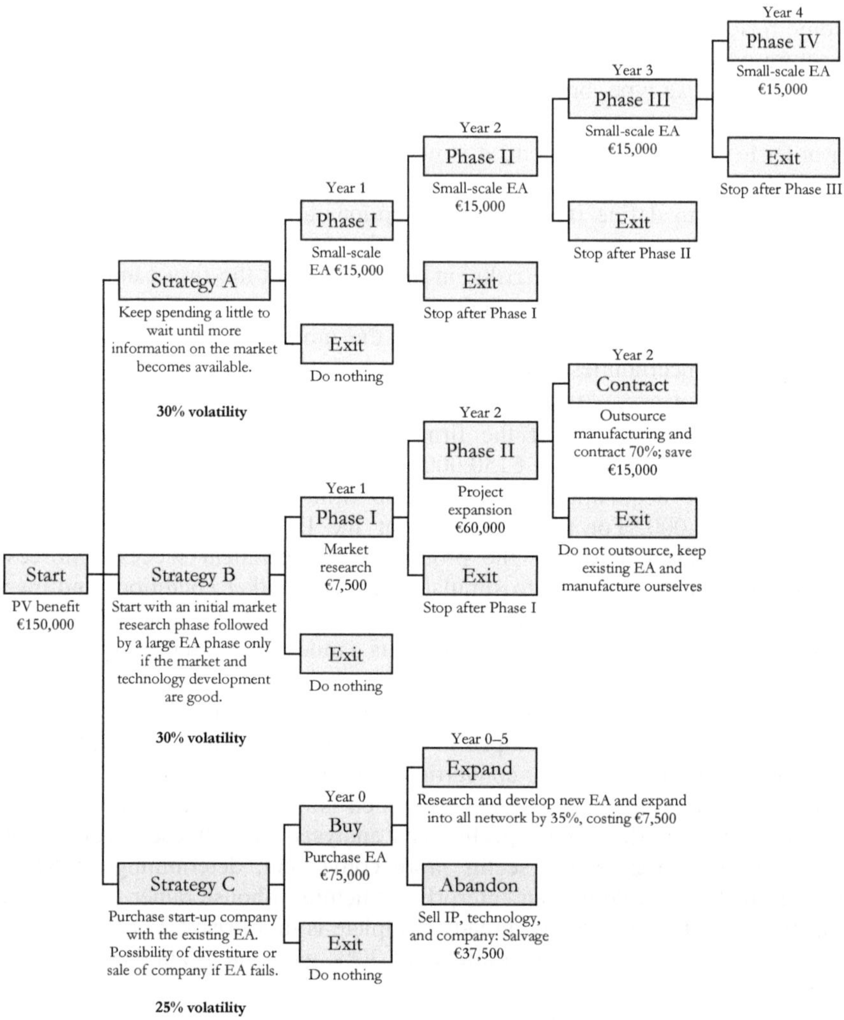

Figure 4.9. Strategic tree analysis

4.8.2 The Real Option Results for the Firm Project

By analysing the project, it is possible to individualise four main options that can be useful to minimise the potential risks: (1) mitigating the development risk connected to the hypothesis of self-doing; (2) mitigating the risk of the market; (3) mitigating the risk of failure in the case of the hypothesis of a start-up firm acquisition; and (4) taking advantage of upside risks each time that it is possible. Moreover, the path depending on the strategies (Figure 4.9) must be taken into account.

Strategy A is related to the development of the enterprise architecture in-house, but in this case it is possible to mitigate the development risk through an exploration option that allows managing the investment according to a staged strategy, where the total cost of €60,000 is split into four steps of €15,000 each. In this way, with each stage based on the date related to the outcomes of the current stage, it is possible to decide on the best strategy to be adopted in the next stage. This means that it is possible to limit the losses to the amount registered up to the valuation point. For example, if the firm observes bad results after one year, it may decide to abandon the project by stopping further investment and limiting the damages to only €15,000 instead of the total amount of €60,000 that the firm risks to lose if starting with the entire project.

Strategy B is always applicable to the development of an extended enterprise network in-house but the market risk is mitigated according to a scale-up option. This means that the management has the right to alter operating scale of the investment in the case of good outcomes. More specifically, a preliminary phase I is performed with a prototype project that requires €7,500 in cost. In this preliminary project, it is possible to test the theoretical enterprise structure and define the benefits associated to the project and replaced with a normal scale. Based on the results gained from the market research in phase I, it is possible to decide if the initiative must be executed or not. In phase II, the project is eventually extended to a full scale. Although the market risk is mitigated through the market research, the development risk still exists. In this case, a contraction option can be useful. More specifically, the firm can find a counterparty for managing the manufacturing risks through a two-year contract, whereby, at any time within the next two years, the firm can have this counterparty firm take care of the development of the enterprise architecture. The total costs of €60,000 are assumed by the firm but it ends up with mitigated development risks and also a saving of €15,000 because it does not need to increase its own manufacturing competencies by hiring outside consultants and purchasing new equipment. In this case, the firm has to define the goodness of the strategic path, whether the market research is valuable and, moreover, how much the firm should share its net profits with the counterparty.

Strategy C is related to the acquisition of the start-up firm for a cost of €75,000. However, note that by acquiring the start-up, the firm obtains additional options. If the results obtained from the new enterprise architecture configuration are lower than expected, it is possible to sell the start-up firm, realising – in the first year – a salvage value of €37,500 for its intellectual properties, patents, technology, assets, and buildings, *etc*. If the sale of the start-up firm is done after the first year, a €1,500 increase for the year must be taken into account, reflecting the increased salvage value because the intellectual property of the start-up firm are expected to increase over time. If the results of the start-up firm are successful within five years, the enterprise architecture can be extended to the global network. In this case, an additional cost of €7,500 is required for expending the already existing enterprise architecture in the model. In this chapter, the Multiple Asset SLS software is used to quantify the value of these strategies. The first step required for the application of the real option theory is related to the definition of all the possible stage values that the uncertain variable can assume. The present value of €150,000 with a volatility of 30 percent can be described as shown in Figure 4.10.

UNDERLYING

										273317.82
									257401.03	
								242411.16		242411.16
							228294.23		228294.23	
						214999.41		214999.41		214999.41
					202478.82		202478.82		202478.82	
				190687.37		190687.37		190687.37		190687.37
			179582.60		179582.60		179582.60		179582.60	
		169124.53		169124.53		169124.53		169124.53		169124.53
	159275.48		159275.48		159275.48		159275.48		159275.48	
150000.00		150000.00		150000.00		150000.00		150000.00		150000.00
	141264.68		141264.68		141264.68		141264.68		141264.68	
		133038.07		133038.07		133038.07		133038.07		133038.07
			125290.53		125290.53		125290.53		125290.53	
				117994.18		117994.18		117994.18		117994.18
					111122.73		111122.73		111122.73	
						104651.45		104651.45		104651.45
							98557.02		98557.02	
								92817.51		92817.51
									87412.24	
										82321.75

Figure 4.10. The *underlying* variable

PHASE 4

										260788.77
									244897.01	
								229932.12		229882.11
							215840.13		215790.21	
						202570.19		202520.38		202470.36
					190074.44		190024.72		189974.80	
				178307.78		178258.16		178208.34		178158.32
			167227.75		167178.23		167128.51		167078.59	
		156794.37		156744.94		156695.32		156645.50		156595.48
	146969.97		146920.64		146871.12		146821.40		146771.48	
137719.10		137669.87		137620.44		137570.82		137520.99		137470.97
	128959.23		128909.89		128860.36		128810.63		128760.70	
		120708.02		120658.57		120608.94		120559.11		120509.08
			112935.85		112886.31		112836.56		112786.62	
				105614.84		105565.19		105515.33		105465.29
					98718.73		98668.95		98618.99	
						92222.76		92172.87		92122.79
							86103.66		86053.65	
								80339.51		80289.36
									74909.66	
										69794.68

Figure 4.11. Graphical representation of Phase 4 in Strategy A

The first option under evaluation is a stage–gate investment organised into four stages. In every phase, the management has the option and flexibility either to continue to the next phase if everything goes well, or to terminate the project. Therefore, with the strategic option value of being able to defer and wait before implementing future phases due to the volatility, there is a possibility that the asset value will be significantly higher. Hence, the ability to wait before making an investment decision in the future is the option value or the strategic value of the project less the NPV. Due to the backward induction process used, the analytical convention is to start with the last phase and go all the way back to the first phase. The option results of phase 4 in strategy A are presented in Figure 4.11.

In the last stage, a positive result is achieved: the option holder can continue to invest in the rest stages. In Table 4.3, assumptions for defining the option value in phase 3 are given, where *OptionOpen* represents the opportunity of keeping the European option open in the intermediate step of the binomial lattice. The intermediate equation *Max(Phase4-Cost,OptionOpen)* represents the profit maximisation decision of either executing the option or leaving it open for possible future execution. In the contrary, the terminal equation *Max(Phase4-Cost,0)* represents the profit maximisation decision at maturity of either executing the option if it is in-the-money or allowing it to expire worthless if it is at-the-money or out-of-the-money. The option value of this stage is graphically represented in Figure 4.12.

Table 4.3. Assumptions for phase 3 in strategy A

Phase 3			
Cost	€15,000.00	Dividend	0.00%
Terminal equation	Max(Phase4-Cost, 0)	Risk-free	5.00%
Intermediate equation	Max(Phase4-Cost, OptionOpen)	Steps	75

PHASE 3

										247617.34
									231751.90	
								216813.28		216710.68
							202747.51		202645.11	
						189503.76		189401.56		189298.95
					177034.16		176932.15		176829.74	
				165293.64		165191.82		165089.60		164986.97
			154239.78		154138.13		154036.08		153933.64	
		143832.64		143731.14		143629.25		143526.98		143424.30
	134034.64		133933.27		133831.51		133729.37		133626.85	
124810.44		124709.15		124607.49		124505.45		124403.05		124300.26
	116025.55		115923.93		115821.96		115719.62		115616.92	
		107749.61		107647.63		107545.30		107442.62		107339.60
			99953.13		99850.72		99747.99		99644.93	
				92608.37		92505.48		92402.28		92298.76
					85689.31		85585.84		85482.08	
						79171.50		79067.34		78962.92
							73032.03		72927.05	
								67249.49		67143.52
									61803.88	
										56676.58

Figure 4.12. Graphical representation of phase 3 in strategy A

On the base of the assumptions in Table 4.4, it is possible to quantify the last two steps. Since the option value gained in phase 2 is also worthy (€111,242.36 as shown in Figure 4.13), it is possible to continue towards the last step. In the case of strategy A adopting a 100-step lattice, the strategic value of this first strategy is €96,974.42 (see Figure 4.14).

Table 4.4. Assumptions for phases 2 and 1 in strategy A

Phase 2			
Cost	€15,000.00	Dividend	0.00%
Terminal equation	Max(Phase3-Cost, 0)	Risk-free	5.00%
Intermediate equation	Max(Phase3-Cost, OptionOpen)	Steps	50
Phase 1			
Cost	€15,000.00	Dividend	0.00%
Terminal equation	Max(Phase2-Cost, 0)	Risk-free	5.00%
Intermediate equation	Max(Phase2-Cost, OptionOpen)	Steps	25

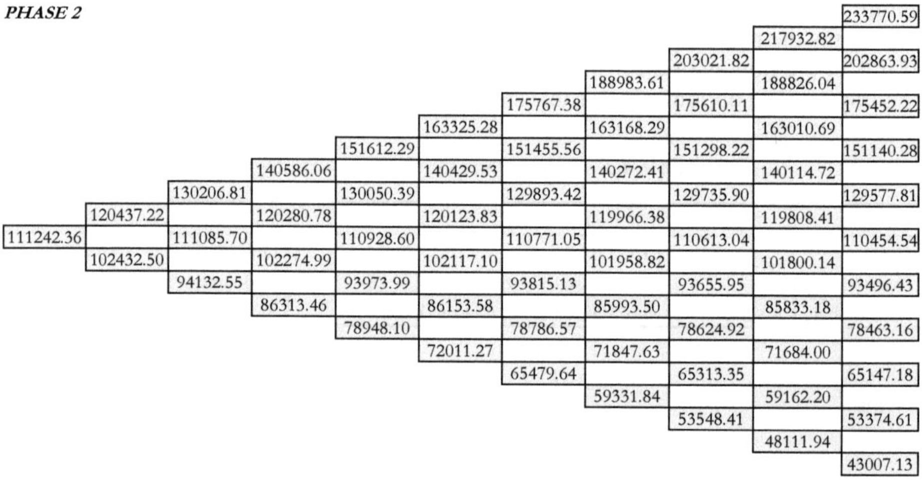

Figure 4.13. Graphical representation of phase 2 in strategy A

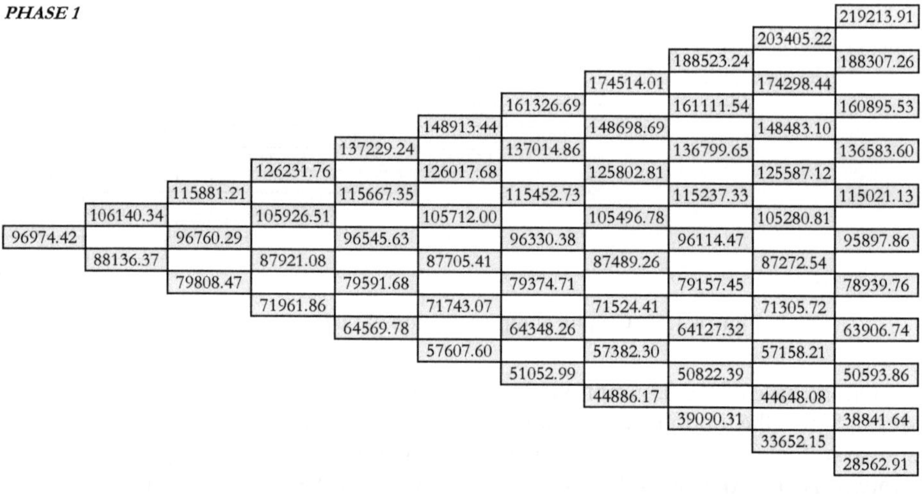

Figure 4.14. Graphical representation of phase 1 in strategy A

As mentioned earlier, the NPV is €90,000. As a consequence, the value of the option is valued at €6,974.42, which means that through the acquisition of the option, it allows the downside risk to be hedged and it is possible to increase the investment value. This result is strongly dependent on the annualised dividend rate; if this rate exceeds 2.5%, the option value becomes zero and the NPV of €90,000 represents the total strategic value of the project. Since in this first strategy the risk of the investment project is mitigated with a stage-gate investment process organised over four years, it is possible to add values to the project every time so that the losses for the year do not exceed the 2.5% of €150,000 (or €3,750).

The total value of strategy B, adopting a 100-step lattice, is of €113,853.17. The assumptions as the base of this strategy valuation are described in Table 4.5.

Table 4.5. Assumptions for phases 2 and 1 in strategy B

Phase 2			
Cost	€60,000.00	Dividend	0.00%
Risk-free	5.00%	Steps	100
Terminal equation	Max(Underlying-Cost, Underlying × Contract + Savings)		
Intermediate equation	Max(OptionOpen, Underlying × Contract + Savings)		
Phase 1			
Cost	€7,500.00	Dividend	0.00%
Risk-free	5.00%	Steps	50
Terminal equation	Max(Phase2-Cost, 0)		
Intermediate equation	Max(Phase2-Cost, OptionOpen)		

As a consequence, after taking into account the costs required for the option acquisition, the NPV of this strategy is €82,500 (equals to 150,000 – 60,000 – 7,500), and the options are valued at €31,353.17. Compared with strategy A, strategy B is a better choice for the management to follow to maximise the economic project value. Without the contraction option, the project value is €88,678.58, where €6,178.58 is generated by the option value and €82,500 by the NPV value. Defining a contraction option with the counterparty is possible to limit the downside technical risk and increase the value of the project for an amount of €25,174.59 (given by €113,853.17 minus €88,678.58). In this case, the value of the project is strongly connected to the contraction factor (*i.e.* how much is allocated to the counterparty) used. For this reason, a sensitivity analysis is performed, as changes in the amount of savings change the contraction factor (Table 4.6).

The expansion option of strategy C values the flexibility of expanding from a current existing enterprise architecture to a larger one organised as a network. The values settled for the quantification of this option are described in Table 4.7.

The last strategy has a total strategic value of €196,659.07. If considering the €75,000 required for the start-up firm acquisition, it is equivalent to a strategic value of €121,659.07. This means that the firm can pay no more than €82,805.9 for the start-up acquisition (*i.e.* the result of 75,000 + 121,659.07 – 113,853.17). If the price required is higher than this amount, the start-up strategy has to be abundant and the management will have to go back to strategy B related to building the enterprise architecture internally (Table 4.8).

Table 4.6. Decision table of savings vs. contraction factors (CF)

	Savings										
	0	5,000	10,000	15,000	20,000	25,000	30,000	35,000	40,000	45,000	50,000
CF	Strategic option values										
0.05	88,717.92	88,830.27	89,002.79	89,248.65	89,575.77	90,000.85	90,579.20	91,262.67	92,085.64	93,100.61	94,201.85
0.10	88,778.14	89,151.38	89,454.91	89,878.53	89,878.53	90,439.01	91,116.63	91,938.58	92,968.86	94,094.42	95,490.99
0.15	88,856.14	89,054.30	89,343.93	89,756.84	90,299.18	90,970.75	91,791.68	92,837.12	94,007.63	95,405.61	96,986.93
0.20	88,974.21	89,246.37	89,635.65	90,159.60	90,825.13	91,644.95	92,705.44	93,920.83	95,321.34	96,984.84	98,754.27
0.25	89,149.34	89,514.90	90,020.43	90,679.82	91,498.75	92,573.80	93,834.04	95,247.06	96,983.69	98,869.51	100,992.99
0.30	89,394.60	89,881.65	90,534.81	91,363.94	92,442.21	93,747.27	95,243.43	96,984.18	98,996.04	101,142.76	103,640.03
0.35	89,743.46	90,390.32	91,233.89	92,310.89	93,660.63	95,239.84	97,020.17	99,125.22	101,444.05	103,959.70	106,771.02
0.40	90,246.60	91,104.41	92,194.37	93,574.32	95,236.38	97,144.03	99,265.25	101,749.59	104,419.68	107,313.89	110,463.64
0.45	90,976.05	92,109.64	93,514.33	95,233.77	97,268.33	99,563.14	102,082.15	104,947.85	108,008.06	111,271.43	114,779.64
0.50	92,026.47	93,512.18	95,307.81	97,398.75	99,863.61	102,599.48	105,562.63	108,800.78	112,273.19	115,887.92	119,751.13
0.55	93,512.02	95,431.72	97,694.27	100,272.31	103,131.38	106,341.16	109,772.10	113,380.58	117,245.17	121,236.52	121,236.52
0.60	95,557.47	97,990.42	100,784.95	103,897.24	107,279.46	110,887.42	114,739.87	118,772.88	122,923.64	127,238.01	131,625.86
0.65	98,486.16	101,478.92	104,801.39	108,401.17	112,236.97	116,309.93	120,523.93	124,849.42	129,307.90	133,824.68	138,406.87
0.70	102,359.42	105,931.83	109,782.81	113,853.17	118,124.94	122,523.98	127,014.18	131,585.52	136,213.51	140,873.38	145,569.01
0.75	107,365.62	111,467.13	115,771.95	120,218.48	124,768.18	129,393.90	134,061.23	138,758.05	143,479.63	148,211.55	152,952.83
0.80	113,479.01	117,973.18	122,578.61	127,250.40	131,956.91	136,683.92	141,423.33	146,170.74	150,921.56	155,674.75	160,429.35
0.85	120,464.34	125,163.02	129,896.44	134,642.43	139,394.04	144,148.24	148,903.44	153,659.15	158,415.12	163,171.18	167,927.28
0.90	127,869.42	132,622.57	137,378.25	142,134.27	146,890.38	151,646.52	156,402.66	161,158.81	165,914.96	170,671.10	175,427.25
0.95	135,365.78	140,121.93	144,878.07	149,634.22	154,390.37	159,146.51	163,902.66	168,658.81	173,414.96	178,171.10	182,927.25

Table 4.7. Assumptions for *expand/abandon* in strategy C

Expand/abandon			
Cost	€7,500.00	Dividend	0.00%
Risk-free	5.00%	Steps	100
Terminal equation	Max(Underlying, Salvage, Underlying × Expansion-Cost)		
Intermediate equation	Max(Salvage, Underlying × Expansion-Cost, OptionOpen)		

Table 4.8. Strategic values of the three strategies

	Strategy A	Strategy B	Strategy C
NPV	90.000	82.500	75.000
Option value	6.974,42	31.353,17	121.659,07
Total strategic value	96.974,42	113.853,17	196.659,07

The real option application of this project is related to the definition and implementation of the best extended enterprise architecture and allows the investment value to be increased. Among the three strategies hypothesised and analysed, the optimal choice is to purchase a start-up firm and, with the further option of abandoning the firm, to gain the opportunity to sell the start-up firm if the results registered are lower than expected.

In the case of not applying the real option theory, the only opportunity available is to develop the enterprise architecture by immediately spending €60,000 and taking an unnecessary risk.

4.9 Conclusions

Effective management of enterprise architectures is a recognised strength of successful enterprises. Enterprise architecture provides a clear and comprehensive view of the structure and operations of an organisation. In this chapter, a real option-based approach is presented to economically manage the investments in extended enterprise architecture, *i.e.* the architecture related not only to a single network but to a network of organisations. The utility of the real option approach derives from several uncertainties associated with the enterprise architecture investment. In contrast to traditional DCF-based methods, the proposed approach manages the uncertainties and allows investments to be configured accordingly. Incorporating managerial flexibility involves understanding and acknowledging the existence of temporal aspects in the investment cycle. This allows managers to build investment configurations that best suit their organisation and its implementation scenario.

References

[4.1] Devaraj, S., Krajewski, L. and Wei, J.C., 2007, "Impact of eBusiness technologies on operational performance: the role of production information integration in the supply chain," *Journal of Operations Management*, **25**(6), pp. 1199–1216.

[4.2] Klein, R., 2007, "Customization and real time information access in integrated eBusiness supply chain relationships," *Journal of Operations Management*, **25**(6), pp. 1366–1381.

[4.3] Francasa, D. and Minner, S., 2009, "Manufacturing network configuration in supply chains with product recovery," *Omega*, **37**(4), pp. 757–769.

[4.4] Melo, M.T., Nickel, S. and Saldanha-da-Gama, F., 2009, "Facility location and supply chain management – a review," *European Journal of Operational Research*, **196**(2), pp. 401–412.

[4.5] Meng, Q., Huang, Y. and Cheu, R.L., 2009, "Competitive facility location on decentralized supply chains," *European Journal of Operational Research*, **196**(2), pp. 487–499.

[4.6] Altiparmak, F., Gen, M., Lin, L. and Karaoglan, I., 2009, "A steady-state genetic algorithm for multi-product supply chain network design," *Computers & Industrial Engineering*, **56**(2), pp. 521–537.

[4.7] Amram, M. and Kulatilaka, N., 1999, *Real Options: Managing Strategic Investment in an Uncertain World*, Harvard Business Publishing, Boston.

[4.8] Dixit, A.K. and Pindyck, R.S., 1994, *Investment under Uncertainty*, Princeton University Press, Princeton.

[4.9] Lin, T.T., 2009, "Applying the maximum NPV rule with discounted/growth factors to a flexible production scale model," *European Journal of Operational Research*, **196**(2), pp. 628–634.

[4.10] Yuri, T., 2008, "Can real options explain financing behavior?," *Journal of Financial Economics*, **89**(2), pp. 232–252.

[4.11] Chen, D., Doumeingts, G. and Vernadat, F., 2008, "Architectures for enterprise integration and interoperability: past, present and future," *Computers in Industry*, **59**(7), pp. 647–659.

[4.12] Garga, A., Kazmanb, R. and Hong-Mei, C., 2006, "Interface descriptions for enterprise architecture," *Science of Computer Programming*, **61**(1), pp. 4–15.

[4.13] Williams, T.J., Bernus, P., Brosvic, J., Chen, D., Doumeingts, G., Nemes, L., *et al.*, 1994, "Architectures for integrating manufacturing activities and enterprises," *Computers in Industry*, **24**(2–3), pp. 111–140.

[4.14] Bernus, P., Nemes, L. and Williams, T.J., 1996, *Architectures for Enterprise Integration*, Chapman and Hall, London.

[4.15] Bass, L., Clements, P. and Kazman, R., 2003, *Software Architecture in Practice*, Addison-Wesley Professional, Boston.

[4.16] ESPRIT Consortium AMICE, 1993, *CIMOSA: Open System Architecture for CIM*, Springer-Verlag, London.

[4.17] Dickerhofa, M., Didic, M.M. and Mampel, U., 1999, "Workflow and CIMOSA – background and case study," *Computers in Industry*, **40**(2–3), pp. 197–205.

[4.18] Williams, T.J., 1994, "The Purdue enterprise reference architecture," *Computers in Industry*, **24**(2–3), pp. 141–158.

[4.19] Goethals, F.G., Snoeck, M., Lemahieu, W. and Vandenbulcke, J., 2006, "Management and enterprise architecture click: the FAD(E)E framework," *Information Systems Frontiers*, **8**(2), pp. 67–79.

[4.20] ISO/TC184/SC5/WG1, 1999, "Annex A: GERAM," In *ISO/DIS 15704: Industrial Automation Systems – Requirements for Enterprise-reference Architectures and Methodologies*, ISO-1999 (ed.).

[4.21] Bernus, P., Nemes, L. and Schmidt, G. (eds.), 2003, *Handbook on Enterprise Architecture*, Springer, London.

[4.22] Doumeingts, G., Vallespir, B., Zanettin, M. and Chen, D., 1992, *GRAI-GIM Integrated Methodology – A Methodology for Designing CIM Systems*, University Bordeaux, France.

[4.23] Proper, H., Bosma, H., Hoppenbrouwers, S. and Janssen, R., 2001, "An alignment perspective on architecture-driven information systems engineering," In *Proceedings of the Second National Architecture Congress*, Amsterdam.

[4.24] Jeongsoo, L., Heekwon, C., Cheol-Han, K. and Kwangsoo, K., 2009, "Design of product ontology architecture for collaborative enterprises," *Expert Systems with Applications*, **36**(2), pp. 2300–2309.

[4.25] Clark, T. and Stoddard, D., 1996, "Interorganizational business process redesign: merging technological and process innovation," *Journal of Information Management*, **13**(2), pp. 9–28.

[4.26] Goethals, F., Vandenbulcke, J., Lemahieu, W. and Snoeck, M., 2004, "Structuring the development of inter-organizational systems," In *Proceedings of the Web Information Systems Engineering Conference*, pp. 454–465.

[4.27] Perks, C. and Beveridge, T., 2003, *Guide to Enterprise IT Architecture*, Springer-Verlag, New York.

[4.28] Saha, P., 2006, "A real options perspective to enterprise architecture as an investment activity," *Journal of Enterprise Architecture*, **2**(3), pp. 32–52.

[4.29] Chen, J., 2006, "An analytical theory of project investment: a comparison with real option theory," *International Journal of Managerial Finance*, **2**(4), pp. 354–363.

[4.30] Cucchiella, F. and Gastaldi, M., 2006, "Risk management in supply chain: a real option approach," *Journal of Manufacturing Technology Management*, **17**(6), pp. 700–720.

[4.31] Cucchiella, F. and Gastaldi, M., 2007, "Risk management in a globalized cosmetic firm," *International Journal of Logistics Economics and Globalisation*, **1**(1), pp. 21–33.

[4.32] Kulatilaka, N., Balasubramanian, P. and Strock, J., 1999, "Using real options to frame the IT investment problem," In *Real Options and Business Strategy: Applications to Decision-Making*, Trigeorgis, L. (ed.), Risk Books, London.

5

Enterprise Networks and Information and Communications Technology Standardisation

Elias G. Carayannis[1] and Yiannis Nikolaidis[2]

[1] School of Business, George Washington University
Washington, DC 20052, USA
Email: caraye@gwu.edu

[2] Department of Technology Management, University of Macedonia
59200 Naousa, Greece
Email: nikolai@uom.gr

Abstract
This chapter is part of a book focused on advancing the state of the art in enterprise networks and logistics for agile manufacturing. In other words, infra-technologies and infra-structures (real and virtual), which support and sustain high speed, throughput, reliability and robustness as key attributes of the systems including ICT-enabled systems, must undergird and leverage. Within any enterprise network, different components are being upgraded at different times, in different places and in different ways. The result is a highly complex and continuously changing environment, which makes ICT standardisation highly important. Apart from many other goals, ICT standardisation aims primarily at assuring interoperability between the various systems of an enterprise network. Without ICT standards to ensure interoperability, negative economic and social consequences will be faced. Today, industry groups are developing continuously their own standards in order to support their networks. The Automotive Network Exchange (ANX) created by the US automotive industry a few years ago and the role of standards is used as a case study in this chapter, leading to important insights and conclusions.

5.1 Introduction

This chapter is part of a book focused on advancing the state of the art in enterprise networks and logistics for agile manufacturing. In other words, infra-technologies and infra-structures (real and virtual), which support and sustain high speed, throughput, reliability and robustness as key attributes of the systems including ICT-enabled systems, must undergird and leverage. The role of information and communications technology (ICT) standards, and especially the standardisation of ICT-enabled systems and components to assure robust and versatile interoperability, portability and functionality within and across a variety of human-centric and techno-centric contexts, cannot be over-emphasised in this regard. This role and its significance constitute the raison d'être for this chapter.

The world economy is becoming more and more information- and knowledge-driven [5.1, 5.2]. Increasingly, smart decision-making is crucial to competitiveness and success in business. Consequently, the systems used to access and distribute information, as well as the technologies that drive them, become the focus of both commercial and regulatory concerns. In addition, new ICT applications, such as mobile communications, personal computers, navigation systems, *etc.*, are radically changing the way that people work and enjoy their leisure time. However, these applications and especially enterprise networking will not reach their full potential unless both they and their supporting infrastructures are fully interoperable. One also needs to consider the rich vs. thin client emerging architectures dilemma (such as web services) and the resulting explosion of design approaches and related impact on standards.

The effectiveness of ICT, especially in enterprise networking, is determined by the ability of its component parts to interoperate, namely to have the ability to work with other systems or products without special effort on the part of the user. Without this, the use of ICT products and services is restricted. Similarly, the use of many applications depends on the ability of products from different manufacturers to interoperate. Two straightforward examples illustrate how interoperable systems can lead to great effectiveness [5.3]. The first one is the *Internet* per se, which can be seen as the ultimate interoperable design to which more and more non-interoperable networks and systems have converged. The second example is *e-mail.* Neither e-mail protocols nor the concept of e-mail were restricted to a limited set of players, and their designs were broadly interoperable. The results of this situation are extraordinary in each instance.

At this point, the role of standards becomes crucial. A standard is an agreement between the parties involved, such as manufacturers, sellers, purchasers, users and regulators (*e.g.* all those participating in an enterprise network) of a particular product, process or service. It contains a technical specification or other precise criteria designed to be used consistently as a rule, guideline or definition. Its adoption ensures for all operators a clear reference in terms of technical specifications, quality, performance and reliability. Its objective is to ensure that products and services are suitable for their purpose and they are comparable and compatible. Standards are summaries of best practices and their creation arises from the experience and expertise of all interested parties.

More specifically, ICT standardisation aims at ensuring that the necessary technologies are properly defined and that interoperability between various systems of an enterprise network is assured. Without ICT standards ensuring interoperability, an opportunity will be lost, with negative economic and social consequences for everyone. In this regard, a standard serves as a risk management and technology roadmap guideline, as well as a strategic technology option that enables the implementation of strategic technology plans, as it provides a substantial forward-looking perspective as to the direction and nature of technology and market dynamics. Standardisation, however, implies a trade-off as well; sometimes, it can be considered to constitute a constraint on creativity and on maximising the added value of technology in many regards as well as on the capacity to capture the full extent of its value-adding potential (one manifestation of this may be the information technology 'productivity paradox' that we discuss below).

Brynjolfsson [5.4] and Brynjolfsson and Hitt [5.5] mention that an important question that has been debated for many years is whether computers and ICT contribute to productivity growth. The former identifies four reasons for the 'productivity paradox', *i.e.* the stability of the productivity despite the increased computing power all over the world, which can also occur in enterprise networking. Among them one can find the mismanagement of information and technology, which can be, indirectly at least, connected to the interoperability between various ICT systems in an enterprise network.

While the first wave of studies sought to document the relationship between investments in computers and potential increases in productivity, recent research is focusing on how to make more computerisation effective in enterprise networks, as Brynjolfsson and Hitt [5.5] pointed out. Computerisation does not automatically increase productivity, but it is an essential component of a broader system of organisational changes that does increase productivity. At this point, standardisation can play an important role.

Within any enterprise network, different components are being upgraded at different times, in different places and in different ways. The result is a highly complex and continuously changing environment, which makes standardisation highly important. The extended possibilities and opportunities for business offered by new networking technologies created an information-based e-economy. Many new companies were born at the end of the twentieth century, but soon a great part of them disappeared. Among the obstacles to their success were a lack of interoperability and too much replicated work, which were problems that could have been overcome through greater standardisation.

Overall, ICT standardisation is beneficial for all parts of society, namely *individuals*, *public administrations* and *enterprises*. Individuals benefit from the additional choices and the lower costs: when standards are used to allow greater ease of access to more systems, the result is extra competition between manufacturers and service providers.

Public administrations benefit from having an instrument for securing policy initiatives:

- ICT standards are vital for the development of interoperable applications, which are important to future economic growth.
- ICT standards provide a measure to judge bids for public procurement tenders. European and national legislation has increasingly referred to European industrial standards to set and demonstrate conformance with health, safety and environmental demands.

Enterprises benefit from:

- Economies of scale: with standardisation, industry can reach a critical mass more quickly, and achieve a return on R&D costs. Standardisation initiatives can open up at least the European market and, if connected to international initiatives, the global market.

- Higher consumer confidence in products or services bought from companies operating according to industry standards.
- Higher sales: interoperable products are more attractive to clients.

5.2 ICT Standards Setting

Technical standards are basic to the exploitation of all technologies. Moreover, introducing standards at the right time is as important as choosing the right technology. While industry takes the lead in ICT evolution, governments usually provide the legal framework for ICT products and services.

For almost 100 years, national and international standards development organisations (SDOs), like CEN (European Committee for Standardisation), CENELEC (European Committee for Electrotechnical Standardisation), ETSI (European Telecommunications Standards Institute) *etc.*, have developed voluntary, consensus-based standards and reduced the need for government-dominated standardisation and state regulation [5.6]. In effect, during the last decades, EU members have assigned – through the European Commission – SDOs with the task of developing certain standards to support ICT. The specific, traditional system for producing formal European standards is rigorous. It ensures consistent quality and guarantees openness and transparency, which are particularly important. However, those traditional approaches to standardisation have often proved to be too slow for ICT, even if the SDOs have introduced a new flexibility to their working methods; such standards take at least two years from conception to adoption. By this time, the market in ICT may have developed further or moved in a different direction.

At the beginning of the twenty-first century, a new standardisation trend has emerged: *market-driven standardisation.* Consortia are often seen as the standardisation organisations that best practice market-driven standardisation (actually, technical standardisation consortia emerged in the 1980s and nowadays, there are more than 400 consortia globally active in ICT). They are part of the expansion of non-governmental organisations that utilise some leverage to by-pass the authority of international organisations and nations. Consortia-driven standardisation is a growing challenge to the heretofore insular community of SDOs that pioneered voluntary, consensus-based standardisation. Consortia are emerging and achieving significant success in providing standardisation services to the same markets and technologies as the SDOs address. They are usually distinguished from SDOs by their lack of accreditation from an independent government-related body. However, another distinction is also true: SDOs represent one or more nations, consortia do not [5.6].

Today, individual commercial companies are the drivers in the development of compatibility standards. When two or more commercial companies support different technologies for a specific standard under development in a nation, a national SDO may not be able to reach consensus. Thus, the national SDO might not bring a unified position to the international SDOs. Consortia on the other hand can gather like-minded companies together to present a unified position wherever they wish. An example of such a case is the lack of a single US position on third generation cellular communications technology. Nevertheless, European companies, with a tradition of respect for standardisation, have developed a common European position supporting GSM cellular communications. In markets with enhanced levels of self-reinforcing effects, the European tradition of respect for standardisation appears more effective than the US desire for market determination. On the contrary, in markets with less enhanced levels of self-reinforcing effects, the US process seems to be more successful.

Consortia support the promotion of a specific commercial agenda on common goals as a requirement for consortium membership. If the goals are clearly stated and acceptable to the significant companies in the desired market or technology, then their successful completion is quite likely. However, the acceptability of a consortium's goals is often a coerced decision for a lot of SMEs and consumer groups that are very unlikely to be represented, not least because they are unable to afford the high participation cost. Besides, when leading industries form a consortium, they may identify a set of goals that are not always in the best interests of other companies in the industry. Consequently, the remaining industries have little choice but to accept the goals presented by the leaders. Resistance would be unproductive, expensive and possibly damaging to business relationships with the industry leaders. Such coercion represents the most socially undesirable aspect of the rise in consortia standardisation.

Two simple reasons are often given to explain the rapid growth of consortia producing standards in the form of technical specifications: consortia have the ability to keep pace with rapid market change, while SDOs need extra time to achieve the consensus necessary for the acceptance of SDO-developed standards. Overall, the advantage of consortia-driven standardisation is that the product can be available quickly, so satisfying commercial needs. Working in closed consortia also gives companies greater control over the release of information that may be commercially sensitive. On the other hand, there are also important problems associated with consortia products. They may be biased, selective and not transparent enough to serve the public interest or they may not be fully competitive. Furthermore, they may risk not meeting European competition law or World Trade Organisation guidelines. The significant differences between consortia and SDOs are identified by Krechmer [5.6] in Table 5.1.

The drawbacks of both aforementioned approaches have led to market demand for a middle way, *i.e.* an open process that combines the tried and tested backing of

Table 5.1. Significant differences between consortia and SDOs

Issue	Consortia	Formal SDOs
Funding source	Often project or product line	Often overhead
Standards development	Varies widely	Trained and well defined
Intellectual property	Negotiation often required	Identified, but not negotiated
National focus	Multi-national	Often regional or national
Brand identification	Not well known	Well known
Standards promotion	Promotion is often funded	Promotion is usually not funded
Compatibility testing	May be offered	Usually not offered
synergy	Legal risks not well tested	Legal risks well tested

the formal standardisation process with a fast, market-driven approach. The SDOs have demonstrated this flexibility with Project Teams of paid individuals to draft documents on a very rapid basis, matched by fast approval procedures.

5.3 Significant References to ICT Standardisation

People, culture and technology are the key building blocks of enterprise networks, supply chains and logistics infra-structures and technologies. The way that the three elements co-exist, co-operate, co-evolve and co-specialise via their interactions, conflicts and synergies determines to a great extent the challenges and opportunities underlying the design and implementation of ICT standards and the process of selecting and upgrading them. The intent of course is to establish and leverage ICT standards that facilitate and enable rather than impede and discourage connectivity, communication and co-operation across organisational, technological and geographic domains. The key is for said standards to be truly supporting and suppressing collaboration, creativity and innovation, meaning that they need to allow for open, flexible, transparent and robust communication that encompasses the requisite information and knowledge sharing. ICT standards can indeed serve as 'bridges' across cultural, technological and organisational divides – if properly chosen, designed and implemented, or as 'separators' and 'buffers' of communities of interest and practice – if not.

As Carayannis and Alexander [5.7] mentioned, ICT standardisation can be considered as a threat on broadband communications. Delivery of specific interactive multimedia applications requires that the carrier support the various standards used to construct those applications [5.8]. As a minimum, all carriers now must offer IP service so as to be serious contenders in broadband communications. But more sophisticated applications, such as virtual reality, will be based upon a dazzling array of standards covering quality of service, compression, data formats and transmission protocols. In addition, industry groups are developing their own standards. For example, the US automotive industry has created ANX, which now certifies which carriers provide IP services that meet the needs of the industry (see below for extensive reference on ANX). Moreover, specific standards are emerging in chemicals, logistics and electronics. Satellite providers will need to track all of these standards developments and ensure that they can in fact support the emerging dominant standards, or they will be shut out of important end-user markets.

Carayannis and Sagi [5.9] discussed an interesting issue that can be connected to the role and the importance of ICT standards on enterprise networks, supply chains and logistics. To be more precise, they investigated the effects of national culture on international ICT systems and networks. Global and multinational corporations are increasingly relying upon information systems that are developed and operated across a multicultural environment. Teams often consist of professionals who vary greatly in their cultural dimensions. However, national cultural differences may contribute to the failure (or success) of these systems.

ICT standards may be the basis to clear the negative consequences of cultural differences and can contribute so that the latter play a role in the design and use of global information systems and networks. For example, Cairncross [5.10] believes that English will be the 'standard' language – consequently a simple form of an ICT standard – with which to cross-communicate in various networks.

Carayannis *et al.* [5.11] mentioned that proponents of increased spending on information technology (*e.g.* Schwartz and Leyden [5.12]) argue that the spread of computing power throughout any business, combined with the ability of individuals and computing devices to communicate through enterprise networks, will lead to very different forms of business transactions and relationships. Note that communication through enterprise networks demands interoperability, which in turn demands ICT standards, as mentioned previously.

In addition, Carayannis *et al.* [5.11] refer to *Envera*, which was founded by Ethyl Corporation, a mid-sized player in the chemicals industry, in August 1999. The official site was developed based on Ethyl's own experiences in building a B2B extranet for its customers. Ethyl then developed the idea to open the extranet to a larger number of chemical producers, while maintaining the focus of the system on selling to core chemical purchasers. The key technical development was the creation of a clearinghouse for all transactions, which *Envera* refers to as the 'single point of contact', using XML to enable standardised exchange of industry data. Now an independent entity, *Envera* refers to its system as 'B4B' (business for business) site, since it is building the basic infrastructure to enable B2B electronic commerce.

Chang and Shaw [5.13] believe that ICT standards have become increasingly important in enhancing supply chain management. Firms that create successful ICT standards can seize new opportunities in industrial collaboration, while firms that are locked out of standardisation processes face difficulties. However, the implementation of ICT standards also becomes critical, while corporate supply chains have become more networked and complex. Chang and Shaw [5.13] have developed an evaluation methodology for measuring the costs and benefits of implementing ICT standards. This methodology is able to evaluate the value from the perspectives of business process, products, information infrastructure, customers and trading partners. At the same time, it simultaneously keeps track of both the internal and the external business environments that influence the success or failure of ICT standards.

Regarding the EU e-procurement experience and supply chain for central and eastern European countries, Carayannis and Popescu [5.14] mentioned that besides preparing legislation and encouraging standardisation at the European international level, the European Commission has launched the SIMAP public-procurement information system, which aims 'at supporting an effective single market by encouraging suppliers and contracting entities to adopt best practices and use electronic commerce and information technology to provide all the information needed to deliver value for money in public procurement'. SIMAP is the EU website on electronic procurement (http://simap.europa.eu/index_en.html), and has been launched in order to promote the use of information technologies for public procurement in EU. The website–network became the opportunity to move towards the use of ICT in the public procurement procedures. Initially, SIMAP was designed to address the provision of information about the EU procurement opportunities to all interested suppliers, with the longer term goal of addressing the whole procurement process, including bids, award of contracts, delivery, invoicing and payment.

Overall, one can argue that the EU e-procurement pilots (*e.g.* SIMAP) have largely achieved its objective to create a common system that can be used by any government in the EU to post procurement notices. Now, millions of suppliers in EU can have on-line access to public procurement information. Therefore, Carayannis and Popescu [5.14] believe that in the relatively short period since SIMAP was established, creation of a single open information market has increased transparency and reduced transaction costs. These are the evident results of its capacity to standardise forms, create a contracting entities database, agree upon a common procurement vocabulary and support the notification function by an efficient information system. The e-procurement pilots improve current practice and quality as well as the information flow among the key actors of the EU public procurement system. However, while EU activities were mainly focused on electronic transmission and dissemination of notices, it seems that there is significant support for systems to handle the entire procurement process electronically.

McAfee [5.15] claims that there are three categories of ICT, each of which provides different organisational capabilities and requires very different kinds of management interventions. Specifically, *function* ICT encompasses technologies – such as spreadsheet and word-processing applications – that streamline individual tasks; *network* ICT includes capabilities like e-mail, instant messaging, and blogs and helps people and enterprises communicate with one another; and *enterprise* ICT is actually all enterprise networks that allow approaches such as customer resource management and supply chain management, and let companies create interactions between groups of workers or with business partners. Moreover, enterprise ICT helps standardise and monitor work, operating similarly to an ICT standard.

Hans *et al.* [5.16] stated that the evolution of today's business towards enterprise networks has led to increasing customer expectations regarding the performance of logistics systems that have to be simultaneously reliable, robust and cost-effective. In order to fulfil these needs, the application of new technologies (*e.g.* radio frequency identification or RFID, which will replace manual identification and bar code technology in many logistics applications during the coming years) is an absolute must. Consequently, they propose a service-oriented approach towards the integration of logistics data, which aims at combining existing systems and standards, thus overcoming today's data and information barriers.

5.4 ICT Standardisation – Why the Best Does Not Always Win

Passell [5.17], who writes about economics for *The Times*, talks about Apple Computer, Inc. At that time, the company that brought the extremely friendly Macintosh was staring at bankruptcy. This was neither a bad break for Apple nor a rare exception to the Darwinian rules in which the best products win the hearts and dollars of consumers. Economists were finally beginning to acknowledge what others had long suspected: *the best does not always win.* Just as biologists are challenging the idea that natural selection drives evolution along 'efficient' and predictable paths, economists are discovering the disorder of their simple, elegant models of capitalist progress. It seems that superior technology will not always survive in the free market.

Recent wisdom on this subject dates back to 1985, the year that P.A. David, an economic historian at Stanford University, published an article about the QWERTY standard [5.18]. Q-W-E-R-T-Y are the first six letters on the upper left of the typewriter keyboard; the universal standard since the 1890s. This layout has prevailed over half a dozen other keyboard layouts that are said to permit faster typing. David [5.18] considers that this happened because QWERTY was the solution to a fleeting technological problem, an arrangement that would minimise the jamming of keys in primitive typewriters. While this explanation has since been challenged, what matters is that one keyboard, chosen for reasons long irrelevant, remains the standard. Competing designs have made about as much headway against QWERTY as Esperanto has made against English! That is because a standardised layout allows typists to learn just one keyboard in order to use all. Once thousands of people had learned to type using QWERTY's merely adequate layout, the technology was effectively locked in. Keyboard design is thus the classic example of 'path dependence', the idea that small, random events at critical moments can determine choices in technology that are extremely difficult and expensive to change.

In the typical path-dependence scenario, producers or consumers see one technology as slightly superior. This edge quickly snowballs into clear economic advantage: production costs fall with greater experience in manufacturing and consumer acceptance grows with greater familiarity. And along the way, the weight of numbers makes the leading product more valuable than one based on competing technologies.

Free marketers fear that 'path dependence' will become a rationale for 'bigger' government; if competitive markets do not guarantee that the best technologies survive, then governments will be more tempted to try to pick winners. However, a world haunted by 'path dependence' asks for more than a government to be the referee who makes everyone play by the same impartial rules. First, any government should slow down and think twice before setting hard-to-reverse technological standards. Besides, the more controversial issue is antitrust – consider Microsoft. It often pays an individual company to set a standard by flexing its own marketing muscle long before a clear winner has emerged. Government will no doubt be called on to take a stand on some looming path-dependence battles: all-purpose personal computers versus cheaper, appliance-like network computers that do one thing well, wireless personal communications versus high-capacity cable, Internet software built around web browsers versus software that piggy-backs on the Microsoft Network, *etc.*

Several years later, Lohr [5.19] gave a different explanation about the fact that old (and usually not the best) technologies are still kicking. He first reminds that Stewart Alsop, the editor of InfoWorld and a thoughtful observer of industry trends, predicted in 1991 that the last mainframe computer would be unplugged by 1996. In February 2008, IBM introduced the latest version of its mainframe, the aged yet remarkably resilient warhorse of computing.

Today, mainframe sales are a tiny fraction of the personal computer market. But with the mainframe facing extinction, IBM retooled the technology, cut prices and revamped its strategy. A result is that mainframe technology – hardware, software and services – remains a large and lucrative business for IBM, and mainframes are still the back-office engines behind the world's financial markets and much of global commerce.

The mainframe stands as a telling case in the larger story of survivor technologies and markets. The demise of the old technology is confidently predicted, and indeed it may lose ground to the insurgent, as mainframes did to the personal computer. But the old technology or business often finds a sustainable, profitable life. Television, for example, was supposed to kill radio, and movies, for that matter. Cars, trucks and planes spelled the death of railways. A current death-knell forecast is that the Web will kill print media.

What are the common traits of survivor technologies? First, it seems that there is a core technology requirement: there must be some enduring advantage in the old technology that is not entirely supplanted by the new. But beyond that, it is the business decisions that matter most: investing to retool the traditional technology, adopting a new business model and nurturing a support network of loyal customers, industry partners and skilled workers. Experts say that the unfulfilled predictions of demise tend to overestimate the importance of pure technical innovation and underestimate the role of business judgment.

To survive, technologies must evolve, much as animal species do in nature. Indeed, John Steele Gordon, a business historian and author, observes that there are striking similarities in the evolutionary process of markets and biological ecosystems. Dinosaurs, he notes, may be long gone, victims of a change in climate that better suited mammals. But smaller reptiles evolved and survived, and today there are more than 8,000 species of reptiles compared with around 5,400 species of mammals.

As a media technology, radio is an evolutionary survivor. Its time as the entertainment hub of American households in the 1930s and 1940s gave way to the rise of television. TV replaced radio as the box that families gathered around in their living rooms. Instead, radio adopted shorter programming formats and became the background music and chat while people ride in cars or do other things at home; 'audio wallpaper' as Paul Saffo, a technology forecaster in Silicon Valley, puts it.

While television did pose a threat to movies, it also served as a prod to innovation, including failures like Smell-O-Vision but also wide-screen, rich-colour technologies like Cinerama and CinemaScope. The idea was to give viewers a more vivid, immersive experience than they could possibly have with television.

Today movies, like other traditional media, face the digital challenge of the Internet. Paul Saffo is betting that after a period of adjustment and experimentation, they will make another life-prolonging adaptation. 'Technologies want to survive, and they reinvent themselves to go on', he said.

The survivors also build on their own technical foundations as well as the human legacy of people skilled in the use of a technology and the business culture and habits that surround it. A change in the economic environment can sometimes lead to the renaissance of an older technology. Railroads, for example, have enjoyed a revival of investment recently as rising fuel costs and road congestion have prompted shippers to move from trucks to trains; some travellers, too, have opted for railways, along routes like the Boston–New York–Washington corridor.

The weight of legacy is underestimated, according to John Staudenmaier, editor of the journal Technology and Culture, because innovation is often portrayed as a bold break with the past. A few stories of technological achievement fit that mould, but they are rare indeed.

5.5 Automotive Network Exchange: an Excellent Example of an Enterprise Network

Gooch *et al.* [5.20] mentioned that the automotive and finance industries have been among the first to realise the potential benefits of virtual private network (VPN) technologies. US motor manufacturers quickly recognised IP enterprise networks as a more efficient way of communicating with partners, suppliers, customers, and attracting new customers. The result was the US ANX, the world's first commercial secure IP enterprise network. ANX has extended into Europe and Asia/Pacific, and has begun to offer its services to customers outside the automotive industry.

Overall, ANX is a TCP/IP enterprise network comprised of trading partner subscribers, certified service providers (CSPs) and network exchange points allowing for efficient and secure electronic communications among subscribers, with only a single connection. The key components of ANX architecture are:

- standards;
- CSPs;
- certified network exchange points operated by certified exchange point operators;
- central monitoring and administration by the ANX Overseer;
- a PKI (public key infrastructure) provided by the ANX certificate authority service providers;
- the public Internet comprised of Internet service providers connected via Internet exchange points.

In addition to ANX, the following networks have been formed (see Table 5.2 for details): the Australian Network (AANX); the Japanese Network (JNX); the European Network (ENX); and the Korean Network (KNX). All these networks are to be connected through ANXGlobal (Figure 5.1).

Table 5.2. Basic information about various ANX networks

Country	Name of network	Starting date	Major participants
USA	ANX	1996	General Motors, Ford and Chrysler
Korea	KNX	1998	Hyundai-Kia, Daewoo and Samsung – Ministry of Commerce Industry and Energy
Australia	AANX	1999	Ford, Holden, Mitsubishi and Toyota – FCAI, FAPM and MTAA
Japan	JNX	2000	JAMA and JAPIA
Europe	ENX	2000	Audi, BMW, Bosch, DaimlerChrysler, DGA, Ford, Karmann, Porsche, PSA Peugeot-Citroën, Renault, Siemens VDO Automotive, Smart GmbH, Volkswagen, ANFAC, GALIA, SMMT and VDA

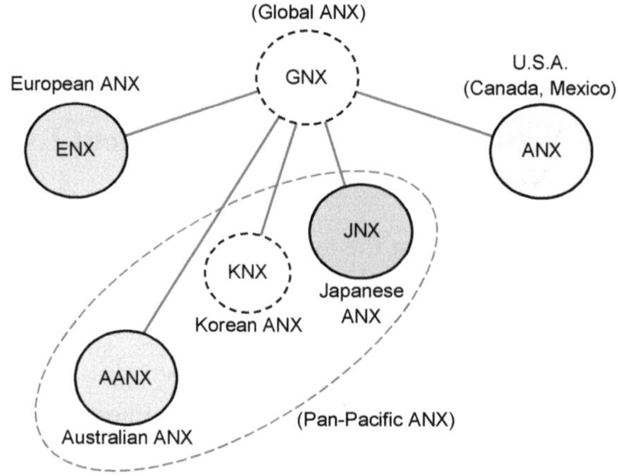

Figure 5.1. The structure of ANXGlobal

5.5.1 The US ANX

The US ANX is a private enterprise network that was initially set up and maintained by the big three automakers through the Automotive Industry Action Group (AIAG), namely General Motors, Ford and Chrysler. It was built around 1996 to provide consistent, reliable speed and guaranteed security for data transmissions between the automakers and the companies that they do business with. A few years later (1999), AIAG sold the ANX enterprise network assets and operations to Science Applications International Corporation (SAIC), which formed ANXeBusiness[1] to grow the network and support the ANX Community. SAIC spun off ANXeBusiness, which is now a subsidiary company of One Equity Partners. Since its introduction, over 4,000 companies have joined the ANX enterprise network.

According to the information provided at the official website of ANXeBusiness (http://www.anx.com), nowadays, the ANX enterprise network is the world's largest multi-provider-managed private network for business (Figure 5.2). Offering guaranteed availability, security and bandwidth, it is a global 'Internet for business' on which some of the world's largest companies depend to communicate with their suppliers, customers, partners and remote locations. Since its inception, the ANX enterprise network has become the trusted foundation for some of the world's most sophisticated and mission-critical e-business. Today, companies with more than $1 trillion in total revenues use the network for their business communication needs. ANX-connected companies use the network to achieve a variety of critical business communications, ranging from reliable file transfer, secure e-mail, Electronic Data Interchange (EDI) transactions (such as parts ordering and shipping notices), financial transactions, complex collaborative engineering, network-based videoconferencing and many more. ANX customers also make hundreds of their applications available to their customers and extended supply chains over the ANX Network.

---

[1] Nowadays, ANXeBusiness manages and operates the network.

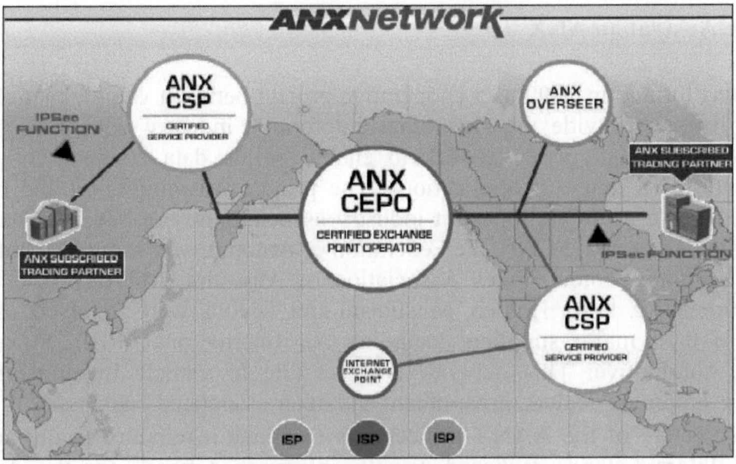

Figure 5.2. ANX enterprise network (from http://www.anx.com/anx_network.jsp)

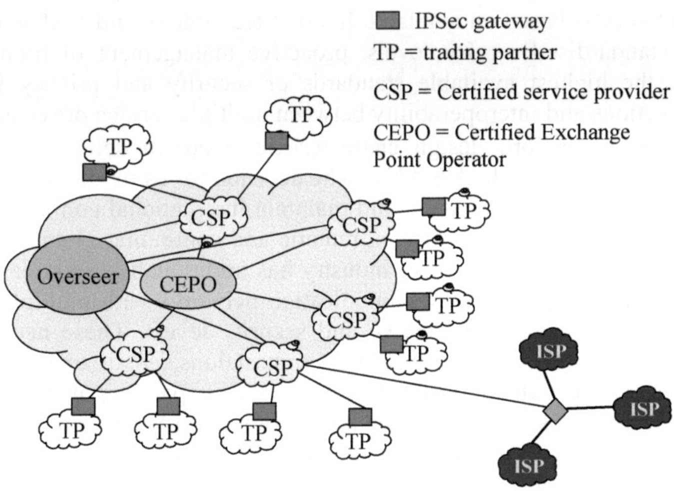

Figure 5.3. Conceptual design of ANX [5.21]

To become a member of the ANX Network, a company needs to subscribe via a CSP. Currently, there are only five CSPs (Figure 5.3) due to the strict regulations and high specifications, namely SBC, Bell Canada, Verizon Business, AT&T and LDMI Telecommunications. The ANX certification process includes a rigorous set of over 120 service quality metrics broken into the following categories: network service, interoperability, performance, reliability, business continuity, security, customer care and trouble handling. Internet service providers wishing to provide ANX services must meet 100% of the service quality metrics in order to become ANX CSPs. Once certified, the CSPs must verify that they continue to meet ANX service quality requirements on a regular basis.

5.5.2 The Australian ANX

AANX was formed in 1999 as a cooperative project between vehicle manufacturers and suppliers to provide the Australian automotive industry with a single, cost-effective private enterprise network to enable online data connectivity between participants for a range of applications. The project was supported and run by a committee made up of the relevant industry associations: the Federal Chamber of Automotive Industries (FCAI), the Federation of Automotive Product Manufacturers (FAPM) and the Motor Trades Association of Australia (MTAA). The four car manufacturers, *i.e.* Ford, Holden, Mitsubishi and Toyota, were involved, as were a number of their major suppliers including Air International, PBR, Plexicor and Tenneco Automotive. This network is accessible to vehicle manufacturers and importers, suppliers, dealers, government and other associated businesses.

The objective of the AANX project was to create a reliable, secure and well-managed Internet-standards based private enterprise network for the Australian automotive industry and its constituents, to provide a platform for conducting domestic and international B2B e-commerce activities. It allows for the timely transmission and exchange of confidential data and business critical messages. This enterprise network is based on available Internet technology and is characterised by agreed and standardised service levels, proactive management of trading partner connections, the highest available standards of security and privacy for trading partner transactions and interoperability between multiple service providers.

The enterprise network design characteristics were modelled largely on the specification developed by the US ANX. The decision for using the latter as a model was motivated by the need to create and maintain international communication and security interoperability in light of the dramatic and continuing globalisation of the automotive industry. The automotive industry has traditionally used a large number of legacy computer systems and communication networks with multiple protocols, multiple links and inconsistent service and security levels. These networks often support only one application, such as EDI transactions, email or computer-aided design file exchange, which means that two trading partners may have several different electronic links with associated duplication of costs and infrastructure.

As far as the AANX structure is concerned, it should be noted that it gives its connected trading partners a choice of communication service providers to enable greater levels of network redundancy for business critical applications. This decision recognises the need to support interoperability between data networks to benefit users and the communication industry. AANX is a multi-provider, virtual private enterprise network where the service providers compete for customers, but comply with common service quality requirements, including security. All trading partners share the same physical infrastructure of the AANX. Within this framework, each electronic conversation occurs via a secure, private logical connection between the two trading partners involved.

Connect Internet Solutions and *Equant* provide the communication services for the network. *KeyTrust* (the trading name of Network Designers Australia) acts as the certificate authority and vendor for managed IPSec security services. High levels of reliability and performance are essential for business transactions carried out between automotive companies. *Connect Internet Solutions* provides communication carriage and network management for AANX connected trading partners, with the managed service of customer premises equipment facilitating proactive monitoring of trading partner connections. *Equant* also acts as a communication carriage and network management service provider for AANX. It carries network traffic over its private network backbone and provides the exchange point operator for AANX, which creates a defined demarcation point between communication service providers. *KeyTrust*, as the network's certificate authority services provider, is responsible for the management of all e-security services and the PKI. It provides the AANX enterprise network with four key security elements:

- Secure data transmission: this is achieved through the use of the IPSec protocol operated under the *KeyTrust* managed services program. IPSec is an industry standard for secure communication over both public and private data networks. In the case of AANX, this is implemented through hardware encryption gateways for permanent connections or client-based encryption software for dialup connections, which automatically encrypt and authenticate all transmissions traversing the network.
- PKI digital certificates: PKI-based digital certificates are used within the AANX enterprise network so that all participants can experience a high level of confidence when transacting over the network. This is achieved through the authentication and identification of all parties taking part in any secure communication session.
- AANX community directory: it is the central policy repository used by security gateways when establishing sessions between trading partners. It contains a *map* of all participants along with their *electronic relationships* and access privileges.
- *KeyTrust* professional managed services: *KeyTrust* monitors and manages the security gateways on a 24×7 basis through its Secure Network Control Centre in Melbourne.

The AANX project covered a specialised area of technological capability that is normally left to information technology and communication experts within each company, but can have significant impact on the competitive standing of each company and the industry as a whole. Implementing the AANX enterprise network enables the local automotive industry to move some of its core business processes online. The AANX has standardised network and security platform, reducing the need for bilateral network design and implementation efforts between each new pair of trading partners and each new application. When a critical mass is achieved, the ability to support multiple applications between multiple trading partners over a single network connection will provide ongoing cost benefits to all participants.

The task of developing an agreed standard that could support the industry's specific application requirements was left to a recognised, respected and non-competitive body accepted by the industry. The task of implementing this standard and making it available as a product from a number of vendors and service providers (in this case communication carriers) was helped by gaining credibility and support from relevant industry associations, major industry participants and the Federal Government.

Specific business knowledge and understanding of key business processes within the industry was required to determine a suitable solution that had sufficient depth and breadth of capabilities to be useful. Personal networking with a wide range of contacts in the automotive and communication industries was required to reach consensus on the requirements and acquire the necessary goodwill and resources for testing and implementing solutions in the pre-commercial phase.

In the early stages of such a venture, the commercial viability is questionable and delicate; a situation not dissimilar to experiences gained from many other technologically-oriented projects and ventures, locally and globally. The project needed long-term support from participants to achieve its desired outcome.

5.5.3 The Japanese ANX

Over the past decades, carmakers in Japan, like Toyota and Honda, won their market positions in part by developing highly efficient, geographically clustered supplier networks [5.22]. For instance, almost all Toyota's factories and suppliers are located within an hour's drive of Nagoya city, allowing suppliers to deliver requested parts to a factory going one way and carry new orders heading the other. Of course, eventually Toyota started faxing and emailing the orders instead of physically delivering them.

This procurement system is very efficient, but it does not permit the carmaker to get optimal prices. Taking also into consideration the advances in networking technology and the competitive pressures brought on by rapid globalisation, one can see several reasons for busting up automotive alliances. After all, why should a carmaker order parts from a supplier just because it is within a one-hour drive when, using a leased-line EDI network, it can order parts just as quickly, and at much lower prices, from a supplier many miles away? However, EDI systems are proprietary in nature; consequently, each of the 13 major carmakers has a different system. For example, Denso, the nation's largest parts supplier, does business with all 13, so it has to deal with 13 different computer systems and 13 different leased lines! And Denso is not alone.

To eliminate such inefficiency, Japan Automobile Manufacturers Association (JAMA) and Japan Auto Parts Industries Association (JAPIA), *i.e.* two big industry associations of Japan, jointly promote standardisation. The first step in their quest is the Japanese ANX (JNX), which standardises information exchange among all suppliers and carmakers in Japan. JNX has been launched in October 2000, after having been tested by 25 suppliers and eight carmakers.

JNX is a standard enterprise network for the automotive industry in Japan and utilises a standard communication technology used by the Internet. This helps suppliers to connect and communicate with all automobile manufacturers with a single link and a single protocol. It reduces communication and operation costs. JNX is designed as a reliable, secured, and high performance infrastructure for the supply chain management, which improves the information flow and reduces the time to the market.

In order to establish a standard communication network, JNX has standardised legacy communication protocols into single communication protocol (TCP/IP) to build an open industrial extranet. In addition, high performance, reliable, and secured service is provided by CSPs. Those service levels is always monitored and managed by JANX overseer.

5.5.4 The European ANX

ENX is the communications enterprise network of the European automotive industry. By using ENX, manufacturers and their suppliers are able to securely communicate and exchange even the most sensitive data via a range of accesses. The high level of security is maintained through ENX's CSPs, which operate the networks' infrastructure separately from the public Internet. It was founded in June 2000. ENX is an association of manufacturers, suppliers and associations from the European automotive industry and it is based on French law. Its members are Audi, BMW, Bosch, DaimlerChrysler, DGA, Ford, Karmann, Porsche, PSA Peugeot-Citroën, Renault, SiemensVDO Automotive, Smart GmbH, Volkswagen, ANFAC (Spain), GALIA (France), SMMT (UK) and VDA (Germany).

5.5.5 The Korean ANX

KNX is an automotive industrial enterprise network launched in November 1998. The Ministry of Commerce, Industry and Energy as well as the Korean 'big three' (Hyundai-Kia, Daewoo, and Samsung) were initial promoters of the project. It has begun operation with some 350 Hyundai and Kia business partners as its first subscribers. From March 2002, this network is used by four carmakers, namely Hyundai, Kia, Renault Samsung and GM Daewoo, as well as 1,200 parts makers, to exchange information such as material procurement and design (about 250,000 cases a day).

KNX is a system that integrates network systems that were operated separately by each firm. In actuality, it is an automobile business network, which combines secure transmission of the private network and convenience of the Internet. It has become a user-oriented enterprise network that guarantees a service-level agreement authenticating the quality of telecommunication service defined in the KNX specification. KNX provides a better service than either the Internet or Extranet in guaranteed bandwidths, encryption (VPN), cost, network operation, flexibility (set up, changes, scalability) and application support activity.

5.6 Conclusions

Standards establish a bridge between research results and the implementation of innovative products. Therefore, standardisation is an essential component for boosting innovation. However, timing is essential for standardisation: an early start provides better chances for being successful. Moreover, the current pace of technological development pushes standardisation and research to proceed in parallel.

As regards ANX in particular, it could be said that although it has started as a way to standardise EDI in the automotive industry, it has evolved into a supply chain network that is not bound to any industry or country. ANX has already gone beyond the automotive industry, and there are many plans to continue that growth. A step in that growth is the global aspect of ANX, and making this an international network, which it is planned to be done with ANXGlobal.

The cases that we reviewed and others that manifest the role that the nature, scope, comprehensiveness and timeliness, as well as the degree of adoption by technology developers and users alike of ICT standards, makes them into key triggers and drivers of as well as impediments to innovation. In particular, innovation in agile manufacturing, where both the scale and scope of the innovation's impact as well as the speed of change are critical, can be greatly impacted by the quality and functionality of the underlying ICT standards. If issues of backward/forward compatibility, interoperability or portability arise, for instance, this could seriously hamper the proper speed and acceleration of the diffusion and adoption of ICT innovations – potentially with disruptive capacity – resulting in a hollowing out of the competitiveness of the business ecosystem that relies on this particular set of ICT standards, including logistic chains and other business transactions-enabling modalities.

References

[5.1] ICT Standards Board, http://www.ictsb.org.

[5.2] Carayannis, E.G. and Sipp, C., 2006, *e-Development toward the Knowledge Economy: Leveraging Technology, Innovation and Entrepreneurship for 'Smart Development'*, MacMillan, Oxford.

[5.3] Gasser, U. and Palfrey, J., 2007, *Breaking Down Digital Barriers – When and How ICT Interoperability Drives Innovation*, Berkman Publication Series, The Berkman Center for Internet & Society, Harvard University, Boston, MA.

[5.4] Brynjolfsson, E., 1993, "The productivity paradox of information technology: review and assessment," *Communications of the ACM*, **36**(12), pp. 67–77.

[5.5] Brynjolfsson, E. and Hitt, L.M., 1998, "Beyond the productivity paradox: computers are the catalyst for bigger changes," *Communications of the ACM*, **41**(8), pp. 49–55.

[5.6] Krechmer, K., 2000, "Market driven standardization: everybody can win," *Standards Engineering*, **52**(4), pp. 15–19.

[5.7] Carayannis, E.G. and Alexander, J., 2001, "Virtual, wireless mannah: a co-opetitive analysis of the broadband satellite industry," *Technovation*, **21**, pp. 759–766.

[5.8] Price, D., 1999, "Deciphering the hieroglyphics of multimedia by satellite," *Satellite Communications*, **23**(5), pp. 34–42.

[5.9] Carayannis, E.G. and Sagi, J.R., 2001, "Dissecting the professional culture: insights from inside the I.T. 'Black Box'," *Technovation*, **21**(2), pp. 91–98.

[5.10] Cairncross, F., 1997, *The Death of Distance: How the Communications Revolution Will Change Our Lives*, Harvard Business School Press, Boston, MA.

[5.11] Carayannis, E.G., Alexander, J. and Geraghty, J., 2001, "Service sector productivity: B2B electronic commerce as a strategic driver," *Journal of Technology Transfer*, **26**(4), pp. 337–350.

[5.12] Schwartz, P. and Leyden, P., 1997, "The long boom," *Wired*, July, pp. 115–172.

[5.13] Chang, H. and Shaw, M., 2004 "Evaluating the economic impacts of IT-enabled supply chain collaboration," In *Proceedings of the Eighth Pacific-Asia Conference on Information Systems*, Shanghai, China.

[5.14] Carayannis, E.G. and Popescu, D., 2005, "Profiling a methodology for economic growth and convergence: learning from the EU e-procurement experience for central and eastern European countries," *Technovation*, **25**, pp. 1–14.

[5.15] McAfee, A., 2006, "Mastering the three worlds of information technology," *Harvard Business Review*, November.

[5.16] Hans, C., Hribernik, K.A. and Thoben, K.-D., 2008, "An approach for the integration of data within complex logistics systems," In *Dynamics in Logistics*, H.-D. Haasis *et al.* (eds.), Springer, Heidelberg, pp. 381–390.

[5.17] Passell, P., 1996, "Why the best doesn't always win," *The New York Times*, May 5.

[5.18] David, P.A., 1985, "Clio and the economics of QWERTY," *American Economic Review*, **75**(2), pp. 332–337.

[5.19] Lohr, S., 2008, "Why old technologies are still kicking," *The New York Times*, March 23.

[5.20] Gooch, D.J., Hubbard, S.D., Moore, M.W. and Hill, J., 2001, "Firewalls – evolve or die," *BT Technology Journal*, **19**(3), pp. 89–98.

[5.21] Les Cottrel, Automotive Network eXchange (ANX), 1998, [online] Slides 1 of 19, Presentation at ICFA-NTF at CERN, retrieved from: http://www.slac.stanford.edu/grp/scs/net/talk/icfa-anx/icfa-anx.PPT.

[5.22] Tachino, K., 2000, "Short cuts: B2B in Japan gets net-worked," *Japan Inc Magazine*, March.

6

Collaborative Demand Planning: Creating Value Through Demand Signals

Karine Evrard Samuel

Centre of Studies and Research in Management
University of Grenoble
UMR 5820 CNRS-UPMF, 150 rue de la Chimie, BP 47
38040 Grenoble Cedex 9, France
Email: karine.samuel@upmf-grenoble.fr

Abstract
This chapter focuses on collaborative demand planning particularly when information is shared in the downstream supply chain between manufacturer and retailer. The use of collaborative practices is transforming traditional supply chain models toward demand-driven supply chains but implies deep organisational changes to prompt vertical alignment. The analysis of practices that permit efficient collaboration between manufacturers and retailers shows that information sharing on demand signals in supply chains is one of the keys to responding to retail demand with greater agility. This chapter aims to show how a manufacturing supply chain needs to be aligned with the retail supply chain in order to create value for the trading partners and for the final consumer. Through the analysis of three case studies, it is attempted to identify which practices allow efficient collaborative demand planning. Regarding the findings, different types of demand signals are identified through the planning process and allow one to highlight some breaking points that prevent the alignment and the optimisation of the retail chain. Research implications are the identification of four steps in the demand planning process that will help managers to better understand which actions should be taken to improve their collaboration practices. The originality of this chapter lies in the fact that it goes beyond historical demand figures analysis and focuses, rather, on information sharing concerning demand signals within supply chains as one of the keys to responding to retail demand with greater agility.

6.1 Introduction

One of the main challenges faced by firms in the current environment is improving customer satisfaction, service and competitiveness at a worldwide level. The retail supply chain plays an important role in achieving this goal since it includes the consumer in the supply chain planning process.

Consumer demand has dramatically changed over the past ten years. The consumer can now buy what he/she wants at the price he/she wishes to pay. If the product is not available at one store, he/she would not hesitate to go to another or to buy the product on-line. This transformation of demand creates more complexity and, of course, more costs along supply chains. Supply chain models have to adapt to complex retail strategies in order to better meet consumer needs. According to the first retail barometer launched by IDC and Microsoft Retail in May 2007, the retail industry has invested more than 4.7 billion euro in information systems to improve customer relationship management at the store level, in supply chain management, and in replenishment planning. In Europe, retailers face major challenges: reduction of profit margins, increased regulation pressures to protect local 'corner shops' and lower consumer purchasing power. There is also greater competition (hard discounts, city centre shops and e-commerce) as well as an increasing difficulty in securing customer loyalty. Retailers are developing new generations of shops and more than never, are working on the improvement of demand forecasting to better anticipate consumer behaviour. On the other hand, more and more evolved information systems are necessary to develop multi-channel distribution systems.

Even if retailers have become increasingly dominant in the fight for a share of the customer's wallet, they still need vendors to help them create additional value for the end consumer. However, manufacturers and retailers do not see the value of collaboration equally. According to a recent study from Forrester Research [6.1], manufacturers want to focus on demand planning and trade promotion management, whereas retailers are more interested in assortment planning and inventory management.

These differences of perception, concerning what collaboration should be, create a breaking point between industrial firms and retailers. Industrial firms are looking for long-term competitive advantage and improving their ROI[1]. They want to build fast, responsive and low-cost supply chains. On the other hand, retailers tend to develop multi-channel strategies to increase their own performance (gross margin, GMROI[2], sales per square foot, turns, stock-to-sales, *etc.*). If the consumer is the most important supply chain participant, he/she sometimes seems to be forgotten in this continuous conflict of interests.

This chapter aims to show how the manufacturing supply chain needs to be aligned with the retail supply chain in order to create value for the trading partners (manufacturer and retailer) and of course, for the final consumer. In the first part, a brief literature review will enable us to analyse current collaborative demand planning initiatives. The use of collaborative practices transforms traditional supply chain models toward demand-driven supply chains but implies deep organisational changes to prompt vertical alignment. We analyse how collaborative demand planning contributes to the improvement of value creation. Three case studies will then be presented and will allow us to highlight some breaking points within supply chains that prevent the alignment and optimisation of the retail chain, and consequently damage the quality of customer service.

---

[1] Return on investment.

[2] Gross margin return on investment (GMROI) is a measure that helps the investor, or management, see the average amount that the inventory returns above its cost. A ratio higher than 1 means the firm is selling the merchandise for more than it costs to acquire it.

In the last part, cross-case analysis allows us to propose a four-phase framework for the demand planning process that identifies demand signals and technologies used to share information. The analysis of practices that permits efficient collaboration between manufacturers and retailers shows that information sharing on demand signals in supply chains is one of the keys to responding to retail demand with greater agility. The objective of this research is to discover what kind of information should be exchanged, in addition to historical demand figures, and how and when during the planning process this information should be shared.

6.2 Creating Value by Implementing Demand-driven Supply Chains (DDSC)

Traditionally the supply chain has been driven from the back, by producers and manufacturers who drive products to markets. In a traditional supply chain, products are pushed downstream towards end consumers. This model is linear in its approach. Businesses in the supply chain merely accept demands resulting from orders received from businesses in front of them. They rarely have any vision of true market demand for a product. To maintain downstream momentum in order to reduce inventory investments, upstream businesses have to constantly exert pressure on the downstream businesses to place orders. In this environment, demand can often be erratic and therefore hard to predict. Items can go from a situation of being under-stocked to being over-stocked in a very short period of time, and businesses across the supply chain do not have timely or accurate information that would allow them to balance the turbulence. The value creation process is rather slow and extremely divided up within the supply chain.

Recent advances in fact-based methods for supply chain management have opened opportunities for its coordination with demand management [6.2]. Demand-driven supply chains are driven from the front by customer demand. Instead of products being pushed to market, they are pulled to market by customers. This model of supply chain requires that companies in a supply chain work more closely to shape market demand by collaborating and sharing information. By doing so, they will have greater visibility and more timely information concerning the demand. The aim of this collaboration is to better position all of the actors, to improve their ability to follow market demands more closely and together produce what the market wants.

The collaboration concept is often badly defined despite the fact that it is frequently used in research in many areas like sociology, economics, information theory, and more recently supply chain management [6.3–6.5]. In the past, two groups of researchers have concentrated their work on the problem of coordination: sociologists on one hand and information systems' theoreticians on the other [6.6–6.11].

The first sociological research, by Fayol [6.12] and Gulick and Urwick [6.13], concentrated mainly on the question of formal authority and direct supervision within organisations. Assets were clearly positioned and controlled inside one firm which theoretically formed a consistent organisation coordinated by distinct tasks resulting from the division of work. Thompson [6.14] proposed a detailed typology of the coordination mechanisms, which has been useful in understanding interdependence modes: reciprocal, sequential or attached to the community. Tushman and Nadler [6.15] pursued this reasoning and proposed the idea according to which information selection should correspond to the environment. In fact, the theoretical foundations of information systems' theory stem from these sociological studies. In a way, coordination can be considered as a joint decision taken in a context where several choices are possible [6.16].

In a supply chain context, coordination does not take place inside one firm but between several organisations interacting with each other. For a long time, transaction costs theory (TCT) has been used to explain governance structures beyond organisational boundaries. Williamson [6.17] justified the existence of firms through the opposition between markets and hierarchies, where hierarchies have a coordination role within organisations. However, TCT has never explained how organisations create value by coordinating their actions and why they need to collaborate.

A traditional supply chain focuses on optimising the internal system. In a demand-driven supply chain, participants are all able to take part in the process of shaping demand as opposed to merely accepting data such as warehouse withdrawal or store receipts. Where businesses traditionally had little or latent insight into market demand, the collaborative technologies employed in implementing a demand-driven supply chain have the overall effect of reducing and even eliminating the gap between upstream businesses and the end consumer. This gives them more accurate and timely information on market trends and enables them to increase the accuracy of their forecasts and hence their ability to interpret and respond to demand fluctuation. This type of market intelligence impacts more than just a business' ability to plan operations; it translates directly into reduced inventory holdings across the supply chain, which, in turn, means an overall reduction in the amount of capital invested therein and the associated risks. Value is jointly created by all partners and theoretically, shared between them.

Research indicates that participation in demand-driven supply chains can be directly translated into improved business performance [6.18, 6.19]. If demand information can be communicated throughout the entire supply chain, each trading partner is able to know how much product needs to be available and when. As a result, lower inventory is needed as a hedge against uncertainty, lead-times can be shortened and sales increased because the right amount of product is available at the right points of consumption. The main impact of demand-driven supply chain participation is in the critical area of demand forecast accuracy, which directly impacts key metrics such as perfect-order fulfilment, supply chain cost and cash-to-cash cycle time. A recent study showed that improvements in demand forecast accuracy increase levels of responsiveness and cut costs for those members of a supply chain who participate in a demand-driven supply chain. Companies that are highly effective at demand forecasting average 15% less inventory, 17% stronger perfect-order fulfilment, and 35% shorter cash-to-cash cycle times, while having one tenth of the stock-outs of their peers [6.20].

Despite these undeniable advantages, it seems that few companies manage to implement effective demand planning processes. Although many industry executives embrace the consumer-driven supply chain concept, they are often not sure where to begin. The first companies that made efforts regarding demand collaboration (Wal-Mart was the pioneer) began these changes at the end of the nineties in the consumer goods industry [6.21]. In 1998, the VICS (Voluntary Inter-industry Commerce Standards) Association formalised the principles of CPFR (collaborative planning, forecasting and replenishment) in a guideline that promoted best practices in order to develop demand collaboration. Despite the existence of this detailed and comprehensive model, and despite the experiences of companies like Wal-Mart or Warner Lambert's which have gained much from CPFR practices, most companies are still looking for a way to implement efficient collaborative planning, in particular to integrate the front end of their supply chain. According to Crum and Palmatier [6.18], suppliers argue that their customers' approach in a demand collaboration relationship is not win–win. For instance, when a customer has spent time and resources to communicate demand information to their suppliers, they expect the products to be available when they said they wanted them. Another point of dissension is that suppliers contend that customers communicate demand information but then do not buy in the volume and timing that was communicated. This leaves the supplier with the excess inventory or they are forced to absorb the additional cost of filling last-minute orders to compensate for unplanned demand [6.22].

The existing literature on CPFR is mainly based on the VICS process model [6.23, 6.24]. CPFR is defined as a business practice that improves accuracy by combining the intelligence of multiple trading participants in the planning and fulfilment of customer demand [6.25]. However, the different experiences analysed by authors show that very few companies systematically implement the model [6.26–6.28]. Using data from seven case studies, Danese [6.29] identified the macro-building blocks upon which CPFR is based and sought to establish relationships between them. This research shows that CPFR is characterised by two dimensions, one that is based on technologies used by the partners to communicate with each other and one based on the organisational concept of liaison devices introduced by Mintzberg [6.30][3].

No direct link is made in the literature between CPFR issues and DDSC. Yet, both imply strong relationships within supply chains and require changes not only in the nature of the relationship with the retailers, but also in the way each participant conducts business. In a recent article, Cederlund *et al.* [6.31] analysed how Motorola turned to CPFR to improve sell-through performance with its retailers. This initiative is described by these authors as a 'time-consuming, painstaking endeavour'. This collaboration required Motorola to develop new business processes, redesign its organisation and adopt systems to support real-time information.

Many CPFR projects fail due to lack of executive support, but also to lack of collaboration rigor or because of unclear objectives at the outset. In particular, the need for a system of technologies and processes to sense demand and react to it in

---

[3] According to Mintzberg, liaison positions are jobs created to directly coordinate the work of two units without having to pass through managerial channels.

real-time, across a network of linked customers, suppliers and employees, is very often underestimated by partners and can lead into dramatic financial results. Starting from the customer means a business process change for many companies, which first requires them to break down the vertical internal functional boundaries within the organisation so that they can collaborate between sales, logistics, procurement and operations. The issue is to know how to move toward a company with multi-tier integration, which supports demand visibility and supply chain collaboration.

Demand collaboration presupposes that information on sales is exchanged in real-time from downstream to upstream partners. Without this information, partners throughout the supply chain can experience a bullwhip effect, in which disruptions intensify throughout the chain. This can negatively affect cash-to-cash cycle time. Thus, integrating customers and suppliers in the supply chain means not only understanding business processes, but also aligning supply chain processes (see Figure 6.1).

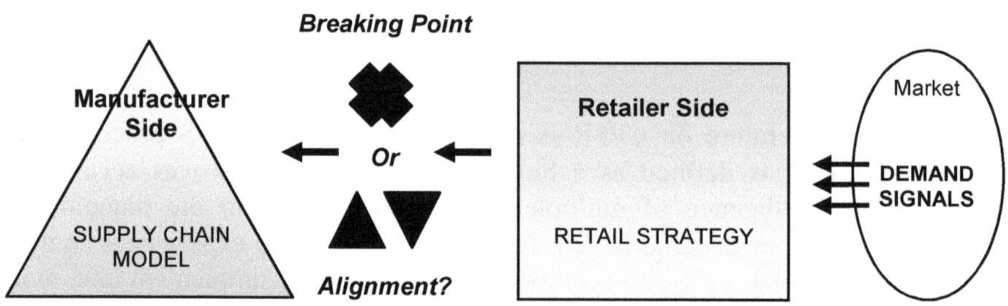

Figure 6.1. Alignment of processes between manufacturer and retailer

In this alignment of processes, information sharing is at the heart of the solution. Software and tools necessarily support the transition, and the choice of technology is essential to elaborate an architecture that is flexible and adaptable. Several types of information sharing in supply chains have been analysed in the literature [6.32–6.34]. Research generally suggests that information sharing can improve supply chain performance but to our knowledge, there is no work that studies how value is created during the demand planning process through the use of inter-organisational information systems. Although researchers have studied the value of information sharing, they have generally considered the upstream share of historical demand figures [6.35].

This chapter proposes a different approach since we consider that the share of demand information encompasses the transfer of 'demand signals' and that value creation within a supply chain is effective when these demand signals are incorporated in joint collaboration planning actions.

Through the comparison of three case studies, the next section aims at showing how collaborative demand planning can increase value within supply chains and which technologies should be used to facilitate the exchange of demand signals in a two-echelon supply chain.

6.3 Using Demand Signals to Develop Collaborative Demand Planning Practices

Generally speaking, demand signals are streams of data that supply chain members may be privy to on an ongoing basis. They include advance information on demand, other than past demand realisations, that correlate with current (real-time) and future demand. The first step of our empirical research was to identify these demand signals. We use a multiple-case study method to investigate the research question. The literature review on CPFR and DDSC guided the selection of the cases. Our empirical analysis focuses on the food and consumer packaged goods (F&CPG) industry. The F&CPG industry provides fertile ground for exploring the impact of manufacturer-retailer partnerships because this industry is highly competitive and characterised by relatively small profit margins. All cases describe a two-level supply chain consisting of one manufacturer (an SME or a large firm) and one retailer (generalist or specialist), where demand planning is a central element in the value creation process for the final consumer (see Table 6.1).

Table 6.1. Presentation of the cases

Manufacturer	Retailer	Product	Demand variability	Demand collaboration mode
Délifruit	Casino	Fruit juice	High	Vendor managed inventory
La Normandise	Casino	Pet food	Low	Advanced inventory
Tefal	Carrefour	Cooking appliance and cookware	Medium	Category management

All data was gathered during company visits between 2006 and 2007, using different sources: semi-structured interviews, company documentation and direct observations. Data analysis involves within-case and cross-case analysis.

6.3.1 Case 1: Délifruit/Casino

The beginning of this relationship was a result of the retailer's initiative. The manufacturer employs 300 people, and the company is held by an American group. It produces fruit juices mainly under retailer brands (70% of its activity). The relationship with the retailer began several years ago, but the replenishment was not optimum from the retailer's point of view, and the manufacturer faced major production planning issues. Transportation costs were high for both parties and the service rate level was low, leading to frequent 'out of stock' incidents in the different points of sale and high inventory levels. The seasonal nature of products, the necessity of regular promotional activity and new product introductions increased complexity and uncertainty within the supply chain. Both parties had a real interest in finding solutions to improve demand planning in order to achieve better results. The main issue was to change the approach of demand planning from a top-down approach to a bottom-up forecasting system. That meant that the retailer had to find a way to transmit demand signals to the manufacturer without giving information that was considered as strategic, for instance data that would give information about competitors' offers or about their own retail strategy. From the retailer's point of view, it was impossible to give real-time information about its sales. Thus, the actors decided to use the distribution centre's inventory out as a reference to organise replenishment. Information was exchanged through Excel files connected with the vendor's EDI[4] system.

As shown in Figure 6.2, a breaking point exists in the information flow between the real stock holdings in stores (based on real consumption) and the replenishment information given to the manufacturer by the distribution centre. This breaking point prevents full *distribution activity planning* for the manufacturer. However, a replenishment system was set up to enable the manufacturer to take responsibility for volumes delivered to the distribution centre. An agreement system exists whereby the manufacturer can make an order proposal with or without confirmation from the retailer. In practice, the system works without confirmation in order to gain time and avoid complex exchanges of information, but the consequence is that the manufacturer has full control over the retailer's inventory and thus the retailer is dependent on the manufacturer's decisions. Even if performance indicators show that this system is efficient (increased service rate, better forecast accuracy, shortened delivery time, improved turnaround), the retailer considers that he loses power in the relationship because of this relative transparency in the transfer of demand information.

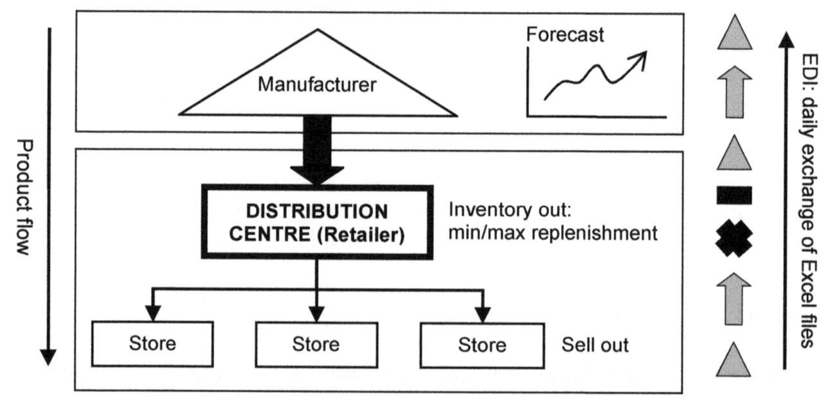

Figure 6.2. Product and information flows in case 1

6.3.2 Case 2: La Normandise/Casino

The manufacturer is a medium-size enterprise of 280 employees, specialised in the production of pet food. It owns a manufacturing plant and a new warehouse and 55% of the turnover is in France. As a result of a decision taken by the CEO, no

---

[4] Electronic data interchange.

more than 5% of sales are realised with one single customer. Retailer brands represent 80% of the vendor's sales and the firm supports the retailers' brand strategies by developing innovative products and by frequent introduction of new products and changes in packaging. The project was initiated by one of the retailers who wished to implement an advanced inventory project. In this particular project, the vendor owns the inventory in a primary warehouse (managed by a third-party logistics provider – 3PL) until the products are delivered to the final consumers (different points of sales). This agreement presents many cost reduction advantages for the retailer, roughly 20 million euro of potential savings on warehouse inventory if the 140 suppliers participate in the project. The project also permits the retailer to ensure its supplies.

From the vendor's point of view, the stakes are difficult to calculate. Of course, they may improve their service rate and reduce the risk of delay penalties because they have information regarding stock levels at the primary warehouse. The supply proposals made by the retailer give information on demand and transportation costs that can be optimised since only one warehouse is used instead of several warehouses in different locations throughout the country (see Figure 6.3).

In this case, power relationships are clearly unbalanced. Collaboration is imposed by the retailer for whom financial gains are at stake. The vendor cannot take the risk of losing a customer like Casino who is one of the leading retailers in France and who represents 5% of its sales. Demand information is held by the retailer who elaborates the forecasts that he transmits to the manufacturer. This practice allows the manufacturer to better control its production planning and to better organise its own supplies. However, inadequate information on demand generates poor inventory management and high stock levels. The vendor also lacks information on promotions because supply proposals are sent four weeks before delivery to the primary warehouse, instead of eight weeks before the beginning of the project. Finally, transportation costs are not controlled because the prices announced by the logistics provider are not respected. The retailer also encountered some difficulties managing its transportation subsidiary, which suffered from many quality issues related in respect of time and delays.

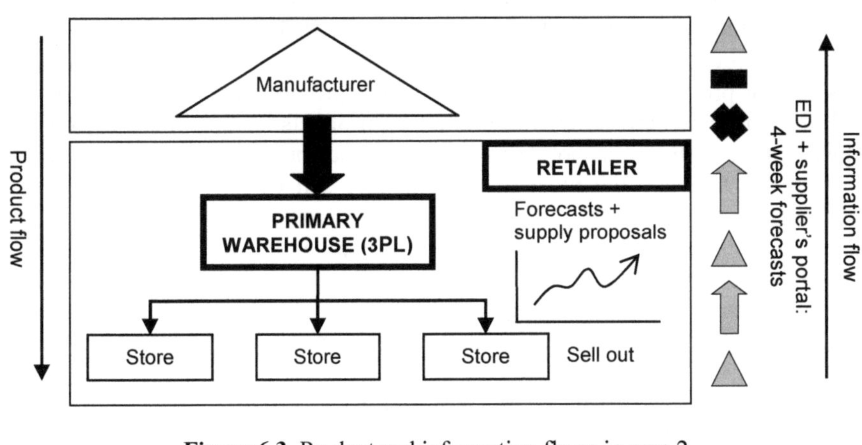

Figure 6.3. Product and information flows in case 2

This system of advanced inventory is to the retailers' advantage. The benefits of such a project can be better shared if the information given to the manufacturer is more precise and regular. Indeed, the vendor needs complete information on demand in order to organise its manufacturing activity. In this case, the retailer claimed to know more about turnovers within the primary warehouse and above all, asked for a daily report on logistics invoicing (number of pallets in stock per day and number of received pallets per day).

Relationships between the retailer's supply department and the manufacturer's negotiation team led to an improved transmission of information. Both parties were satisfied with the project and the experience was seen as globally positive thanks to the pooling of product flows in a single warehouse.

6.3.3 Case 3: Tefal/Carrefour

One of the roles of a retailer is to receive new products on a daily basis (retailer brand, national brand, or low-price products) and to coordinate the assortment and the allocation of space in relation to a particular product. In its basic form, category management entails a leading supplier playing a proactive role in consumer data analysis and computer-aided space planning. In this particular type of partnership, the leading manufacturer in an industry is made 'category captain' and takes on the organisation of the category itself for the retailer [6.36]. It is this special role that Tefal took on with several retailers who accepted to work with the category management concept to organise their points of sale. The products concerned are culinary utensils and small home equipment goods. Working in concert, Tefal and Carrefour (one of the main retailers in France) developed a detailed category plan for the supermarket that set out which SKUs (stock-keeping units) it will carry, their retail prices, promotional programs, and a 'planogram' displaying the layout, space and format or the offer[5]. The term 'category management' represents an array of collaborative manufacturer-retailer relationships. Particularly in this case, we study the demand planning process through observation of techniques that facilitate collaborative relations. These include open-book costing, joint performance measurement, joint forecasting, pricing guidance, consumer profitability analysis, cost analysis and formal profit-sharing arrangements [6.37].

Tefal has developed a software called *Assortman* to optimise layout organisation. This software provides advice about pricing and promotions and attempts to build the ideal assortment across the whole range of products by detailing the number of products, those with strong potential and the layout presentation. The objective of this project is to make the most of data, mainly sell outs, to better optimise the assortment.

Data exchanges concern sales details at a reference level, in terms of value and volume (sell outs). This data is considered as highly strategic by the retailer and one of its objectives is to promote the quality of the offering. The retailer estimates that the supermarkets do not have the internal expertise required to price and display all

---

[5] A *planogram* is a type of architectural drawing of the space each item will occupy on the store's fixtures.

of their thousands of SKUs. Tefal, as a category captain, is able to calculate profitability at category, account and product levels.

In addition to data given by the retailer, Tefal uses other information obtained from consumer survey data (provided by GfK[6]). By comparing both data sources, Tefal is able to analyse sales and evaluate the positioning of the retailer on the market. It is also possible to observe which brand is over or under represented, which is absent, and which product is not proposed in layouts when there is a good turnover.

As shown in Figure 6.4, information is fairly exchanged along the entire supply chain and demand data analysed at a store level can be seen as a source of competitive advantage, to be shared by manufacturer and retailer. It implies a reduction of risk for both partners and lower transaction costs that contribute largely to the success of this inter-organisational arrangement.

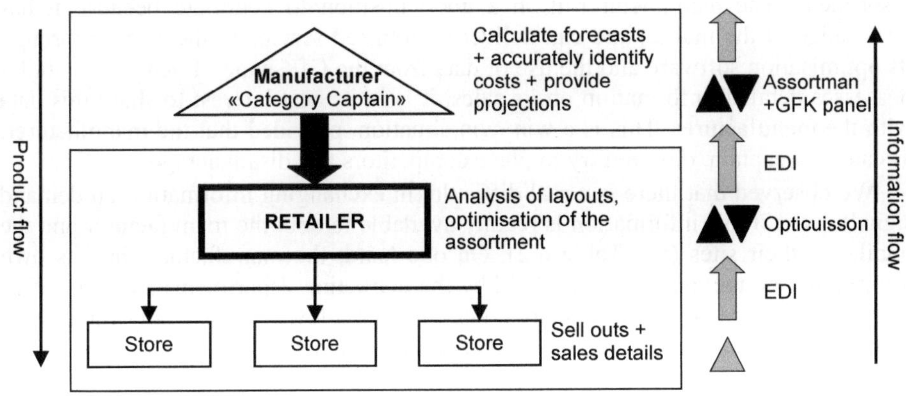

Figure 6.4. Product and information flows in case 3

6.4 Cross-case Analysis and Discussion

Case 1 and case 2 permit us to observe the relationship between an SME and a major retailer. The retailer's experience can help manufacturers to better organise their downstream supply chains, but in both cases, collaboration was undertaken under duress and started by the retailer.

The Vendor Managed Inventory (VMI) project with Casino led Délifruit to start reflecting on the implementation of an advanced planning system (APS) to better manage its planning process. Even if the retailer took the initiative in developing a partnership, the manufacturer has gained agility in its internal supply chain and has improved its logistic reactivity. It probably would have been less active in structuring its supply chain if the retailer had not set up this project.

---

[6] The GfK Group is one of the largest market research companies and the number five in the world. GfK delivers market research services, from data collection and analysis to consulting, in all major consumer, pharmaceutical, media and service sector market segments.

The Advanced Inventory project was more complicated because the power relationships were highly imbalanced. The manufacturer had no possible tactic to bring the negotiation around to its advantage and was progressively convinced by the retailer to accept the project. However, the simplification of the logistic flows toward one primary warehouse (instead of seven previously) led to significant gains on logistic costs for the manufacturer and the retailer reached its strategic objective of inventory reduction.

Regarding the transmission of demand signals by the retailer, we can observe that there is a breaking point in the upward information flow, thanks to these two cases. Manufacturers' supply chain models are 'push' models where forecasting is statistically calculated by vendors. There is no compatibility between such supply chain models and the retail strategy of the retailer who has no real interest in sharing strategic information, such as that related to as sell outs, for instance.

Case 3 is different because the retailer faces the necessity to better optimise its assortment. The manufacturer is in a good position to negotiate because it has knowledge of the market and can deliver customised service to the retailer through its optimisation software and the use of data from the GfK panel. Even if the retailer possesses detailed information on its sales, it is in its best interest to share this data with the manufacturer. This is a win–win situation, provided that the manufacturer, as category captain, does not try to place competitors at a disadvantage.

We observed that there is a real difficulty in exchanging information on demand signals, even if this information is readily available to both the manufacturer and the retailer at their sites (see Table 6.2). On one hand, the manufacturer has its own information on the market (partly held by the marketing department) and very often spends a lot of money for information systems that allow to statistically forecasting demand (*e.g.* advanced planning and scheduling systems). Of course, it is not ready to share this analysis with a specific retailer. On the other hand, the retailer has information on final consumer behaviour through the sales for each retail point and is able to sense demand variability in real time. Very often, the retailer does not work actively with this data or considers that information on its sales cannot be shared with its suppliers because it is strategic and confidential. This point is particularly true in Europe, probably due to cultural restraints and lack of trust among trading partners. The result of this situation is that the manufacturer calculates forecasts without a real time view of what is happening on the market, and the retailer does not develop any forecasting abilities.

Table 6.2. Identification of demand signals

Demand signals held by the manufacturer	Demand signals held by the retailer
• Sales forecasts (statistically calculated) • Inventory levels by location • Available to promise (ATP) • Customers' orders to date and historical sales figures • Marketing data (product evolution and calendar of new product introductions, pricing changes, promotions)	• Sell outs by product, by SKU (item level) • Stocking locations (stores and distribution centres) • Planned promotions (catalogues) • Sales history, including end-user or consumer sales data

If sharing sales data is not possible between a manufacturer and a retailer, demand signals cannot go upstream along the supply chain and demand planning will be poor at both levels. According to the analyst firm AMR Research, in the consumer packaged goods (CPG) industry, 56% of companies take more than two weeks to sense a demand. While 70% of CPG companies gather downstream sales data, less than 3% are able to use this information to sense demand, much less to react to it.

The third case study shows that when demand information is shared directly between store and manufacturer, it is possible to create a database that can be used to gather and integrate point-of-sale information and demand insight data. As a result, the data can be put into a meaningful format for business users in sales, marketing, finance, supply-chain planning and R&D. The use of category management inspires firms to share data on the market and improve the demand planning process. This observation can be compared to the 'flowcasting' concept, recently introduced by Martin *et al.* [6.38]. Flowcasting uses the same time-tested approach at the retail level that has been used in distribution (DRP) and manufacturing (MRPII) for years: starting with a forecast of sales by product at the shelf, each store calculates what it will need to bring in as simulation based on its current on-hand balances and ordering rules. The sum of the stores' planned arrivals represents a stream of planned withdrawals from the retail distribution centre and implicitly from the manufacturing plant, and the chain reaction of demand throughout the entire supply chain is recalculated on a daily basis as market conditions change.

However, despite the benefits reaped by store-level forecasting (or 'bottom-up' forecasting), the implementation of this method requires an appropriate planning system and a well designed business process tailored to the retail environment. At the moment, it seems that very few industrial firms can or have established this narrow connection with the store level, in as much as the distribution centre represents a breaking point in the flow of information.

As a result for this research, we propose to analyse four different stages in the development of demand planning practices (see Table 6.3).

Table 6.3. Identification of the four stages in the demand planning process

		Demand signals	Technology used	Frequency of data exchange
STAGE 1	Static demand planning	Historical demand figures	Excel files + APS	Monthly
STAGE 2	Demand sensing	Historical demand figures + inventory outs	Excel files + APS + DRP (EDI)	Weekly
STAGE 3	Demand shaping	Sell outs	Excel files + APS + DRP + S&OP	Daily rolling forecast
STAGE 4	Knowledge sharing	Sell outs + consumer sales data + demand analysis	Excel files + EDI + supplier portal + workflow tools	Real time

At stage 1, the demand planning process is mainly based on historical demand figures. Forecasts are generally calculated by the manufacturer and there is no alignment between the supply chain model of the vendor and the retail strategy of the distributor. Data is exchanged once a month and there is no visibility along the downstream supply chain.

At stage 2, firms develop an ability to read demand signals from the market but the translation of these signals still leads to calculated forecasts, generally updated on a weekly basis. Manufacturers and retailers use actionable operating insights but the unification of data is not effective because of lack of shared information systems.

At stage 3, manufacturers and retailers develop a capacity to shape demand. Their understanding of demand signals is improved and they can react rapidly to a change of trend. Supportive sales and marketing tactics can be shared to develop a proactive supply chain closely connected with the market.

At stage 4, information is shared between manufacturer and retailer in order to develop profitable demand response, *i.e.* trading partners share not only data but knowledge. This allows real-time forecasting and thus a better visibility along the downstream supply chain.

The three cases show that information is mainly exchanged through Excel files, EDI, and supplier portals. Collaborative technologies are a permanent area of concern between retailers and manufacturers, particularly when there is a battle to know who runs the implementation. Even a simple EDI system can be transformed into a two-year project if the compatibility of information systems between partners is not well controlled. IT tools can offer better visibility within the extended supply chain only if retailer and manufacturer think the same way about sharing data. Supply chain tools interact with ERP systems but the problem of data compatibility is recurrent. Enterprise resource planning vendors are developing more and more solutions to support information exchange among companies across the supply chain but despite these solutions, coordination between manufacturer and retailer is, in most cases, not deep enough to synchronise downstream supply chain with customer demand [6.39].

Finally, another problem is the dependence link created by a joint IT project. If manufacturer and retailer invest time and money to develop a shared tool, the possibility of choosing another supplier or retailer is lower. The fear of becoming subordinated to a trading partner can create a real brake to collaboration because neither the manufacturer nor the retailer wants to be committed to a relationship that could prevent the possibility of improving its profit margins. This point probably explains why Excel files are the most common way to exchange information between manufacturer and retailer. Beyond sophisticated information systems, the use of a spreadsheet is a simple way to treat information and to exchange it individually.

6.5 Conclusions

To effectively manage demand, collaboration needs to occur between manufacturers and retailers. True demand management is not just about sales forecasting, it is also about having a global vision of information and manufacturing flow from plant capacity and raw materials through to the finished goods in the warehouse – all of these assets are needed to effectively respond to customer demand. Using data from the three case studies, this research underscores the necessity for a manufacturer to have the right planning tools within its organisation in order to synchronise the shop floor with manufacturing and scheduling options. This is the only way to quickly respond to changing demand signals. The difficulty, however, is to overcome the breaking point in the upstream information flow due to the challenges of demand signals exchange with the retailer. Better alignment of supply and demand could be obtained by carring out volume and variability demand profiling, but this activity should be shared by manufacturers and retailers.

This research allows us better understanding the demand planning process. We suppose that value can be created both for manufacturers and retailers if information on demand is better shared between them. As the collaboration intensity increases, it becomes possible for the partners to create shared knowledge. Thus, the four-stage demand planning process proposed after the cross-case analysis (see Table 6.3) can be set in a more global framework in connection with the collaboration intensity level that characterises a relationship between a manufacturer and a retailer (see Figure 6.5).

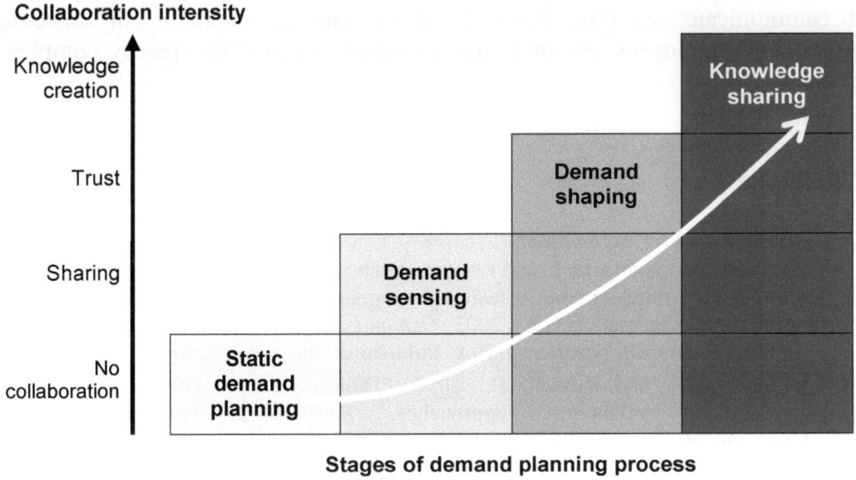

Figure 6.5. Proposition of a framework to analyse collaborative planning practices

From an academic point of view, this research can contribute to advancing theory on demand planning in two ways: (1) by proposing a model to explain the evolution of demand planning practices toward more collaborative relationships between manufacturers and retailers; and (2) by giving more details on how to share demand signals in order to improve relationships between manufacturers and retailers. However, the case studies that were analysed here are limited to a small sample and only to the F&CPG industry. A wider sample of networks in several industries should be the subject of other research in order to confirm our initial results.

The theoretical analysis of the demand planning process also has practical implications that can help managers to better understand the critical role played by information sharing. Of course, evolving from traditional approaches to supply chain management to value-network orchestration will take many years and will imply deep changes in business practices. This is particularly true for the relationship between vendors and retailers. This research shows that demand management plays an essential role in improving the ability to coordinate an extended network because it contributes to increased visibility along the supply chain. Thus, sharing demand signals provides the basis for shared performance measurements within the downstream supply chain.

Future research needs to be conducted to identify types of demand planning strategies that correspond to a specific alignment between the supply chain model and retail strategies. As retail demand planning really impacts retail business profitability, there are major stakes for retailers to better analyse and simulate demand (reduction of out-of-stocks). Cultural restraints observed concerning information sharing should also be overcome, particularly in Europe. From the manufacturers' point of view, the change from a 'push' supply chain model toward a 'pull' one should permit a better integration of demand management and supply management processes, and as a result to substitute demand information for inventory. To effectively manage demand, technologies and processes that can sense and communicate real-time demand, taking into consideration of customers, suppliers and employees, are of course necessary, but not necessarily complex to implement.

References

[6.1] Baird, N., Leaver, S., Bradner, L., Overby, C.S., Stromberg, C. and Gaynor, E., 2006, "The state of manufacturer and retailer collaboration," *Forrester Research*, available at http://www.forrester.com/Research/Document/Excerpt/0,7211,40694,00.html.

[6.2] Crum, C. and Palmatier, G.E., 2003, *Demand Management Best Practices: Process, Principles and Collaboration*, J. Ross Publishing, Inc., Boca Raton, FL.

[6.3] Whipple, J.M. and Russell, D., 2007, "Building supply chain collaboration: a typology of collaborative approaches," *International Journal of Logistics Management*, **18**(2), pp. 174–196.

[6.4] Simatupang, T.M. and Sridharan, R., 2005, "An integrative framework for supply chain collaboration," *International Journal of Logistics Management*, **16**(2), pp. 257–274.

[6.5] Min, S., Roath, A.S., Daugherty, P.J., Genchev, S.E., Chen, H., Arndt, A.D. and Richey, R.G., 2005, "Supply chain collaboration: what's happening?" *International Journal of Logistics Management*, **16**(2), pp. 237–256.

[6.6] Boulding, K., 1956, *The Image*, University of Michigan, Ann Arbor, MI.

[6.7] Wiener, N., 1961, *Cybernetics: Or Control and Communication in the Animal and the Machine* (2nd edn.), MIT Press, Cambridge, MA.

[6.8] Barr, A. and Feigenbaum, E. (eds.), 1981, *The Handbook of Artificial Intelligence* (Vol. 1), William Kaufmann, Los Altos, CA.

[6.9] Hewitt, C., 1986, "Offices as open systems," *ACM Transactions on Office Systems*, **4**(3), pp. 271–278.

[6.10] Miller, M. and Drexler, K., 1988, "Market and computation: agoric open systems," *The Ecology of Computation*, North-Holland, Amsterdam, pp. 133–176.
[6.11] Huberman, B., 1988, *The Ecology of Computation*, North-Holland, Amsterdam.
[6.12] Fayol, H., 1949, *General Industrial Management*, Pitman, London.
[6.13] Gulick, L. and Urwick, L. (eds.), 1937, *Papers on the Science of Organizations*, Institute of Public Administration, Columbia University, New York.
[6.14] Thompson, J., 1967, *Organizations in Action: Social Science Bases of Administrative Theory*, McGraw Hill, New York.
[6.15] Tushman, M. and Nadler, D., 1978, "Information processing as an integrating concept in organization design," *Academy of Management Review*, (3), pp. 613–624.
[6.16] Radner, R., 1992, "Hierarchy: the economics of managing," *Journal of Economic Literature*, **30**(3), pp. 1382–1415.
[6.17] Williamson, O.E., 1985, *The Economic Institutions of Capitalism: Firms, Markets, Relational Contracting*, Free Press, New York.
[6.18] Crum, C. and Palmatier, G.E., 2004, "Demand collaboration: what's holding back?" *Supply Chain Management Review*, January/February, pp. 54–61.
[6.19] Hau, L.L., 2004, "The triple-A supply chain," *Harvard Business Review*, October, pp. 102–112.
[6.20] O'Marah, K. and Souza, J., 2004, *21st Century Supply Chain: The Demand-Driven Supply Network*, http://www.amrresearch.com/Content/View.aspx?compURI=tcm:7-17058.
[6.21] Ireland, R. and Bruce, R., 2000, "CPFR: only the beginning of collaboration," *Supply Chain Management Review*, September/October, pp. 80–88.
[6.22] Aviv, Y., 2007, "On the benefits of collaborative forecasting partnerships between retailers and manufacturers," *Management Science*, **53**(5), pp. 777–794.
[6.23] Harrington, L., 2003, "9 Steps to success with CPFR", *Transportation and Distribution*, April, pp. 50–52.
[6.24] Barrat, M. and Oliveira, A., 2001, "Exploring the experiences of collaborative planning initiatives," *International Journal of Logistics Management*, **31**(4), pp. 266–289.
[6.25] VICS CPFR Committee, *CPFR Guidelines*, http://www.vics.org/.
[6.26] Attaran, M. and Attaran, S., 2007, "Collaborative supply chain management: the most promising practice for building efficient and sustainable supply chains," *Business Process Management Journal*, **13**(3), pp. 390–404.
[6.27] Samuel, K.E. and Spalanzani, A., 2006, "Developing collaborative competencies within supply chains," In *Proceedings of the 4th International Conference on Supply Chain Management and Information Systems*, Taichung, Taiwan, July 5–7.
[6.28] Skjoett-Larsen, T., Thernoe, C. and Andersen, C., 2003, "Supply chain collaboration," *International Journal of Physical Distribution & Logistics Management*, **33**(6), pp. 531–549.
[6.29] Danese, P., 2006, "Collaboration forms, information and communication technologies and coordination mechanisms in CPFR," *International Journal of Production Research*, **44**(16), pp. 3207–3226.
[6.30] Mintzberg, H., 1979, *The Structuring of Organizations*, Prentice Hall, New York.
[6.31] Cederlund, J.P., Kohli, R., Sherer, S.A. and Yuliang, Y., 2007, "How Motorola put CPFR into action," *Supply Chain Management Review*, **11**(7), pp. 28–35.
[6.32] Hadaya, P. and Cassivi, L., 2007, "The role of joint collaboration planning actions in a demand-driven supply chain," *Industrial Management and Data Systems*, **107**(7), pp. 954–978.
[6.33] Mishra, B., Raghunathan, S. and Yue, X., 2006, "Demand forecast sharing in supply chains," Working Paper, University of Texas at Dallas, Richardson, TX.

[6.34] Kaipia, R. and Hartiala, H., 2006, "Information sharing in supply chains: five proposals on how to proceed," *International Journal of Logistics Management*, **17**(3), pp. 377–393.

[6.35] Simchi-Levi, D. and Zhao, Y., 2003, "The value of information sharing in a two-stage supply chain with production capacity constraints," *Naval Research Logistics*, **50**, pp. 888–916.

[6.36] Institute of Grocery Distribution (IGD), 1999, *Category Management in Action*, IGD Business Publications, London.

[6.37] Free, C., 2007, "Accounting practices in the UK retail sector: enabling or coercing collaboration?" *Contemporary Accounting Research*, **24**(3), pp. 897–933.

[6.38] Martin, A., Doherty, M. and Harrop, J., 2007, *Flowcasting, the Retail Supply Chain*, Factory 2 Shelf Inc, http://www.flowcastingbook.com.

[6.39] Hillman, M. and Hochman, S., 2007, *Supply Chain Technology Landscape Has Radically Changed for Everyone*, http://www.supplychainbrain.com/content/nc/world-regions/canada/single-article-page/article/supply-chain-technology-landscape-has-radically-changed-for-everyone/, January.

7

Value Creation and Supplier Selection: an Empirical Analysis

Blandine Ageron and Alain Spalanzani

University of Grenoble, 51, rue B. de Laffemas, BP 29
26901 Valence Cedex 9, France
Emails: blandine.ageron@upmf-grenoble.fr; alain.spalanzani@upmf-grenoble.fr

Abstract
In order to increase their competitiveness amid a growing internationalisation, many companies outsourced a part of their activities by the late 1980s and 1990s. This outsourcing process has transformed organisational boundaries and created supply chains where suppliers and sub-contractors are essential parts of these chains. There is a growing tendency to select these partners and, consequently, they are fewer in number and tend to be found farther and farther from the network leader. This geographical distance between the company and its suppliers affects the organisational density of the network and raises the problem of cooperation–coordination in the buyer–supplier relationships. The objective of this chapter is to examine the criteria used in the suppliers' selection process and thereby in the supply chain. More specifically, in the context of distributed supply sources and partnership objectives, the geographical distance between the company and its suppliers raises the problem of organisational density and increases the need for partnership cooperation–coordination in the supply chain.

7.1 Introduction

The process of selecting suppliers compels all companies to focus on the 'make or buy' dilemma. As [7.1, 7.2] noted, companies are streamlining their operations from a vertical integration (hierarchy) towards a more external contractualisation of key activities (market). In recent years, many firms have changed their relationships from traditional arm's length relations toward new arrangements based on cooperation [7.3]. Several factors can explain this trend: access to lower production costs, value-adding partnerships, stock reduction, development of agility and flexibility *etc.*, and the emergence of IT (information technology). Firms that outsource some of their activities have to decide which specific assets they should keep or develop, in order to concentrate on their core business [7.4]. The choice of partners (suppliers or sub-contractors) and supply chain coordination are two fundamental core competencies essential for companies engaged in network business.

Therefore, company boundaries now reflect a new economic and cognitive rationality that is dependent on the on-going debate on the duality of differentiation–coordination. The buyer is responsible for the quality of the suppliers' portfolio, that is to say the differentiation; thereby the supply chain manager is in charge of the coordination of the network constituting this portfolio. Even if the two 'roles', that of the buyer and the supply chain manager, are different, the internal cooperation linking them is fundamental. They both influence each other: the size and the quality of the suppliers' network built by the buyers will be essential for the order capability developed by the supply chain manager. Consequently, the criteria for selecting suppliers will have to reflect the preferences of the latter, because his/her job consists of optimising the flows (evaluated by the order–fill ratio criterion) from upstream suppliers to downstream customers.

In this chapter, we attempt to study the major criteria used in selecting suppliers within the supply chain. With the help of empirical evidence from selected companies in France, the aim of this chapter is to:

- Characterise the supplier selection process. We outline the importance of trust and long-term advantage in the relationships set up by companies with their suppliers. Considering this process, we observe that the number of suppliers, which network leaders are engaged with, influences the characteristics of the supplier selection.

- Illustrate the supplier selection according to classical criteria and 'secondary' ones. These results confirm that IT is a selection criterion in the upstream supply chain and contribute in the supplier selection in two ways. First, the study identifies the tools used in the supplier selection process. For instance, we examine different tools according to their capability to exchange, collaborate or decide in the upstream supply chain. Second, it questions the impact of these different tools on the relationships constructed by network leaders with their suppliers. We observe that collaborative tools can modify the type of relationships established.

- Highlight the value creation for supply chain partners through the selection process. We discuss value creation in upstream supply chain in regards of competencies and performance. Concerning competencies, we investigate more precisely IT competences and the way suppliers acquire or develop these capabilities. Moreover, we investigate network leaders' attitudes towards their suppliers and the responses given by suppliers in return of companies' attitudes. Performance is examined concerning the management of suppliers and the definition of key performance indicator (KPI) in order to evaluate the success of their selection process. Value creation is finally assessed by addressing difficulties and interrogations confronting network leaders.

In the second section, the supplier selection process is discussed. In Section 7.3, methods and materials are developed. Results concerning supplier selection process, criteria and value creation are presented in Section 7.4. Finally, concluding remarks are presented in Section 7.5.

7.2 Supplier Selection

According to [7.5–7.8], the question of strategic purchasing in upstream supply chain construction is associated with the development of the supplier selection process. Indeed, this selection phase is certainly the most important one in the buying process. Consistent with the fact that the buyer needs to periodically evaluate the performance of its suppliers, the authors of [7.5] argue that the more rigorous and methodical the selection, the higher the performance.

In this context of supplier selection, lots of models focusing on the selection and evaluation process of suppliers have been developed [7.9–7.11]. Even if they do not predict 'one best way' of selecting suppliers, these models are very important for the decision process of companies. They offer a very structured and rigorous approach that can help evaluating suppliers. If there is a follow-up, it is also possible to evaluate flexibility, reaction capacity, comprehension and reliability in order to minimise risk and maximise value creation.

By the same time, lots of researches have emerged on the supplier selection and the criteria used in this process. Although it can be described as disparate and contentious [7.12], there is a consensus on four main traditional criteria: price, quality, deadlines and services. Apart from these criteria, other studies question criteria such as supplier characteristics (size, reputation, *etc.*), available offer and trust between buyers and suppliers [7.13]. As suggested by [7.5], criteria differ according to whether it concerns upstream suppliers or downstream customers. Hence, in the manufacturing upstream, we can usually identify and study criteria such as quality, cost, deadlines and company technical capacity [7.14, 7.15]. As regards the upstream distribution chain, the criteria that are most often analysed are deadlines, the quality of delivered products, and more generally customer satisfaction. It is argued in [7.16–7.18] that information levels and customer requirements force companies to set up upstream collaboration together with downstream ones. To satisfy their customers, they need to minimise their costs by maintaining an optimal level of competitivity and productivity. In this way, the effect of intrinsic and personal characteristics of buyers on the supplier selection process has also been studied [7.19, 7.20].

The importance of IS (information systems) and IT for the development of new organisational forms, such as strategic partnership or networks has been observed by the late 1990s [7.21–7.23]. This observation was confirmed by empirical studies [7.24, 7.25]. The impact of IT on the upstream supply chain is nowadays critical because companies have to deal simultaneously with suppliers that are culturally and geographically more distant and customers that are demanding a high level of satisfaction. As a result, they are at a competitive advantage [7.22, 7.26–7.28] and network leaders have changed their behaviour as they begin to develop more distant relationships and to increase externalised activities [7.29]. At the same time, suppliers benefit from IT and have gained more negotiation power over customers. This is highlighted in [7.30], which demonstrates that the use of IT enables deeper, more stable and more relevant relationships, thanks to the benefits distributed between all stakeholders. Thus, the deployment of IT provides substantial benefits through lowered transaction costs in the field of invoicing, payment settlement, inventory management and the development of new products [7.31, 7.32].

7.3 Methods and Materials

7.3.1 Questionnaire

We developed a questionnaire structured in three parts:

- The first part is introductive and presents the theme, the objectives and the aim of our research. Moreover, confidentiality and anonymity are addressed.
- The second part relates the selection process, namely the relations established by companies with their suppliers.
- The last part concerns the identification characteristics of the interviewee (sex, age, function, *etc.*).

7.3.2 Data Collection

Our data were collected with a face-to-face administration methodology. Given the complexity and length of the questionnaire, we chose this method in order to help and support respondents who were faced with difficulty in understanding. Nonetheless, because some respondents were geographically at a distance or very busy, we chose electronic administration (email) as an alternative methodology. This mail administration was an opportunity because 30% of respondents were able to be reached thanks to it. Out of 110 questionnaires, 20 were inadequate for analysis, because of a lack of information or incomplete answers. The good survey results (90 out of 110 questionnaires were completed) are certainly due to the fact that we chose the face-to-face administration, even though it is more difficult to organise.

7.3.3 Companies Sampled

Companies were not chosen according to their size, geographic location or activity sector. In order to get a representative sample of French companies, we decided to establish the quality of the respondent as the selection criterion. The target respondent has to be in charge of global purchasing (small and medium sized companies) or product purchasing (bigger companies). Conscious of the bias introduced by a single informant, we decided to set up a second condition. He or she has to be also directly in charge of the supplier selection process within the company. These two conditions reduced the number of potential respondents and often left us with little choice as to the 'eligible' respondents. This method was extremely profitable and a lot of companies were involved in our study: SMEs, multinationals, industrial or services companies.

7.4 Results

7.4.1 Typology of Companies

To begin with, our findings concern the typology of companies (Figure 7.1). We observed that about 50% of companies are linked to the manufacturing sector.

Moreover, our sample is mainly made up of large companies: 57% employ more than 1000 workers (51 companies), 20% between 200 and 1000 (13 companies) and finally 28% of companies employ less than 200 workers (26 companies). It is nevertheless interesting to note that the presence of a buyer within a company is related to company size. Only companies with over 1000 employees have created and developed this role. For smaller companies, purchasing is often managed by the CEO (20%) or the operations manager (20%).

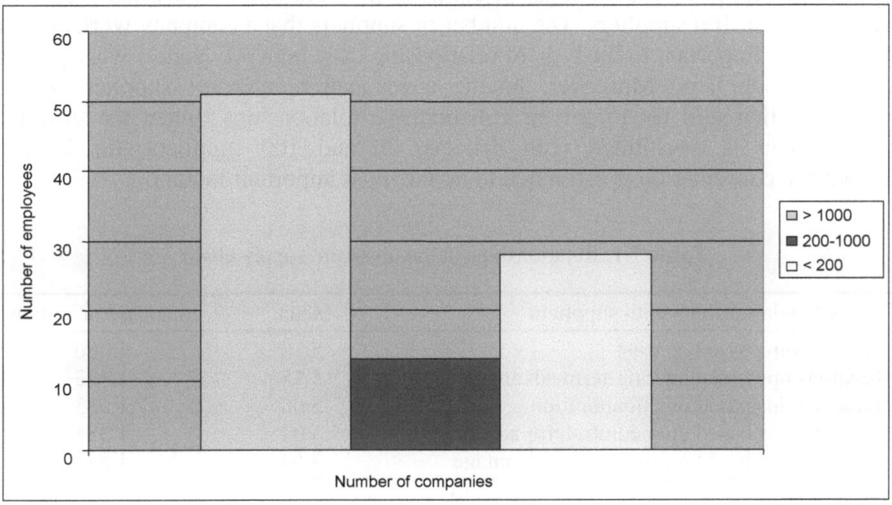

Figure 7.1. Typology of companies

7.4.2 Characteristics of Supplier Selection

In product costs, purchasing represents one of the most important elements [7.33]. This fact combined with the reduction of supplying and buying sources and the geographic distance of suppliers, has led companies to manage their stakeholder relationships more profitably [7.13]. In this way, companies are looking for and building long-term relationships that are organised and based on a relationship of trust (Table 7.1). Therefore, companies that tend to reduce the number of their suppliers have to find new ways to deal with risk [7.13, 7.34]. With more relationships built on trust, companies gear themselves towards risk management and, as a result, suppliers are more loyal and the upstream supply chain is denser. Nonetheless, it is important to note that trust is not in itself a selection criterion but is simply an important characteristic in sustainable collaborative relationships. We can affirm, thereby, that the supplier selection process is a strategic decision, linking companies to one another on a long term basis. For instance, our research outlines an average two years relationship between suppliers and network leaders.

Meanwhile, our study points out the importance of relationships based on long-term advantage in comparison with collaboration. Companies prefer this type of relationships certainly because it requires less implication from partners. Long-term advantage can be seen as the first step in supply chain associations in which suppliers and network leaders work together in an 'extended network'.

However many suppliers involved with companies, the way of managing the different partners in the upstream supply chain is essential. Even if companies do work with more suppliers, trust and a long-term advantage are foundations for partnership relations (Figure 7.2). Nevertheless, network leaders are keener to build 'value added' relationships with a narrow basis of suppliers. About 25% of companies engaged with less than 100 suppliers aimed to set up relationships based on trust and long-term advantage. In contrast, they are only 10% when they work with more than 100 suppliers. The number of suppliers that a company works with is in some way important to the type of relationships that network leaders want to build with their suppliers. Moreover, having a reduced number of suppliers affects companies that will try to set up collaborative relationships (when the upstream supply chain is constituted with between 20 and 100 suppliers, for 87% of companies, coordination is estimated to be the most important factor).

Table 7.1. Relationships in the upstream supply chain

Types of relationships with suppliers	Mean*	Standard deviation
Relationships based on trust	5.18	1.480
Relationships based on long-term advantage	5.18	1.472
Relationships based on collaboration	5.08	1.533
Relationships based on medium-term advantage	5.04	1.381
Relationships based on short-term advantage	3.54	1.874

*Lickert scale: 1–7 (1 = no accordance, 7 = complete accordance)

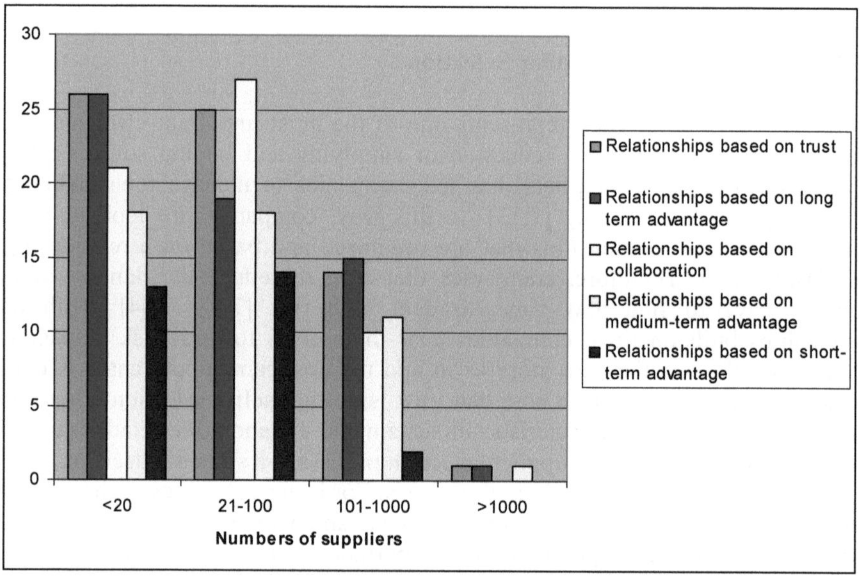

Figure 7.2. Impact of number of suppliers on upstream supply chain relationships

7.4.3 Selection Criteria

7.4.3.1 Classical Criteria

Traditional criteria related to the supplier selection process have been studied. We confirm that price, quality, flexibility represent the three fundamental criteria in the supplier selection process. In this way, more than 80% of companies consider price (85%) and quality (80%) to be essential in their selection process. While the geographical market does not seem to have a significant impact on these essential criteria [7.35, 7.36], similar conclusions can be drawn about personal relationships (Table 7.2).

Table 7.2. Importance of criteria in supplier selection process

Choice of the importance of criteria in supplier selection process	Mean*	Standard deviation
Price	2.36	2.053
Quality	2.36	1.860
Flexibility	2.36	2.411
Size	2.36	2.754
Inter-operational capacity of internal IS	4.35	4.078
Trust	4.78	3.033
Long-term relationships	5.58	3.058
Mastering of internal IS	6.13	3.679
Geographic proximity	6.77	4.442
Collaboration thanks to IS	7.19	4.210
Personal relationships	8.07	5.524

*Scale: 1–13 (1 = most important, 13 = least important)

Meanwhile, we observe that long-term relationships are not essential in the supplier selection process. This observation is contrary to the previous results and the characteristics of supplier selection process where we outline the importance of long-term advantage as a basis of upstream supply chain relationships. The same observation can be made about trust that is considered to be a subsidiary criterion in the supplier selection process. We argue that network leaders engaged in upstream partnership, do not evaluate theirs suppliers on these criteria as they consider that these two criteria are a prerogative.

The size of network leaders influences the ranking only when less important criteria are taken into consideration (Table 7.3). Consequently, companies with less than 50 employees consider trust, personal relationships and geographic proximity as essential criteria. Meanwhile, companies with more than 1000 employees continue to consider price and quality as the most important criteria. The asymmetric relationships can certainly explain this result.

90% of companies agree with the fact that price and quality 'exceed' all other criteria. We observe that other criteria such as size, trust and long-term relationships are becoming more important in the upstream selection process. It is interesting to note, however, that IT and more particularly inter-operational capacity and internal

IS are seen as important compared to other traditional criteria such as geographic proximity or personal relationships (Table 7.3). By this way, even if traditional criteria are more valued, IT can be considered as a real selection criterion in the supplier selection process.

Table 7.3. Impact of size of network leaders on the ranking of selection criteria

Selection criterion \ Size of network leaders†	<50	50–200	200–1000	>1000	Total
Price	15*	11	12	45	83
Quality	14	11	12	43	80
Flexibility	6	7	8	30	51
Size	5	2	3	24	34
Inter-operational capacity of internal IS	7	3	4	20	34
Trust	6	3	5	20	34
Long-term relationships	4	2	6	18	30
Mastering of internal IS	8	4	2	12	26
Geographic proximity	7	2	3	18	30
Collaboration thanks to IS	4	4	3	12	23
Personal relationships	8	4	3	10	25

† 14 companies have fewer than 50 employees, 12 companies have 50–200 employees, 13 companies have 200–1000 employees, and 51 companies have more than 1000 employees.
* Number of companies ranking this criterion as 'extremely important'

7.4.3.2 IT Criterion

Critical to upstream partnerships is the flow of information. Therefore, establishing and managing a knowledge-sharing network between the buying and supplying organisations is vital. Even if the importance of IT in collaborative relationships has been established for a long time, their integration at a strategic level is more recent [7.23]. For many companies, however, IT represents a real opportunity as new boundaries are drawn for new organisational forms.

As mentioned in Table 7.3, IT is a significant selection criterion in upstream supply chain. The capacity of suppliers to master their internal IS is obviously essential and reflects the innovative behaviour of suppliers. It is a positive 'signal' for network leaders engaged in deep and sustainable relationships with theirs suppliers. Collaboration thanks to IS became easier and more profitable for all the partners. These results are consistent with Figure 7.3 and the fact that only 52% of companies consider that IT is a significant criterion even if only 31.9% check that their suppliers are able to use their IT. As IT represents a huge investment, one can argue that many companies, notably small ones, cannot afford them without financial help [7.36]. The smaller the suppliers are, the more difficult for them it is to acquire and develop IT. Moreover, specialised relation-specific investments made for the partnership is often of little value outside the related partnership. This necessary resulting dependence can hinder the supplier to invest in specific IT. In these contexts, leading companies are keen on selecting suppliers who do not yet totally master IT but constitute a potential of acquisition and development, notably if they are supported and assisted.

Concerning the categories of tools mobilised and preferred in the selection process, new tendencies can be outlined. Transactional tools are significant in the supplier selection and choice processes even if internal and external collaborative ones are getting more attentions according to the network leaders (Table 7.4).

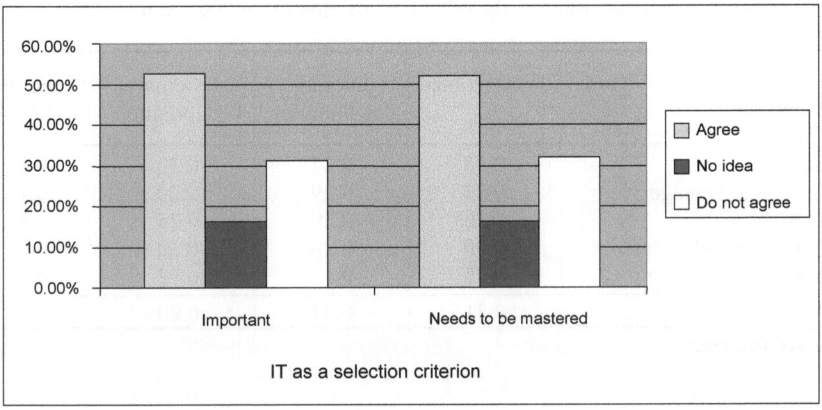

Figure 7.3. IT characteristics in the supplier selection process

Table 7.4. Tools used in the supplier selection process

Tools used in the supplier selection process	Mean*	Standard deviation
Use of transactional tools	4.24	1.85
Use of internal collaborative tools	3.59	1.84
Use of external collaborative tools	3.49	1.79
Use of decision-making tools	2.83	1.44

*Lickert scale: 1–7 (1 = no accordance, 7 = complete accordance)

The tools used in the supplier selection process have a weak impact on the type of relationship that a company establishes with its suppliers (Table 7.5). Even if we note that internal and external collaborative tools are increasingly becoming more popular, transitional tools stay significant in the supplier selection process.

However, relationships based on collaboration become more significant than other types of relationships. Trust and long-term advantage tend to be less important. Development of IT between partners strengthens the relationships and engages network leaders and their suppliers in a more coordinated supply chain.

Surprisingly, if transactional tools do not impact the relationships characteristics, concerning collaboration tools, we observe that collaboration is becoming more important. Companies are paying more attention on deep and long-term relationships. This is consistent with the fact that companies have changed their relationships from traditional arm's length relations towards new arrangements based on cooperation and trust [7.3]. Supply chain coordination is a fundamental core competency essential for companies engaged in networked business.

Finally, there is a difference between external and internal collaborative tools since external ones are less developed. We argue that the maturity of tools is probably one explanation. Whereas internal collaborative tools are widely accepted by companies, external ones need more time to be acquired.

Table 7.5. Tools according to the selection criterion of the upstream supply chain

Tools / Relationships based on	Transactional	Internal collaboration	External collaboration	Decision making
Trust	0.33*	0.27	0.20	0.11
Long-term advantage	0.33	0.29	0.20	0.11
Collaboration	0.36	0.29	0.25	0.10
Medium-term advantage	0.30	0.20	0.20	0.09
Short-term advantage	0.15	0.13	0.12	0.05
Total	0.31	0.24	0.20	0.11

* Observed frequency

7.4.4 Supplier Selection and Value Creation

7.4.4.1 Competencies Acquisition and Development

The capacity of suppliers to develop competencies in order to meet the needs of their customers is essential. The goal of the assessment of future capabilities is to be sure that the supplier will continue to add value over time. More precisely and related to IT competences, one can argue that it is through influence and help rather than more aggressive forms of collaboration that IT competencies are acquired and developed (Figure 7.4). We observe that 19 companies out of 33 influence their suppliers, and only 6 constraint them. This assisting or influencing attitude is more relevant in the specific context of ERP (enterprise resource planning) and EDI (electronic data interchange) tools. These IT essentials for network leaders need to be mastered by suppliers. In this context, and in order to help suppliers to acquire and develop them, companies assist (15 out of 33) and influence (6 out of 33) rather than constrain (3 out of 33) them. These results are largely explained by the type of relationships that are built between networks leaders and suppliers. It appears to be difficult for companies to set up long-term advantage and trust relationships by coercive decisions and actions.

Tools, such as EDI, ERP and sourcing are used in all the three forms of acquisition and/or development of competencies. A portal is preferred for assistance to tracing which was imposed by network leaders (forced). The significant place of EDI and ERP is once again confirmed. These tools, important in the supplier selection process, are preferred in terms of competencies, whatever their acquisition and/or development form.

Even if few network leaders try to deal actively with their suppliers (only 33 companies out of 90), those who do, observe that their suppliers tend to react positively. This reaction has leveraged positive relationships whatever the acquisition or development mode. Companies that have constrained their suppliers to obtain and use IS found their suppliers reacted positively. Among these network leaders, only one decided to cease relations with its supplier because of a refusal (Figure 7.5).

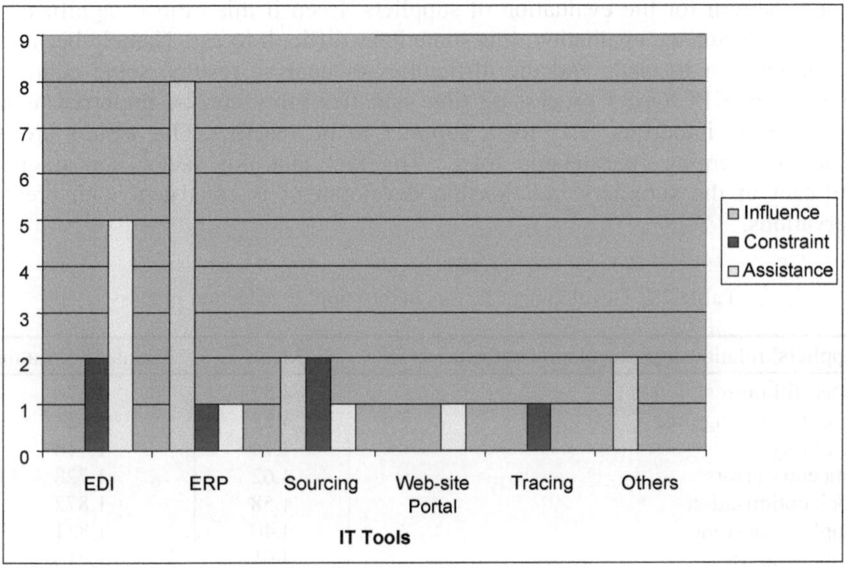

Figure 7.4. Network leaders' attitudes towards their suppliers

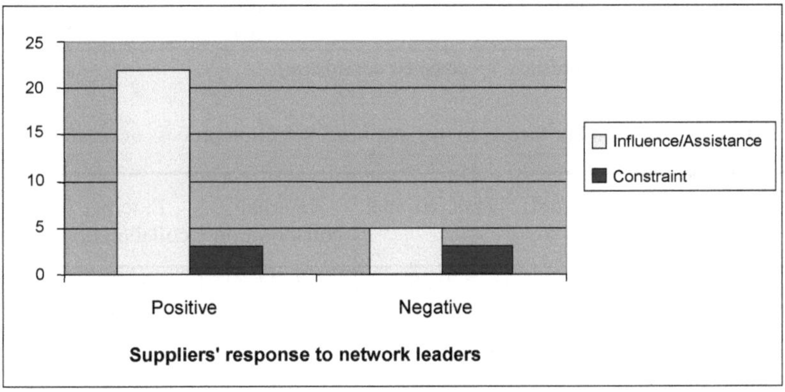

Figure 7.5. Suppliers' responses to their network leaders

7.4.4.2 Performance

The more rigorous and methodical the selection, the higher the performance [7.5]. The evaluation of supplier selection has to take place once the supplier selected and the buying process engaged. Indeed, once a collaborative supplier has been selected and a contract awarded, the next issue is the management of that supplier. However this constraint, network leaders need to define KPI in order to evaluate the success of their selection process. In this context, a lot of companies assess their suppliers on their ability to respond to the order–fill ratio (Table 7.6). This ratio, which sums up simultaneously the process time, quantity and quality of an order, represents an essential pattern for the evaluation of suppliers. Even if this ratio is significant for supplier performance evaluation, it is sometimes difficult to use. Namely because of the combination it needs and the difficulties of analyse results, some companies prefer others KPI. Order processing time and flexibility can be preferred as they highlight how important it is for a supplier to be reactive. Our results are also relevant concerning 'partnership trust'. The fact that this factor appears to be significant in the suppliers' relationship development is consistent with previous observations.

Table 7.6. Development factors in the supplier selection process

Suppliers' relationships development thanks to IT	Mean*	Standard deviation
Order–fill ratio	4.82	1.670
Order processing time	4.82	1.624
Flexibility	4.74	1.716
Data entry errors	4.62	1.728
Stock optimisation	4.58	1.872
Supplier lead-time	4.40	1.831
Partnership trust	4.05	1.710
Quality	4.00	1.866
Purchasing price	3.98	1.846
Quality problem solving	3.88	1.853
Technical innovations' benefit	3.81	1.842
New product set-up	3.77	1.811

*Lickert scale: 1–7 (1 = no accordance, 7 = complete accordance)

Table 7.7. Development factors in the suppliers' selection process according to IT

IT tools / Relationships development of supplier	Transactional	Internal collaboration	External collaboration	Decision making
Order–fill ratio	0.39*	0.27	0.23	0.11
Flexibility	0.39	0.32	0.28	0.08
Data entry errors	0.38	0.28	0.19	0.09
Stock optimisation	0.34	0.26	0.20	0.09
Supplier lead-time	0.35	0.27	0.27	0.08
Partnership trust	0.28	0.19	0.12	0.08
Quality	0.30	0.24	0.26	0.09
Purchasing price	0.29	0.26	0.19	0.10
Quality problem solving	0.31	0.21	0.21	0.08
Technical innovations' benefit	0.20	0.20	0.20	0.09
New product set-up	0.23	0.21	0.18	0.08
Total	0.32	0.25	0.21	0.09

*Observed frequency

Our study demonstrates IT leveraged supplier relationships, according to past experiences and used evaluation criteria (Table 7.7). Value creation and performance are different according to the tools. Transactional tools are more essential than other tools and their impact is more significant on the order processing time and less so on quality and flexibility. This observation can be explained by the nature of the transactional tools themselves. In contrast, characteristics such as the order–fill ratio, supplier lead-time or quality might benefit more precisely from external collaborative relationships. Finally, the largest homogeneity observed in internal collaborative tools shows the importance of cohesion, when it comes to decision making, for network leaders.

7.4.4.3 Difficulties

The importance of IT in the supplier selection process and their deployment confronts companies with many difficulties (Table 7.8). Even if it is financial problems that are the most serious, the acquisition and deployment of competencies and/or IT are supposed to be a big investment for suppliers [7.37]. This predominant financial impact can be explained by the medium size of suppliers (59% of suppliers have fewer than 100 employees and 75% fewer than 500) and the asymmetric relationships between network leaders and their suppliers. In order to build relationships based on cooperation, network leaders need to support their suppliers. In this way, some companies delegate human resources in the suppliers' organisation. The benefit of this delegation is for the network leader to be able to anticipate or react as soon as a difficulty has been highlighted by the supplier; for the supplying company, this collaboration enables competencies transfer and value creation.

Table 7.8. Difficulties resulting from IT deployment in upstream supply chain

Difficulties resulting from IT deployment with suppliers	Mean*	Standard deviation
Financial cost	4.96	1.589
Return on investment	4.39	1.765
Human capabilities	4.38	1.803
Supplier size	4.36	1.678
Systems incompatibility	4.17	1.852
Supplier rigidity	4.05	1.782
Top managers commitment	4.03	1.722
Confidentiality	4.00	1.782
Supplier dependence	3.88	1.880
Supplier's material capabilities	3.84	1.752
Network leader's material capabilities	3.74	1.739
Security	3.60	1.831
Network leader previous experiences	3.34	1.556

*Lickert scale: 1–7 (1 = no agreement, 7 = complete agreement)

Even if these financial problems are the most serious, the acquisition and deployment of competencies concerning IT, in particular human capabilities, are supposed to be a 'complex' investment for suppliers [7.37]. To conclude, we can note that material capabilities from the suppliers or the network leaders are insignificant against the network leaders' previous experiences. These results can be explained by the fact that IT deployment does not need huge and different material investment from other than information technologies.

7.5 Conclusions

The objective of this chapter is to examine the criteria used in the suppliers' selection process and thereby in the supply chain. The empirical evidence from selected companies confirms that the suppliers' selection is a strategic decision because companies tend to reduce the number of suppliers in order to set up sustainable collaborative relationships. These relationships are mainly based on long-term advantage and trust, due to the degree of the closeness and information sharing involved in these upstream partnerships. The impact of the number of suppliers is significant according to the type of relationships that network leaders want to set up with their upstream partners. We observed that companies engaged with a reduced basis of suppliers seek to set up relationships based on trust and long-term advantage.

Moreover, this chapter highlights that traditional criteria such as price, quality and flexibility play a major role in the supplier selection process. Nonetheless, we observed that the size of the network leaders is significant in regards of this ranking. Less important criteria such as trust, personal relationships and geographic proximity become more important as the size of the network leaders decrease. In this way, companies under 20 employees establish geographic proximity and personal relationships as the third and the fourth criteria. Above these traditional criteria, more technical ones are confirmed. The technological issues of potential suppliers appear to be evaluated, particularly in relation with IT and the capabilities of suppliers to acquire and develop specific tools in response of network leaders' needs.

By the same time, we outlined new companies' tendencies concerning the importance of information systems and types of tools mobilised and preferred in this process [7.20]. Even if transactional tools are more appropriate for collaborative relationships, we observed that internal and external collaborative tools are getting more attention according to the network leader. Meanwhile, internal tools that are more accepted by companies demonstrate the importance of internal cohesion as a condition for the choice and the use of more developed tools in the selection process (such as external tools).

In the companies studied, value creation in the selection process is related to competencies acquisition and development and level of performance. According to the first point, one can argue that network leaders play a positive role in value creation because they influence and help their suppliers rather than constrain them. This positive attitude has a direct impact on suppliers' attitude as they mostly react positively. The partnerships built on long-term advantage and the trust between all the members of the upstream supply chain allow new arrangements and explain this situation. The main KPI, the order–fill ratio, results in a simultaneous combination of process time, quantity and quality of an order. Suppliers are evaluated on this ratio but also on others indicators such as supplier lead-time and stock optimisation. In such a context of suppliers' performance, one can note that purchasing price is relatively insignificant.

To conclude with our results, even if IT is a significant criterion in the supplier selection process, their acquisition and deployment confront companies with a lot of problems and interrogations. Financial cost and return on investment emerge as the two main difficulties confronting suppliers. The human capabilities needed for such investments appear to be also a difficulty that suppliers have to deal with. The size of the suppliers is nevertheless a problem that network leaders have to take into account in their supplier selection. We confirm previous observations and the fact that suppliers' size is a fundamental element that can hinder some partnerships. In contrast, material capabilities do not seem to be a significant problem that suppliers or network leaders are facing.

Concluding remark concerns limitations of our research. Particularly, we have not discussed the arbitration of criteria that is preferred in the selection process, nor have we examined the influence of each criterion, even if many companies are now beginning to develop multi-criteria models.

References

[7.1] Williamson, O., 1975, *Markets and Hierarchies: Analysis and Antitrust Implications*, Free Press, New York.

[7.2] Williamson, O., 1981, *The Economic Institutions of Capitalism*, Free Press, New York.

[7.3] Spalanzani, A., 2007, "Organisation de la Gestion Industrielle : Un Siècle d'Innovation Continue," In *Regards sur la Recherche en Gestion, Contributions Grenobloises*, l'Harmattan (ed.), pp. 323–349

[7.4] Oberoi, J.S. and Khamba, J.S., 2005, "Strategically managed buyer-supplier relationships across supply chain: an exploratory study," *Human Systems Management*, **24**(4), pp. 275–283.

[7.5] Pearson, J. and Ellram, L., 1995, "Supplier selection and evaluation in small versus large electronics firms," *Journal of Small Business Management*, **33**(4), pp. 53–65.

[7.6] De Boer, L., Der Wegen, L.V. and Telgen, J., 1998, "Outranking methods in support of supplier selection," *European Journal of Purchasing & Supply Management*, **4**(2–3), pp. 109–118.

[7.7] Verma, R. and Pullman, M.E., 1998, "An analysis of the supplier selection process," *Omega*, **26**(6), pp. 739–750.

[7.8] Chan, F.T.S., Chan, H.K. and Lau, H.C.W., 2007, "A decision support system for supplier selection in the airline industry," *Journal of Engineering Manufacture*, **221**(4), pp. 741–758.

[7.9] De Boer, L., Labro, E. and Morlacchi, P., 2001, "A review of methods supporting supplier selection," *European Journal of Purchasing & Supply Management*, **7**(2), pp. 75–89.

[7.10] Sarkis, J. and Talluri, S., 2002, "A model for strategic supplier selection," *Journal of Supply Chain Management*, **38**(1), pp. 18–28.

[7.11] Chan, F.T.S., 2003, "Interactive selection model for supplier selection process: an analytical hierarchy process approach," *International Journal of Production Research*, **41**(15), pp. 3549–3579.

[7.12] Cheraghi, S.H., Hossein, S., Dadashzadeh, M. and Subramanian, M., 2004, "Critical success factors for supplier selection: an update," *Journal of Applied Business Research*, **20**(2), pp. 91–108.

[7.13] Donney, P.M. and Cannon, J.P., 1997, "An examination of the nature of trust in buyer–seller relationship," *Journal of Marketing*, **61**(2), pp. 35–51.

[7.14] Dempsey, W.A., 1978, "Vendor selection and the buying process," *Industrial Marketing Management*, **7**, pp. 257–267.

[7.15] Dickson, G.W., 1966, "An analysis of supplier selection systems and decisions," *Journal of Purchasing*, **2**(1), pp. 5–17.

[7.16] Berens, J., 1972, "A decision matrix approach to supplier selection," *Journal of Retailing*, **47**(4), pp. 47–53.

[7.17] Shipley, D., 1985, "Resellers' supplier selection criteria for different consumer products," *European Journal of Marketing*, **19**(7), pp. 26–36.

[7.18] Ellram, L. and Carr, A., 1994, "Strategic purchasing: a history and review of the literature," *International Journal of Physical Distribution and Logistics Management*, **30**(2), pp. 9–18.

[7.19] Brown, G., Boya, U.O., Humphresy, N. and Widing, R.E., 1993, "Attributes and behaviours of sales people preferred by buyers: high socializing vs. low socializing industrial buyers," *Journal of Personal Selling and Sales Management*, **13**(1), pp. 25–30.

[7.20] Swift, C.O. and Gruben, K.H., 2000, "Gender differences in weighting of supplier selection criteria," *Journal of Managerial Issues*, **12**(4), pp. 502–512.

[7.21] Ellram, L., 1990, "The supplier selection decision in strategic partnership," *Journal of Purchasing and Materials Management*, **26**(4), pp. 8–14.

[7.22] Bakos, J.Y. and Brynjolfsson, E., 1993, "Information technology incentives and the optimal number of suppliers," *Journal of Management Information System*, **10**(2) pp. 37–53.

[7.23] Mentzer, J.T., Min, S. and Zacharia, Z.G., 2000, "The nature of interfirm partnering in supply chain management," *Journal of Retailing*, **76**(4), pp. 549–568.

[7.24] Hart, P.J. and Saunders, C.S., 1998, "Emerging electronic partnerships: antecedents and dimensions of EDI use from a supplier's perspective," *Journal of Management Information System*, **14**(4), pp. 87–111.

[7.25] Lee, H.S. and Kim, S.K., 2001, "Supplier selection and management system considering relationships in supply chain management," *IEEE Transactions on Engineering Management*, **48**(3), pp. 307–319.

[7.26] Sirkka, L.J. and Blakes, I., "The global network organization of the future: information management opportunities and challenges," *Journal of Management Information Systems*, **10**(4), pp. 25–57.

[7.27] Dyer, J. and Singh, H. 1998, "The relational view: cooperative strategy and sources of inter-organizational competitive advantage," *Academy of Management Review*, **23**(4), pp. 660–679.

[7.28] Cash, J.I. and Konsynski, B.R., 1985, "IS redraws competitive boundaries," *Harvard Business Review*, March–April, pp. 134–142.

[7.29] Clemonds, E. and Row, M., 1993, "Limits to interfirm coordination through information technology: results of a field study in consumer packaged goods distribution," *Journal of Management Information Systems*, **10**(1), pp. 73–95.

[7.30] Subramani, M., 2004, "How do suppliers benefit from information technology use in supply chain relationships," *MIS Quarterly*, **28**(1), pp. 47–73.

[7.31] Ghosh, M. and John, G., 1999, "Governance value analysis and marketing strategy," *Journal of Marketing*, **63**, pp. 131–145.

[7.32] Mukhopadhyay, T. and Kekre, S., 2002, "Strategic and operational benefit of electronic integration in B2B procurement processes," *Management Science*, **48**(10), pp. 1301–1313.

[7.33] Goffin, K., Szwejczewski, M. and New, C., 1997, "Managing suppliers: when fewer can mean more," *International Journal of Physical Distribution and Logistics Management*, **27**(7), pp. 422–435.

[7.34] Gallivan, M.J. and Depledge, G., 2003, "Trust, control and the role of interorganizational system in electronics partnerships," *Information System Journal*, **13**(2), pp. 159–190.

[7.35] Min, H., 1994, "International supplier selection," *International Journal of Physical Distribution and Logistics Management*, **24**(5), pp. 24–33.

[7.36] Humphreys, P., Mak, K.L. and Yeung, C.M., 1998, "A just-in-time evaluation strategy for international procurement," *Supply Chain Management: an International Journal*, **3**(4), pp. 175–186.

[7.37] Riggins, F.J. and Mukhopadhyay, T., 1994, "Interdependent benefits from interorganizational systems: opportunities for business partner reengineering," *Journal of Management Information System*, **11**(2), pp. 37–57.

8

Supplier Selection in Agile Manufacturing Using Fuzzy Analytic Hierarchy Process

Cengiz Kahraman and İhsan Kaya

Department of Industrial Engineering, Istanbul Technical University
34367 Macka, Istanbul, Turkey
Emails: kahramanc@itu.edu.tr; kayai@itu.edu.tr

Abstract
The focus on competitive supply chains and extended enterprises requires the adoption of agile manufacturing practices demanding their suppliers to have agile attributes. This study designs and implements a procedure for judging the suitability of suppliers for an organisation competing on agile manufacturing characteristics. Quantitative and qualitative factors are used to appraise and select appropriate suppliers to fit within an organisation's agility practices. Under incomplete information from the experts, the fuzzy set theory is used to handle the uncertainty. A fuzzy analytic hierarchy process is used for the selection of the best supplier for agile manufacturing. The selection criteria are determined after a wide literature review. A detailed application is given to illustrate the model.

8.1 Introduction

After the Second World War, Eiji Toyoda of the Toyota Motor Company was developing major initiatives to improve his manufacturing and engineering processes. He made several visits to the Ford 'Rouge' factory in Detroit, USA, which was widely recognised to be the 'flagship' of the Ford plant. His first visit to Rouge was in 1950 and, although Ford was producing more vehicles than Toyota, he believed that there was considerable room for improvement in the Ford methods of production. This led to the method of working that became known as 'lean manufacturing'. Taiichi Ohno, an engineer at Toyota, was instrumental in making many of the efficiency improvements within the manufacturing process. His work with Toyota focused attention on machine utilisation: most notably the set-up and changeover processes and procedures. During a ten-year improvement process, Ohno reduced die changeover times from one day to 3 min. One of the key aspects of the Japanese philosophy has been the drive for workers and resources to do more with less. This has led to the developments in 'lean thinking' and 'agile manufacturing'. Over the last twenty years, there has been an increasing drive for organisations to reduce costs and improve the efficiency of their processes. From a production perspective, this has often centred on the Japanese lean manufacturing and JIT (just-in-time) waste elimination processes. In 1991, the Iaccocca Institute in Bethlehem, Pennsylvania commissioned a report specifically to analyse the changing nature of manufacturing within a turbulent marketplace. The Agile Manufacturing Forum was initiated the following year with the term 'agile manufacturing' being introduced to a business environment specifically to investigate how organisations could respond in a volatile market. The term 'agile' has been used to counteract the diversity and fragmentation of the marketplace and the likelihood of reducing volume requirements. An agile manufacturer can be defined as 'the fastest to market, with the lowest total cost and the greatest ability to meet varied customer requirements – the final measure being the ability to delight the customer' [8.1].

The swift trend towards a multiplicity of finished products with short development and production lead times has led many companies into problems with inventories, overheads, and inefficiencies. They are trying to apply the traditional mass-production approach without realising that the whole environment has changed. Mass production does not apply to products where the customers require small quantities of highly customised, design-to-order products, and where additional services and value-added benefits like product upgrades and future reconfigurations are as important as the product itself. Approaches such as rapid prototyping (RP), rapid tooling (RT), and reverse engineering are helping to solve some of these problems. RP is a relatively new class of technology used for building physical models and prototype parts from three-dimensional (3D) computer-aided design (CAD) data. RT falls into two categories: (1) advanced methods of making tools using RP technology, an additive process, and (2) advanced methods of making tools using milling technology, a subtractive process. Reverse engineering encompasses a variety of approaches to reproducing a physical object with the aid of documentation, drawings, or computer models. In the broadest sense, reverse engineering is whatever it takes, manual or under computer control, to reproduce something [8.2].

Agility is the ability to thrive and prosper in an environment of constant and unpredictable change. Some of the reasons why the manufacturing paradigm is changing from mass production to agile manufacturing include [8.2]:

1. global competition is intensifying;
2. mass markets are fragmenting into niche markets;
3. cooperation among companies is becoming necessary, including companies who are in direct competition with each other;
4. customers expect low volume, high quality, custom products;
5. very short product life-cycles, development time, and production lead times are required; and
6. customers want to be treated as individuals.

Agile manufacturing is a term applied to an organisation that has created the processes, tools and training required to enable it to respond quickly to customer needs and market changes while still controlling costs and quality. It is a response to complexity brought about by constant change. Agility is an overall strategy focused on thriving in an unpredictable environment. Focusing on the individual customer, agile competition has evolved from the unilateral producer-centred customer-responsive companies inspired by the lean manufacturing refinement of mass production to interactive producer–customer relationships.

In a similar sense, some researchers contrast flexible manufacturing systems (FMS) and agile manufacturing systems (AMS) according to the type of adaptation: FMS is reactive adaptation, while AMS is proactive adaptation. Agility enables enterprises to thrive in an environment of continuous and unanticipated change. It is a new, post-mass-production system for the creation and distribution of goods and services. Agile manufacturing requires resources that are beyond the reach of a single company. Sharing resources and technologies among companies becomes necessary. The competitive ability of an enterprise depends on its ability to establish proper relationships, and thus cooperation seems to be the key to possibly complementary relationships. An agile enterprise has the organisational flexibility to adopt for each project the managerial vehicle that will yield the greatest competitive advantage. Sometimes, this will take the form of an internal cross-functional team with participation from suppliers and customers [8.3].

Businesses are restructuring and re-engineering themselves in response to the challenges and demands of the twenty-first century. In this century, businesses will have to overcome the challenges of demanding customers seeking high-quality, low-cost products, responsive to their specific and rapidly changing needs. Agility addresses new ways of running companies to meet these challenges. Agility is about casting off those old ways of doing things that are no longer appropriate – changing pattern of traditional operation. In a changing competitive environment, there is a need to develop organisations and facilities significantly more flexible and responsive than current existing ones [8.4]. Agility is a business-wide capability that embraces organisational structures, information systems, logistics processes and, in particular, mindsets. A key characteristic of an agile organisation is flexibility. Indeed, the origins of agility as a business concept lie in FMS. Initially, it was thought that the route to manufacturing flexibility was through automation to enable rapid change (*i.e.* reduced set-up times) and thus a greater responsiveness to changes in product mix or volume. Later, this idea of manufacturing flexibility was extended into the wider business context and the concept of agility as an organisational orientation was born. Agility should not be confused with *leanness*. Lean is about doing more with less. The term is often used in connection with lean manufacturing to imply a 'zero inventory', just-in-time approach [8.5].

As manufacturing strategies have evolved, the focus has shifted away from being big and stable with complete control, to being small, nimble and more responsive to the market. This evolution reflects the introduction of new technology, new trends and, in particular, new customer behaviour. New markets are up for grabs because being big and stable is no longer a competitive formula. Agility is the small manufacturer's chance to seize the market by responding faster to customer demands. Today's manufacturing world leaders are characterised by their ability to deliver the products that customers want with minimum time-to-market and maximum capability to revamp products to meet market expectations. To become an agile manufacturer, a company must recognise change in the marketplace and then manage and master that change. Today's customers focus on unique products and expect one-to-one marketing. As a result, they are less willing to accept mass-marketed goods and are rejecting one-size-fits-all products. This means that manufacturers must adapt and adopt the make-to-order mentality displayed by Japanese manufacturers in the 1980s in order to become customer-focused and supply customised products and services designed to match a particular customer profile. Agile manufacturing promises high-quality, individually customised, price-competitive products produced on demand [8.6].

The key enablers of agile manufacturing include: (i) virtual enterprise formation tools/metrics; (ii) physically distributed manufacturing architecture and teams; (iii) rapid partnership formation tools/metrics; (iv) concurrent engineering; (v) integrated product/production/business information systems; (vi) rapid prototyping; and (vii) electronic commerce [8.7].

8.1.1 Agile Manufacturing Criteria

Ramesh and Devadasan [8.8] identified twenty criteria for agile manufacturing (AM) based on a literature study:

- flattened organisational structure;
- devolution of authority;
- responsiveness;
- adaptability and flexibility;
- enriching customers;
- rapid increase in productivity;
- knowledge-driven employees;
- fully empowered employees;
- open management;
- robust response to customers;
- short and effective product life cycle;
- short and flexible product service life;
- continuous design improvement;
- shorter manufacturing planning time;
- cost management;
- automation;
- information technology integration;
- change in business and technical processes;
- very efficient time management;
- supply chain management application.

Apart from the twenty AM criteria identified in [8.8], there are another ten criteria that have been used by other authors:

- adaptability;
- virtual corporation;
- reconfiguration;
- long-term gains;
- responsiveness;
- deployment of technology;

- continuous improvement practiced;
- strategically viewed;
- innovative culture;
- customer-integrated process.

Gunasekaran [8.7] developed a conceptual model for the development of AMSs based on the literature survey as shown in Figure 8.1. The model has been developed along the four key dimensions of strategies, technologies, people and systems.

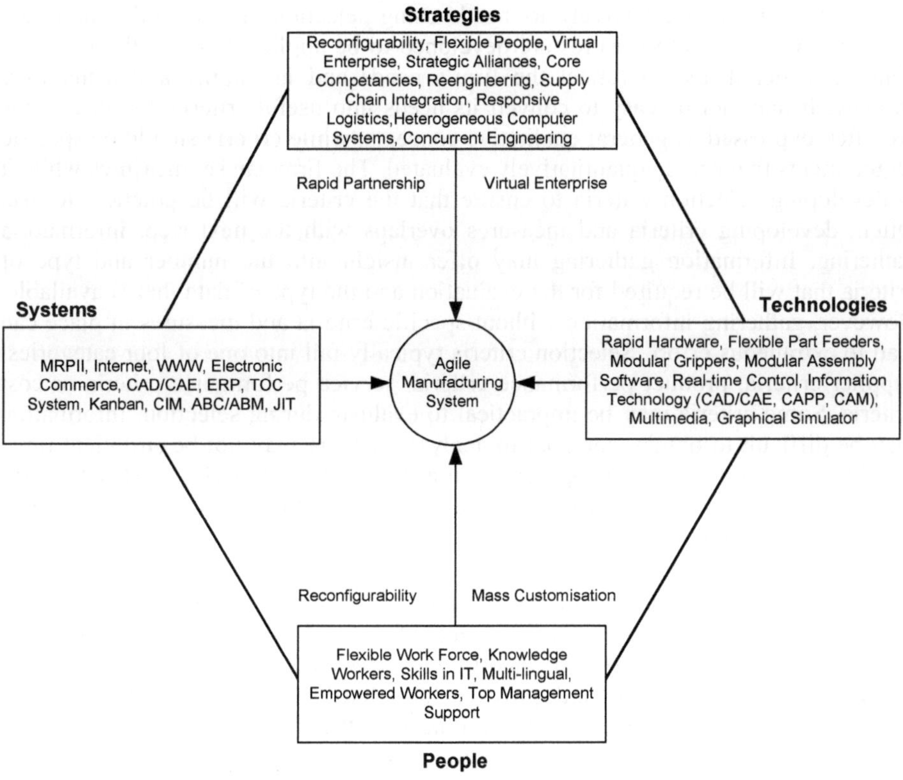

Figure 8.1. Development of an agile manufacturing system

Customers expect personalisation in their supply-chain relationships and best practices from their supply-chain partners. Manufacturers that can offer a more personalised relationship to their customers and confirm their use of world-class practices will survive. Those that cannot will lose their competitive edge and, eventually, lose customers and even whole markets. Even brand awareness, traditionally the linchpin of customer loyalty, is becoming less important than the ability to execute and meet customer needs.

Choosing the right supplier involves much more than scanning a series of price lists. The choice depends on a wide range of factors such as quality, reliability and service. Regardless of how they are weighed up, the importance of these different factors is based on the business' priorities and strategy. Supplier selection is one of the most important activities of organisations, where the short- and long-term success of a buyer's organisation depends upon the proper selection of its suppliers. The current competitive environment has placed increasing competitive pressures on suppliers to match the needs of buyer(s) in terms of quantity, quality, product mix, cost, time and place of delivery, to name only a few performance measures. The focus on competitive supply chains and extended enterprises needs the adoption of agile manufacturing practices requiring their suppliers to have agile attributes.

To select prospective suppliers, the firm judges each supplier's ability to meet consistently and cost-effectively its needs using selection criteria and appropriate measures. Criteria and measures are developed to be applicable to all the suppliers being considered and to reflect the firm's needs and its supply and technology strategy. It may not be easy to convert its needs into useful criteria, because needs are often expressed as general qualitative concepts, while criteria should be specific requirements that can be quantitatively evaluated. The firm can set measures while it is developing selection criteria to ensure that the criteria will be practical to use. Often, developing criteria and measures overlaps with the next step, information gathering. Information gathering may offer insight into the number and type of criteria that will be required for the evaluation and the type of data that is available. However, gathering information without specific criteria and measures in place can lead to extraneous effort. Selection criteria typically fall into one of four categories: supplier criteria, product performance criteria, service performance criteria or cost criteria. Some criteria may be impractical to evaluate during selection. Information may be difficult to obtain, complex to analyse, or there may not be sufficient time. The firm's criteria should be appropriate to its planned level of effort. Also, the firm may initially develop criteria or measures that it eventually finds inapplicable to some suppliers or certain products and services. Applying common criteria to all suppliers makes objective comparisons possible.

Decision-makers usually find that it is more confident to give interval judgments than fixed value judgments. Therefore, most of the evaluation parameters cannot be given precisely. The fuzzy set theory was specifically designed to represent uncertainty and vagueness and provide formalised tools for dealing with the imprecision intrinsic to many problems. Fuzzy set theory is used to model systems that are hard to define precisely. Fuzzy logic is a precise logic of imprecision and approximate reasoning. As a methodology, fuzzy set theory incorporates imprecision and subjectivity into the model formulation and solution process. Fuzzy set theory implements classes or groupings of data with boundaries that are not sharply defined (*i.e.* fuzzy). Any methodology or theory implementing 'crisp' definitions, such as classical set theory, arithmetic and programming, may be *fuzzified* by generalising the concept of a crisp set to a fuzzy set with blurred boundaries. The benefit of extending crisp theory and analysis methods to fuzzy techniques is the strength in solving real-world problems, which inevitably entail some degree of imprecision and noise in the variables and parameters measured and processed for the application. Accordingly, linguistic variables are a critical aspect of some fuzzy logic applications, where general terms such as 'large', 'medium', and 'small' are each used to capture a range of numerical values. Fuzzy set theory encompasses fuzzy logic, fuzzy arithmetic, fuzzy mathematical programming, fuzzy topology, fuzzy graph theory and fuzzy data analysis, although the term 'fuzzy logic' is often used to describe all of these [8.9].

The rest of this chapter is organised as follows. Section 8.2 summarises the literature review on supplier selection. Section 8.3 shows the proposed model for supplier selection in an agile manufacturing system. Section 8.4 gives the supplier selection criteria for agile manufacturing. Section 8.5 presents an application of the proposed model. Finally, Section 8.6 concludes the chapter with suggested future work.

8.2 Literature Review

Supplier selection has gained importance in the literature by some authors. In this section firstly, supplier selection and agile supplier selection studies have been analysed. Choy and Lee [8.10] proposed a case-based supplier management tool (CBSMT) using the case-based reasoning (CBR) technique in the areas of intelligent supplier selection and management that can enhance performance as compared to using the traditional approach. Cebeci and Kahraman [8.11] and Cebeci [8.12] measured customer satisfaction of catering service companies in Turkey by using fuzzy AHP. Ghodsypour and O'Brien [8.13] presented a mixed integer non-linear programming model to solve the multiple sourcing problems, which takes into account the total cost of logistics, including net price, storage, transportation, and ordering costs. Buyer limitations on budget, quality, service, *etc.*, can also be considered in the model. Feng *et al.* [8.14] presented a stochastic integer programming approach for simultaneous selection of tolerances and suppliers based on the quality loss function and process capability indices. Boer *et al.* [8.15] presented a review of decision methods reported in the literature for supporting the supplier selection process. The review is based on an extensive search in the academic literature. Masella and Rangone [8.16] proposed four different vendor selection systems (VSSs) depending on the time frame (short-term versus long-term) and on the content (logistic versus strategic) of the co-operative customer/supplier relationships. Liu *et al.* [8.17] compared suppliers for supplier selection and performance improvement using data envelopment analysis (DEA). Braglia and Petroni [8.18] described a multi-attribute utility theory based on the use of DEA, aiming at helping purchasing managers to formulate viable sourcing strategies in the changing market place. Dowlatshahi [8.19] focused on facilitating an interface and collaboration among designers at three planning horizons: strategic, tactical and operational with respect to supplier relations. To accomplish this interface, nine propositions for all areas of interface at three levels of planning are presented. Motwani *et al.* [8.20] attempted to fill a void in supplier selection research by developing a model for sourcing and purchasing in an international setting, particularly in developing countries. Ittner *et al.* [8.21] examined whether supplier selection and monitoring practices affect the association between supplier strategies and organisational performance. Ganeshan *et al.* [8.22] examined the dynamics of a supply chain that has the option of using two suppliers – one reliable, and the other unreliable. They analysed the cost economics of two suppliers in a broader inventory logistics framework, one that includes intrinsic inventories and transportation costs.

Verma and Pullman [8.23] examined the difference between managers' rating of the perceived importance of different supplier attributes and their actual choice of suppliers in an experimental setting. Boer *et al.* [8.15] showed, by means of a supplier selection example, that an outranking approach may be very well suited as a decision-making tool for initial purchasing decisions.

O'Brien and Ghodsypour [8.24] proposed an integration approach of an analytical hierarchy process and linear programming to consider both tangible and intangible factors in choosing the best suppliers and placing the optimum order quantities among them such that the total value of purchasing becomes maximum. Noci [8.25] designed a conceptual approach that first identifies measures for assessing a supplier's environmental performance and second suggests effective techniques for developing the supplier selection procedure according to an environmental viewpoint. Choi and Hartley [8.26] compared supplier selection practices based on a survey of companies at different levels in the auto industry. Mummalaneni *et al.* [8.27] reported the results of an exploratory study examining the trade-offs made by Chinese purchasing managers among the six attributes identified earlier. Swift [8.28] examined the supplier selection criteria of purchasing managers who have a preference for single sourcing and those who have a preference for multiple sourcing.

Gupta and Nagi [8.29] developed a flexible optimisation framework for partner selection in agile manufacturing. They developed a flexible and interactive decision support system that formally combines fuzzy qualitative information to aid in optimal selection of manufacturing partners for a business initiative in an agile manufacturing environment. Bocks [8.30] developed a data management framework (DMF) that can be defined as the ability of an enterprise to manage and distribute data, information and knowledge as the decisive enabler for core enterprise business process to support agility in manufacturing. The purpose of DMF was to provide a seamless enterprise data management solution in support of the AM environments.

Wang *et al.* [8.31] presented an Internet-assisted manufacturing system for AM practice. This system used the Internet as an interface between a user and a central network server (CNS) and allowed a local user to operate remote machines connected to the Internet. It was consisted of an integrated CNS of computer-aided design/computer-aided process planning/computer-aided manufacturing/computer-aided analysis, which links to local flexible manufacturing systems or computer numerically controlled machines by means of cable connections. Yang [8.32] proposed an object-oriented model of an AMS with a definition of the agile objects at four levels and their features. Meanwhile, it explained the process in which the agile objects, under the stimulation of market demands, get assembled into objects at higher levels and are integrated into agile system by sending information to each other and by accepting information selectively. McMullen [8.33] showed how the philosophies, practices, decision processes, measurements, logistics and systems architectures of the theory of constraints all work together to provide an infrastructure for AM. It was suggested that theory of constraints systems could be moved to a co-standard status with traditional MRP/capacity requirements planning systems, in order to encourage the systems community to provide the MRP II and ERP systems infrastructure required to support the emerging agile manufacturers. Merat *et al.* [8.34] proposed an agile workcell for light mechanical assembly. They used an object-oriented software running under VxWorks, a real-time operating system, for workcell control. In this way, an agile software architecture was developed to allow the rapid introduction of new assemblies through code re-use. Incorporated into the software architecture is a capability for workcell simulation so that the controller software could be written and tested on-line, enabling the rapid introduction of new products and facilitating agility in manufacturing.

Aoyama [8.35] proposed a fundamental redesign of the software development process, called the *agile software process* to meet the conflicting demands of delivering software products faster while simultaneously facilitating their widely distributed development. Monsplaisir [8.36] described the evaluation of two computer supported cooperative work (CSCW) prototypes to aid engineering teams in the design of an AM facility. The CSCWs facilitated consideration of a large number of flexibility and agility criteria associated with the design of manufacturing systems. Kusiak and He [8.37] developed three rules applicable to the design of products for agile assembly from an operational perspective. These rules are intended to support the design of products to meet the requirements of AM. Lee *et al.* [8.38] described an AM database system designed for capturing and manipulating the operational data of a manufacturing cell. Song and Nagi [8.39] proposed a framework for production control in an AM environment in which: (i) information was modelled in a hierarchical fashion using object oriented methodology; (ii) information transactions were specified by the workflow hierarchy consisting of partner workflows; (iii) information flow between partners was controlled by a set of distributed workflow managers interacting with partner knowledge bases, which reflect partner-specific information control rules on internal data exchange, as well as inter-partner mutual protocols for joint partner communications; and (iv) the prototype system was accomplished using the Web based on a client–server architecture.

Gunasekaran [8.4] and Gunasekaran and Yusuf [8.40] summarised a literature survey for the development of a framework for an agile manufacturing system (AMS). They presented a classification of the literature available on AMS and a brief review of each article was presented. Meade and Sarkis [8.41] introduced a decision methodology and structure that allows for the evaluation of alternatives to help organisations become more agile, with a specific objective of improving the manufacturing business processes for manufacturing and organisational agility improvement. They proposed a networked hierarchical analysis model based on the various characteristics of agility to evaluate alternatives that impact the business processes. The proposed evaluation model includes the analytic network process methodology for solving complex and systemic decisions.

Naylor *et al.* [8.42] presented a case study demonstrating how agility and leanness have been combined successfully within one supply chain to meet customer requirements. Wu *et al.* [8.43] presented an integer programming formulation to partner selection in agile manufacturing. Mason-Jones *et al.* [8.44] classified supply chain design and operations according to the lean, agile and leagile paradigms, which enables to match the supply chain type according to marketplace need. They also applied lean, agile and leagile principles according to the real needs of the specific supply chain. Christopher and Towill [8.5] showed how the lean and agile paradigms may be selected according to marketplace requirements. These were distinctly different, since in the first case the market winner was cost, whereas in the second case the market winner was availability. They also emphasised that agile supply chains were required to be market sensitive and hence nimble that the definition of waste was different from that appropriate to lean supply. Their paper concluded by proposing a cyclic migratory model that described the PC supply chain attributes during its evolution from traditional to its present customised 'leagile' operation.

Chan and Zhang [8.45] proposed an object and knowledge-based interval timed petri-net (OKITPN) approach, which provided an object-oriented and modular method of modelling manufacturing activities. It included knowledge, interval time, modular and communication attributes. The features of object-oriented modelling allowed the agile manufacturing systems to be modelled with the properties of classes and objects, and make the concept of software IC possible for rapid modelling of complex agile manufacturing systems. Once all of the interval timed petri-net (ITPN) objects were well defined, the developers need to consider only the interfaces and operations relating to the ITPN objects. In order to demonstrate the capability of the proposed OKITPN, it has been used to model rapidly the agile manufacturing systems that were reconfigured according to requirements.

Aitken *et al.* [8.46] suggested a three-level framework bringing together the various strands that contribute to the agile enterprise. In their integrative model, the key principles that underpin the agile supply chain such as rapid replenishment and postponed fulfilment, the individual programmes such as lean production, organisational agility and quick response, and finally individual actions to be taken are demonstrated in a layered model. Sanchez and Nagi [8.3] reviewed a wide range of literature on agile manufacturing. They reviewed about 73 papers from premier scientific journals and conferences, and proposed a classification scheme to organise these. McCullen and Towill [8.47] considered the effect of an agile manufacturing strategy on a company global supply chain that consists of overseas warehouses, a central finished good warehouse and a UK factory. Elkins *et al.* [8.48] explored agile manufacturing systems for engine and transmission machining applications as a key enabler in an automotive agile manufacturing strategy and discussed two simple decision models that provide initial insights and industry perspective into the business case for investment in agile manufacturing systems in the automotive industry. The models were applied to study the hypothetical decision of whether to invest in a dedicated, agile or flexible manufacturing system for engine and transmission parts machining. They focused on the use of flexible and agile manufacturing systems in the automotive industry for engine and transmission machining applications. They also analysed how automotive engineers perceive the differences between dedicated, agile, and flexible manufacturing equipment for machining applications.

Ip *et al.* [8.49] presented an investigation on the partner selection problems with engineering projects. Firstly, they described the problem as a 0–1 integer programming with non-analytical objective function. It was proven that the partner selection problem is a type of earliness and tardiness production planning problems. They also proposed a branch and bound algorithm with project scheduling to obtain the solution of partner selection. Lou *et al.* [8.50] discussed the concepts and characteristics of an agile supply chain, which was regarded as one of the pivotal technologies of agile manufacture based on dynamic alliance. They also emphasised the importance of coordination in the supply chain and presented a general architecture of agile supply chain management based on a multi-agent theory, in which the supply chain was managed by a set of intelligent agents for one or more activities.

Yusuf *et al.* [8.51] discussed the drivers and emerging patterns of supply chain integration. They developed a conceptual model of supply chain practices as determinants of manufacturing competitiveness and business performance. They also analysed the relationship between the patterns of supply chain and attainment of competitive and business performance. The analysis was based on the data collected from a survey using the standard questionnaire administered to 600 companies in the UK. Gunasekaran and Ngai [8.52] defined a build-to-order supply chain (BOSC) as the configuration of firms and capabilities in the supply chain that created the greatest degree of flexibility and responsiveness to altering market conditions in a cost-effective manner. BOSC incorporated the characteristics of both the lean and agile manufacturing strategies.

Agarwal *et al.* [8.53] explored the relationship among lead-time, cost, quality and service level and the leanness and agility of a case supply chain in fast-moving consumer goods business by using analytic network process (ANP). They presented a framework for modelling the performance of lean, agile and leagile supply chains on the basis of interdependent variables. The framework provided an aid to decision makers in analysing the variables affecting market sensitiveness, process integration, information driver and flexibility in lean, agile and leagile supply chains for the performance improvement of the case supply chain. They evaluated the influence of various performance dimensions on the specified objectives of SC, such as timely response to meet the customer demand. They also considered the influence of the performance determinants on each other. Finally, they concluded with a justification of the framework, which analyses the effect of market winning criteria and market qualifying criteria on the three types of supply chain: lean, agile and leagile.

Cao and Gao [8.54] analysed partner selection problem in agile manufacturing environment. The partner selection problem was described as a 0–1 integer programming model with nonlinear objective function. Its objective was to maximise project success probability within the constraints of time and cost. Because of the complexity and the nonlinearity of the model, it could not be solved by conventional methods easily. They proposed a dynamic and adaptive penalty-guided genetic algorithm approach to solve the problem. Dotoli *et al.* [8.55] analysed the design and optimisation of integrated e-supply chains for agile manufacturing systems. They emphasised that e-supply chains integrate the Internet and Web-based electronic market with promising systems to achieve agility for manufacturing systems, and the integrated e-supply chains (IESCs) is a key issue for the configuration of the partner network. They proposed a single- and multi-objective optimisation model to configure the network of IESCs and used an integer linear programming (ILP) problem solution. The ILP problem solution provided different network structures that allow for improving supply chain flexibility, agility and environmental performance during the design process. Ismail and Sharifi [8.56] analysed the parallel developments in the areas of agile systems and manufacturing, and how supply chain management led to the introduction of the agile supply chains

(ASC) concept. They tried to answer how to achieve agility in supply chains. They proposed a framework for the development of ASC that was based on the integration of existing supply chain analysis and development models and techniques with those of the supply chain design (SCD) and also the design for the supply chain (DfSC).

Krishnamurthy and Yauch [8.57] analysed a case study to determine whether the concept of leagility could be applied to a single corporation with multiple business units and whether a decoupling point would be necessary to distinguish the lean and agile portions of the enterprise. The case study findings were used as the basis for describing a theoretical corporate leagile infrastructure and for stimulating new research questions. Ramesh and Devadasan [8.8] managed a review on the literature and contributed a comprehensive model that would identify the criteria for attaining agility and suggested a procedure to successfully implement it in the manufacturing arena. They studied the literature dealing with AM criteria and derived meaning and definitions of AM.

Srinivasan [8.58] outlined some challenges faced by petroleum refineries that seek to be lean, agile and proactive. He analysed the role of artificial intelligence – software agents, pattern recognition and expert systems – to pave the path toward agility. Forsberg and Towers [8.59] introduced agile merchandising as a new value adding strategy for the European clothing and textile manufacturing industry. They investigated the concept of creating strategic agile supply networks in the textile fashion industry based on a case study involving eleven European textile manufacturers. They suggested that by incorporating the cooperative advantages of European manufacturers into the sourcing process, retailers would be better able to respond to the volatile and unpredictable nature of fashion garment demand.

Demirli and Yimer [8.60] analysed the adaptation of the build-to-order supply chain (BOSC) to become agile in a mass customisation process in order to meet diversified customer requirements for manufacturers of assembled products such as cars, computers, furniture, *etc.* They proposed an integrated production–distribution planning model for a multi-echelon, multi-plant and multi-product supply chain operating in a build-to-order environment for agile manufacturing systems. The uncertainties associated with estimation of the various operational cost parameters were represented by fuzzy numbers. The BOSC scheduling model was constructed as a mixed-integer fuzzy programming problem with the goal of reducing the overall operating costs related to component fabrication, procurement, assembling, inspection, logistics and inventory, while improving customer satisfaction by allowing product customisation and meeting delivery promise dates at each market outlet. They also suggested an efficient compromise solution approach by transforming the problem into an auxiliary multi-objective linear programming model. Hasan *et al.* [8.61] designed and implemented a procedure for judging the suitability of suppliers for an organisation competing on agile manufacturing characteristics. They used quantitative and qualitative factors to appraise and select appropriate suppliers to fit within an organisation's agility practices. They used two techniques, the analytical network process (ANP) and data envelopment analysis (DEA) to select appropriate suppliers in a multi-phased supplier selection approach. In the first stage, ANP was executed to appraise suppliers on their qualitative benefits, generating quantitative data from these qualitative dimensions then DEA was used to synthesise the data to arrive at a ranking of the suppliers.

8.3 Supplier Selection Criteria for Agile Manufacturing

Selection criteria typically fall into one of four categories: supplier criteria, product performance criteria, service performance criteria or cost criteria [8.9].

8.3.1 Supplier Criteria

A firm uses supplier criteria to evaluate whether the supplier fits its supply and technology strategy. These considerations are largely independent of the product or service sought. Supplier criteria are developed to measure important aspects of the supplier's business: financial strength, management approach and capability, technical ability, support resources and quality systems.

- *Financial* – the firm should require its suppliers to have a sound financial position. Financial strength can be a good indicator of the supplier's long-term stability. A solid financial position also helps ensure that performance standards can be maintained and that products and services will continue to be available.
- *Managerial* – to form a good supplier relationship, companies need to have compatible approaches to management, especially for integrated and strategic relationships. Maintaining a good supplier relationship requires management stability. The firm should have confidence in its supplier's management ability to run the company. It is also important that the supplier's management be committed to managing its supply base. The supplier's level of quality, service, and cost are directly affected by the suppliers' ability to meet its needs.
- *Technical* – to provide a consistently high quality product or service, promote successful development efforts, and ensure future improvements, a firm needs competent technical support from its suppliers. This is particularly important when the firm's supply and technology strategy includes development of a new product or technology or access to proprietary technology. Technical criteria may motivate a firm to move into the global marketplace. Sometimes, a desirable technology has been developed overseas and is not available to domestic suppliers.
- *Support resource* – the supplier's resources need to be adequate to support product or service development (if necessary), production and delivery. Criteria need to consider the supplier's facilities, information systems, and provisions for education and training. When considering international suppliers, a firm needs to carefully examine the industrial infrastructure that supports the supplier. With international suppliers, a firm also needs to establish appropriate mechanisms to handle financial transactions and product deliveries, as well as any related legal and regulatory matters. Some form of global customer service may be required to support project implementation and day-to-day operations.
- *Quality systems and process* – the supplier's quality systems and processes that maintain and improve quality and delivery performance are key factors.

Selection criteria may consider the supplier's quality assurance and control procedures, complaint handling procedures, quality manuals, ISO 9000 standard registration status, and internal rating and reporting systems. As the customer, a firm especially wants to examine the supplier's programs or processes for assessing and addressing customer needs.

- *Globalisation and localisation* – a firm's sourcing strategy may recognise definite advantages or disadvantages associated with choosing suppliers in a particular region or country. The firm's risk assessment should have identified potential risks, such as currency fluctuations, shifts in political policy, and the accompanying domestic or international regulatory and market changes.

8.3.2 Product Performance Criteria

A firm can use product performance criteria to examine important functional characteristics and measure the usability of the product being purchased. The exact criteria depend on the type of product being considered. A firm may need to examine conformance to specifications in any of the following areas:

- *end use* – quality, functionality (speed, capacity, *etc.*), reliability, maintainability, compatibility, durability/damage tolerance;
- *handling* – packaging, shelf-life, storage requirements;
- *use in manufacturing (components)* – quality, testability, manufacturability, compatibility, end-use performance;
- *other business considerations* – environmentally-friendly features such as recycled product content, ergonomic features, product availability, stage of the technology life cycle, market trends.

If the product or service is yet to be developed, the firm's supplier criteria needs to examine whether the supplier has the basic management, technical, and quality support necessary to develop the product or service. In the international market, technical standards may vary between countries. The firm either needs to become familiar with manufacturer's standards or test the product using its own standards. Products may have to be reworked to be compatible or interchangeable with domestic products.

8.3.3 Service Performance Criteria

A firm can use service performance criteria to evaluate the benefits provided by supplier services. When considering services, a firm needs to clearly define its expectations since there are few uniform, established service standards to draw upon. Because any purchase involves some degree of service, such as order processing, delivery and support, a firm should always include service criteria in its evaluation. If the supplier provides a solution combining products and services, the firm should be sure to adequately represent its service needs in the selection criteria. The service aspect can easily be lost amid product specifications when purchasing a highly technical product. Some of the concepts employed to judge products also apply to services, however, the terminology is often different, and services require other considerations. When assessing the fitness of services, a firm may need to examine the following areas:

- *customer support* – accessibility, timeliness, responsiveness, dependability;
- *customer satisfiers* – value-added;
- *follow-up* – to keep customer informed, to verify satisfaction;
- *professionalism* – knowledge, accuracy, attitude, reliability;
- *cost criteria* – cost criteria recognise important elements of cost associated with the purchase.

The most obvious costs associated with a product are "out of pocket" expenses, such as purchase price, transportation cost and taxes. These are typically considered during selection. Operational expenses, such as transaction processing and cost of rejects, may also be included, although these require more effort to estimate. Although a firm can express any criteria in terms of estimated cost, in some cases, obtaining reliable estimates may be too involved for the level of analysis in selection. A firm should re-evaluate cost in more detail during qualification. To evaluate suppliers based on a firm's selection criteria it needs to develop measures of supplier performance, product or service performance, and cost. There should be consensus within the team or organisation on the measures, standards, and methods used to rate or compare suppliers. A firm needs to develop effective measures for each of its selection criteria. A firm can evaluate the effectiveness of a measure of quality by determining the degree to which it is related to customer requirements, and developed with inputs from and consensus with work.

Meade and Sarkis [8.41] determined the agile supplier dimension and attributes for their model as follows: leverage people and information, master change and uncertainty, collaborative relationships and enrich customer, cross-functional training, continuous education and training, internalisation of societal values, value-added enterprise metrics, open information/communication policy, competency driven operations, integrated and interactive partner relations, proactive information sharing policies, virtual enterprise partnering, electronic commerce operability, modification, expansion, mix/range, routing, machine, volume flexibilities, individualised products, production-to-order, extra quality standard, and market opportunity pulled production.

Ramesh and Devadasan [8.8] reviewed the literature of AM to identify the criteria that distinguish an AM enterprise from traditional manufacturing company and obtained the followings: organisational structure, devolution of authority, manufacturing set-ups, status of quality, status of productivity, employees' status, employee involvement, nature of management, customer response adoption, product life cycle, product service life, design-improvement, production methodology, manufacturing planning, cost management, automation type, information technology integration, change in business and technical process, time management, and outsourcing. Olsson [8.62] structured a hierarchy for supplier selection in an agile manufacturing system (see Figure 8.2).

In this chapter, the following main and sub-criteria to determine the best supplier for an agile manufacturing firm have been determined after a wide literature review and consulting with experts' opinions. The hierarchical structure is illustrated in Figure 8.3.

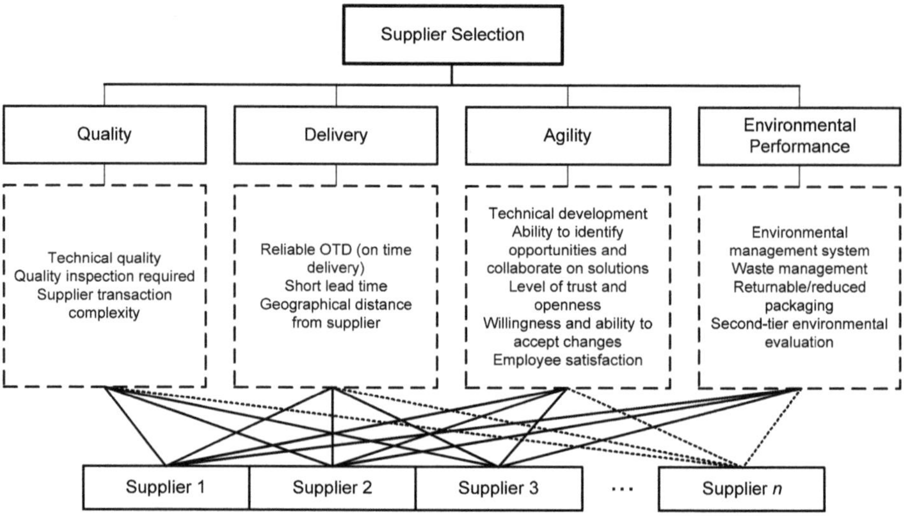

Figure 8.2. The hierarchy for supplier selection of an agile manufacturing firm

1. Quality (C1):
   i. technical quality (C11);
   ii. inspection procedures (C12);
   iii. quality systems (C13).
2. Delivery (C2):
   i. reliability (C21);
   ii. distance (C22);
   iii. lead time (C23);
   iv. transport cost (C24).
3. Agility (C3):
   i. flexibility (C31);
   ii. means of information (C32);
   iii. electronic data interchange (C33);
   iv. workforce (C34).
4. Performance (C4):
   i. new product introduction and development time (C41);
   ii. customer responsiveness (C42);
   iii. concurrent engineering (C43).
5. Management (C5):
   i. financial status (C51);
   ii. organisation structure (C52);
   iii. warranties and claim policies (C53).
6. Service (C6):
   i. lab facility (C61);
   ii. packaging ability (C62);
   iii. tool and processing (C63);
   iv. R&D (C64).

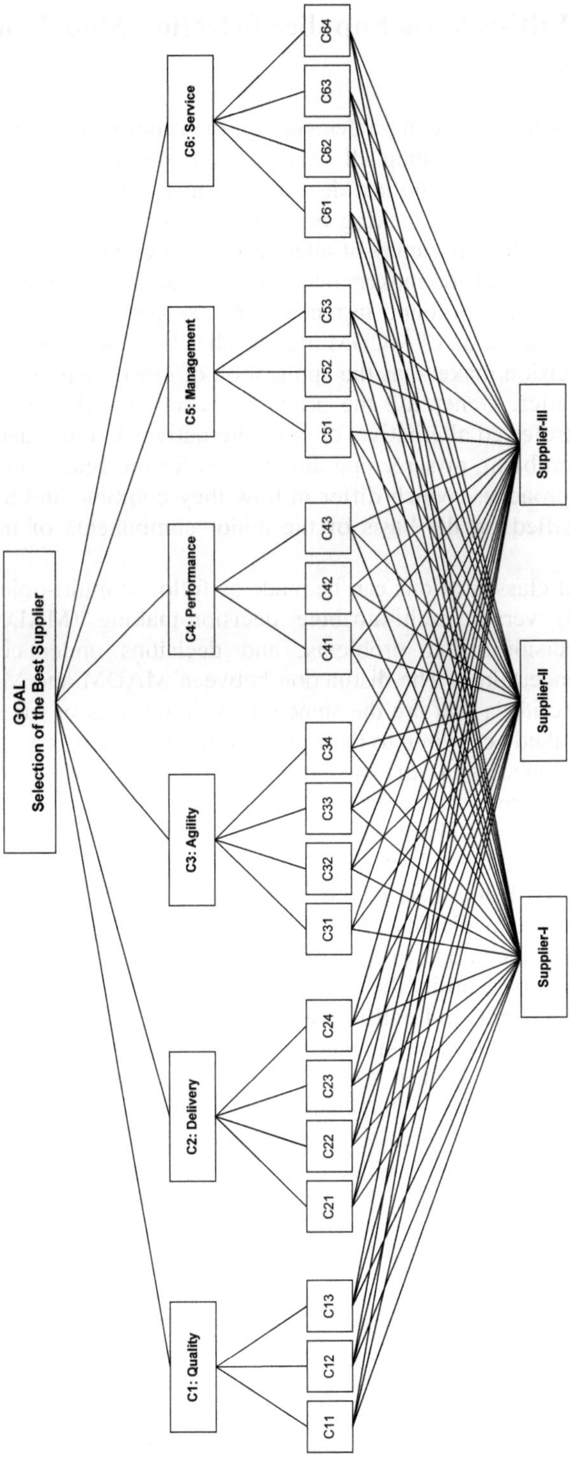

Figure 8.3. The considered hierarchical structure

8.4 A Fuzzy Multi-criteria Supplier Selection Model for Agile Manufacturing

Decision-making, such as making a decision on the order of projects, is the process of finding the best option among all feasible alternatives. It has been one of the fastest growing areas during the last decades owing to the changes in the business sector. In almost all decision-making problems, there exist many criteria and the process of evaluating the efficiency of alternatives is a complicated one. That is, for many such problems, decision makers need to take multiple criteria decision-making (MCDM) techniques into account. In recent years, as computer usage has increased significantly, the application of MCDM methods has become considerably easier for both users and decision makers, as the application of most methods correspond with complex mathematics. Generally, the primary concern of MCDM problems is to choose the most preferred alternative, or rank alternatives in the order of importance for the selection problem, or screening alternatives for the final decision. There are many MCDM approaches, which differ in how they combine and utilise data, and they can be classified on the basis of the major components of multiple criteria decision analysis.

Three different classifications can be made as follows: multi-objective decision-making (MODM) versus multi-attribute decision-making (MADM); individual versus group decision-maker problems; and decisions under certainty versus decisions under uncertainty. The distinction between MADM and MODM is based on the evaluation criteria that are the standards of judgments or rules on which the alternatives are ranked according to their desirability. Criterion is a general term and includes both the concepts of attributes and objectives. An attribute is a measurable quantity whose value reflects the degree to which a particular objective is achieved. An objective is a statement about the desired state of the system under consideration and indicates the direction of attribute improvement. Objectives are functionally related to, or derived from, a set of attributes [8.63]. In the literature, some multi-criteria approaches are available to help the decision-making process, such as AHP, outranking methods, technique for order performance by similarity to ideal solution (TOPSIS), multi-attribute utility theory (MAUT), simple additive weighting method, weighted product method, and multi-objective linear programming (MOLP) and its variants such as multi-objective stochastic integer linear programming, interactive MOLP and mixed 0–1 MOLP. Other multi-criteria approaches include multi-objective goal programming (MOGoP), multi-objective geometric programming (MOGeP), multi-objective nonlinear fractional programming, multi-objective dynamic programming and multi-objective genetic programming.

AHP is a structured approach to decision making developed by Saaty [8.64]. It was introduced for choosing the most suitable alternative, which fulfils the entire set of objectives in a multi-attribute decision-making problem and allows a set of complex issues to be compared with the importance of each issue relative to its impact on the solution to the problem. It is a weighted factor-scoring model and has the ability to detect and incorporate inconsistencies inherent in the decision-making process. Therefore, AHP has been applied to a wide variety of decision-making problems, including the evaluation of alternatives. Traditional AHP needs exact judgments. In addition, due to the complexity and uncertainty involved in real-world decisions, it is sometimes unrealistic or even impossible to perform exact comparisons. It is therefore more natural or realistic that a decision maker is allowed to provide fuzzy judgments instead of precise comparisons. In this chapter, AHP based on fuzzy set theory is applied to the decision-making process. Sometimes a decision maker's judgments cannot be crisp, and it is relatively difficult for them to provide exact numerical values. Therefore, most of the evaluation parameters cannot be given precisely. The evaluation data of an alternative project's suitability for various subjective criteria and the weights of the criteria are usually expressed in linguistic terms by the decision maker. In this case, the fuzzy logic that provides a mathematical strength to capture the uncertainties associated with human cognitive process can be used. A fuzzy multi-criteria decision-making methodology is proposed here to select the best alternative and to avoid traffic congestion [8.65].

Fuzzy set theory was specifically designed to mathematically represent uncertainty and vagueness, and provide formalised tools for dealing with imprecision intrinsic to many problems. The root of fuzzy set theory goes back to 1965 when Zadeh initiated the concept of fuzzy logic [8.66]. It uses approximate information and uncertainty to generate decisions. This is why it looks somewhat similar to human reasoning. Since knowledge can be expressed in a more natural way by using fuzzy sets, many engineering and decision problems can be greatly simplified. Fuzzy set theory implements groupings of data with loosely defined boundaries. Keeping this in mind, any methodology or theory implementing 'crisp' definitions may be 'fuzzified', if needed, by generalising the concept of a crisp set to a fuzzy set with blurred boundaries. The main benefit of extending crisp analysis methods to fuzzy techniques is the strength in solving real-world problems, which has imprecision in the variables and parameters measured and processed for an application. To achieve this benefit, linguistic variables are used as a critical aspect of some fuzzy logic applications. If a variable can take words in natural languages as its value, it is called a linguistic variable, where the words such as *good*, *mediocre* and *bad* are characterised by fuzzy sets defined in the universe of discourse in which the variable is defined. Several geometric mapping functions have been widely adopted, such as triangular, trapezoidal and S-shaped membership functions (MF) where the triangular MF is a special case of the trapezoidal one. In this chapter, standardised trapezoidal fuzzy numbers (STFNs) are used. As shown in Figure 8.4, a trapezoidal fuzzy number, $\tilde{A}=(a, b, c, d)$, is a normal, convex fuzzy set, on the real line, with a piecewise continuous membership function.

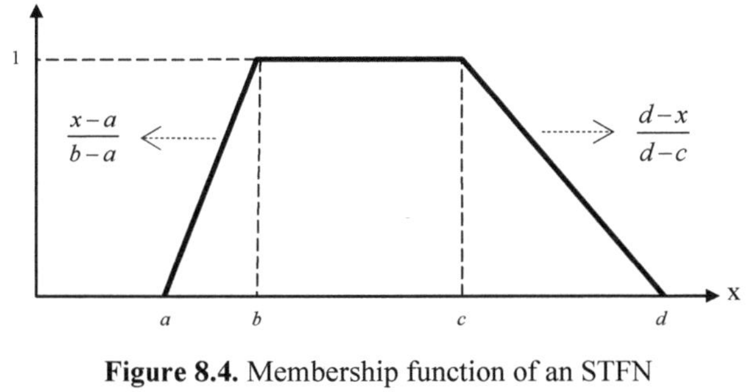

Figure 8.4. Membership function of an STFN

The properties of the four points $a \le b \le c \le d$ and the membership functions are given in Equation 8.1 [8.65]:

1. $\mu(x) = 0$ for every $x \in (-\infty, a) \cup (d, \infty)$
2. μ is increasing on $[a, b]$ and decreasing on $[c, d]$
3. $\mu(a) = \mu(d) = 0$ and $\mu(x) = 1$, for every $x \in [b, c]$

$$\mu_{\tilde{A}}(x) = \begin{cases} \dfrac{(x-a)}{(b-a)}, & \text{for } a \le x \le b \\ 1 \quad , & \text{for } b \le x \le c \\ \dfrac{(d-x)}{(d-c)}, & \text{for } c \le x \le d \\ 0 \quad , & \text{otherwise} \end{cases} \tag{8.1}$$

The cases $a = -\infty$ and $d = \infty$ are admitted, and then the fuzzy number will be, by the left or by the right, asymptotically zero, so its support will not be bounded.

As mentioned in the previous sections, both the benefits and the costs embedded in a very important strategic investment project are mostly intangible. This is the main reason making the investment decision evaluation extremely hard. If sufficient objective data were available, the probability theory would be preferred in such a decision analysis. Unfortunately, decision makers do not have enough information to perform such a decision analysis, since probabilities can never be known with certainty and the decisions about strategic level information technology investments are attributable to many uncertain derivations. In this situation, decision makers should rely on their knowledge in modelling projects, which are the investments to decrease traffic congestion. To deal quantitatively with such an imprecision or uncertainty, the fuzzy set theory is used.

According to Zeng *et al.* [8.67], influential factors can be decomposed by brainstorming or checklist techniques, scored by fuzzy membership functions and weighed by AHP. However, one drawback of the current AHP method is that it can only deal with definite scales in reality. To deal with this drawback, fuzzy AHP is proposed. A fuzzy AHP is an important extension of the typical AHP method, which was first introduced by Laarhoven and Pedrycz [8.68]. Later, a few other fuzzy AHP approaches were developed and applied to some industrial problems (*e.g.* by Buckley [8.69], Chang [8.70] and Kahraman *et al.* [8.71]).

In this chapter, a modified AHP method proposed by Zeng *et al.* [8.67] and combined with a different fuzzy ranking method is used for selecting the traffic congestion project best fitting the municipality's policies. In this method, fuzzy aggregation is used to create group decisions, and then defuzzification is employed to transform the fuzzy scales into crisp scales for the computation of priority weights. The group preference of each factor is then calculated by applying fuzzy aggregation operators, *i.e.* fuzzy multiplication and addition operators. The steps of the methodology [8.65, 8.67] are described below.

Step 1: Measure the factors in the hierarchy

The decision-makers are required to provide their judgments on the basis of their knowledge and expertise for each factor at the bottom level in the hierarchy. As different decision-makers having different perspectives could have different influence on the final decision, a fuzzy eigenvector based weighting is used in the model to calculate the decision-makers' competence. The decision-makers can provide a precise numerical value, a range of numerical values, a linguistic term or a fuzzy number. For m decision-makers in the evaluation group, the ith decision maker is assigned a weight c_i, where $c_i \in [0, 1]$ and $c_1 + c_2 + \cdots + c_m = 1$.

Step 2: Compare the factors using pair-wise comparisons

The attributes having impacts on the selection of a project are listed and classified in a hierarchical structure as can be seen in Figure 8.3. The pair-wise comparison matrixes for the main and sub-criteria are built by a simple Microsoft Excel based evaluation form. A modified fuzzy AHP method is applied to work out the priority weights of selected attributes. In a typical AHP method, decision-makers would have to give a definite number from Table 8.1 to the pair-wise comparison so that the priority vector can be computed.

However, attribute comparisons often involve certain amount of uncertainty and subjectivity. Here are two very important special examples to illustrate this situation:

Table 8.1. Scale of relative importance [8.64]

Intensity of importance	Definition	Explanation
1	Equal importance	Two activities contribute equally to the objective
3	Weak importance of one over another	Experience and judgment slightly favour one activity over another
5	Essential or strong importance	Experience and judgment strongly favour one activity over another
7	Demonstrated importance	An activity is strongly favoured and its dominance demonstrated in practice
9	Absolute importance	The evidence favouring one activity over another is of the highest possible order of affirmation
2, 4, 6, 8	Intermediate values between the two adjacent judgments	When compromise is needed
Reciprocals of above non-zero	If activity *i* has one of the above non-zero numbers assigned to it when compared with activity *j*, then *j* has the reciprocal value when compared with *i*	

(i) Assume that a decision maker is sure that attribute 1 is more important than attribute 2, but he/she cannot give a definite score to the comparison because of being not sure about the degree of importance for attribute 1 over attribute 2. In this case, the decision maker may prefer to provide a range of 3–5 to describe the comparison as, 'attribute 1 is between weakly more important to strongly more important than attribute 2'. (ii) Assume that, the decision-maker cannot compare two attributes due to the lack of adequate information. In this case, a typical AHP method has to be discarded due to the existence of fuzzy or incomplete comparisons.

A modified fuzzy AHP is used in this study to overcome these shortcomings. The decision-makers are required to compare every attribute pair-wise in their corresponding group structured in the hierarchy shown in Figure 8.3, and define scores of the project alternatives against these attributes.

Step 3: Convert preferences into STFNs

As described in steps 1 and 2, because the values of factors provided by experts are crisps, *e.g.* a numerical value, a range of numerical value, a linguistic term or a fuzzy number, the STFN is employed to convert these experts' judgments into a universal format for the composition of group preferences. A crisp numerical value, a range of crisp numerical values and a triangular fuzzy number can be converted to an STFN as follows:

- a crisp number '*n*' is converted to the STFN as $\tilde{A} = (n,n,n,n)$, *i.e.* $a=b=c=d=n$;
- a linguistic term 'about *n*' is converted to the STFN as $\tilde{A} = (n-1, n, n, n+1)$, *i.e.* $a=n-1$, $b=c=n$, $d=n+1$;
- a range, whose scale is likely between (*n*, *m*), is converted to the STFN as $\tilde{A} = (n,n,m,m)$, *i.e.* $a=b=n$, $c=d=m$;
- a triangular fuzzy number, $\tilde{T} = (x, y, z)$, is converted to the STFN as $\tilde{A} = (x, y, y, z)$, *i.e.* $a=x$, $b=c=y$, $d=z$;
- if a decision maker cannot compare any two factors at all, it is represented by $\tilde{A} = (0,0,0,0)$, *i.e.* $a=b=c=d=0$.

The decision makers are encouraged to give fuzzy scales if they are not sure about the exact numerical values or leave some comparisons absent as they cannot compare two attributes at all. In each case, as can be understood from the conversion information in the brackets above, a single standardised trapezoidal fuzzy number (STFN) is employed to convert these decision makers' judgments into a generic format for the composition of group preferences.

Step 4: Aggregate individual STFNs into group STFNs

The aim of this step is to apply an appropriate operator to aggregate the individual preferences made by individual experts into a group preference of each factor. The aggregation of STFN scores is performed by applying the fuzzy weighted trapezoidal averaging operator, which is defined by:

$$\widetilde{S}_i = \widetilde{S}_{i1} \otimes c_1 \oplus \widetilde{S}_{i2} \otimes c_2 \oplus .... \oplus \widetilde{S}_{im} \otimes c_m \tag{8.2}$$

where $\widetilde{S}_i$ is the fuzzy aggregated score of factor F_i; $\widetilde{S}_{i1}, \widetilde{S}_{i2}, ..., \widetilde{S}_{im}$ are the STFN scores of the factor F_i measured by m decision makers DM_1, DM_2, ···, DM_m, respectively; ⊗ and ⊗ denote the fuzzy multiplication and fuzzy addition operators, respectively; and c_1, c_2, ···, c_m are contribution factors allocated to the decision-makers, DM_1, DM_2, ···, DM_m and $c_1 + c_2 + \cdots + c_m = 1$.

Similarly, the aggregation of STFN scales is defined as

$$\widetilde{a}_{ij} = \widetilde{a}_{ij1} \otimes c_1 \oplus \widetilde{a}_{ij2} \otimes c_2 \oplus .... \oplus \widetilde{a}_{ijm} \otimes c_m \tag{8.3}$$

where $\widetilde{a}_{ij}$ is the aggregated fuzzy scale of attribute i comparing to attribute j for $i, j = 1, 2, ..., n$; and $\widetilde{a}_{ij1}, \widetilde{a}_{ij2}, ..., \widetilde{a}_{ijm}$ are the corresponding STFN scales of attribute i compared to attribute j measured by the decision makers DM_1, DM_2, ···, DM_m, respectively.

It should be noted that the aggregation should discard the absent scale while it comes with non-zero scales provided by other decision makers under the same comparison. This process can be defined as in Equation 8.4.

$$\widetilde{a}_{ij} = \frac{\widetilde{a}_{ij1} \otimes c_1 + \widetilde{a}_{ij2} \otimes c_2 + ... + \widetilde{a}_{ijm} \otimes c_m}{1 - \sum c_r} \tag{8.4}$$

where $\sum c_r$ is the total weight of the decision makers who provide zero scales. If none of the decision makers can evaluate a particular comparison, this comparison should be left absent.

Step 5: Defuzzification

After having all the required aggregated STFNs, it is now the time for defuzzification to find the representative crisp numbers. Assume an aggregated STFN, $\widetilde{a}_{ij} = \left(a_{ij}^a, a_{ij}^b, a_{ij}^c, a_{ij}^d\right)$, the representative crisp value a_{ij} can be obtained from Equation 8.5.

$$a_{ij} = \frac{a_{ij}^a + 2\left(a_{ij}^b + a_{ij}^c\right) + a_{ij}^d}{6} \tag{8.5}$$

where $a_{ii} = 1$ and $a_{ji} = 1 / a_{ij}$.

Consequently, all the aggregated fuzzy scales $\widetilde{a}_{ij} \left(i, j = 1,2,...,n\right)$ are transferred into crisp scales a_{ij} within the range of [0, 9].

Step 6: Calculate the priority weights of factors

Let F_1, F_2, ···, F_n be a set of factors in one section, a_{ij} is the defuzzified scale representing the quantified judgment on F_i comparing to F_j. Assuming that A_1, A_2, ···, A_n represent a set of attributes in one group, pair-wise comparisons between A_i and A_j in the same group yield an $n \times n$ matrix as defined in Equation 8.6.

$$A = a_{ij} = \begin{array}{c|cccc} & F_1 & F_2 & \cdots & F_4 \\ F_1 & 1 & a_{12} & \cdots & a_{1n} \\ F_2 & 1/a_{12} & 1 & \cdots & a_{2n} \\ F_3 & \cdots & \cdots & \cdots & \cdots \\ F_4 & 1/a_{1n} & 1/a_{2n} & \cdots & 1 \end{array}, \qquad i, j = 1,2,\ldots,n \tag{8.6}$$

where $a_{ii} = 1$, $a_{ji} = 1/a_{ij}$.

The priority weights of factors in the matrix A can be calculated by using the arithmetic averaging method:

$$w_i = \frac{1}{n}\sum_{j=1}^{n} \frac{a_{ij}}{\sum_{k=1}^{n} a_{kj}} \qquad i, j = 1,2,\ldots,n. \tag{8.7}$$

where w_i is the section weight of F_i. Assume that F_i has t upper sections at different level in the hierarchy, and $w_{section}^{(i)}$ is the section weight of the ith upper section that contains F_i in the hierarchy. The final weight w_i' of F_i can be derived using

$$w_i' = w_i \times \prod_{i=1}^{t} w_{group}^{(i)} \tag{8.8}$$

All individual upper section weights of $w_{group}^{(i)}$ can also be derived by Equation 8.7 to prioritise sections within the corresponding cluster in the hierarchy.

Step 7: Calculate the final fuzzy scores

When the scores and the priority weights of factors are obtained, the final fuzzy scores $\left(\tilde{FS}\right)$ can be calculated by

$$\left(\tilde{FS}\right) = \sum_{i=1}^{n} \tilde{S}_i\, w_i' \qquad i = 1,2,\ldots,n \tag{8.9}$$

Step 8: Compare the $\left(\tilde{FS}\right)$ *values using an outranking method*

This step is added to increase the reliability of the results.

In this chapter, two ranking methods proposed by Yuan [8.72] and Tran and Duckstein [8.73] are used to rank final fuzzy scores. The main principle of the Yuan's ranking method is explained below [8.65].

Let C_i and $C_j \in F(\Re)$ be normal and convex. A fuzzy relation that compares the right spread of C_i with the left spread of C_j is defined as

$$\Delta_{ij} = \int_{c_{i\alpha}^{+} > c_{j\alpha}^{-}} (c_{i\alpha}^{+} - c_{j\alpha}^{-}) d_{\alpha} + \int_{c_{i\alpha}^{-} > c_{j\alpha}^{+}} (c_{i\alpha}^{-} - c_{j\alpha}^{+}) d_{\alpha} \tag{8.10}$$

$$\mu(C_i, C_j) = \frac{\Delta_{ij}}{\Delta_{ij} + \Delta_{ji}} \tag{8.11}$$

where $\mu(C_i, C_j)$ is the degree of largeness of C_i relative to C_j and $C_i, C_j \in F(\Re)$. C_i is larger than C_j if and only if $\mu(C_i, C_j) > 0.5$. C_i and C_j are equal if and only if $\mu(C_i, C_j) = 0.5$.

The second method by Tran and Duckstein [8.73] is based on the comparison of distances from fuzzy numbers (FNs) to some predetermined targets: the crisp maximum (*Max*) and the crisp minimum (*Min*). The idea is that an FN is ranked first if its distance to the crisp maximum (D_{max}) is the smallest but its distance to the crisp minimum (D_{min}) is the greatest. If only one of these conditions is satisfied, the FN might be outranked by others depending upon the context of the problem (*e.g.* the attitude of the decision maker in a decision situation) [8.65, 8.74].

Max and *Min* are chosen as follows:

$$\begin{aligned} Max(I) &\geq \sup\left(\bigcup_{i=1}^{I} s(\tilde{A}_i)\right) \\ Min(I) &\leq \inf\left(\bigcup_{i=1}^{I} s(\tilde{A}_i)\right) \end{aligned} \tag{8.12}$$

where $s(\tilde{A}_i)$ is the support of FNs A_i, $i = 1, \ldots, i$. Then, D_{max} and D_{min} of fuzzy number A can be computed as follows:

$$D^2(\tilde{A}, M) = \begin{cases} \left(\frac{a_2 + a_3}{2} - M\right)^2 + \frac{1}{2}\left(\frac{a_2 + a_3}{2} - M\right) \times [(a_4 - a_3) - (a_2 - a_1)] + \\ \frac{1}{3}\left(\frac{a_3 - a_2}{2}\right)^2 + \frac{1}{6}\left(\frac{a_3 - a_2}{2}\right) \times [(a_4 - a_3) + (a_2 - a_1)] + \\ \frac{1}{9}[(a_4 - a_3)^2 + (a_2 - a_1)^2] - \frac{1}{9}[(a_2 - a_1) \times (a_4 - a_3)] \end{cases} \tag{8.13}$$

where M is either *Max* or *Min* and $\tilde{A}$ is a trapezoidal fuzzy number. Hence,

$$D_{\max} = \sqrt{D^2(\widetilde{A}, Max)} \text{ and}$$

$$D_{\min} = \sqrt{D^2(\widetilde{A}, Min)} \tag{8.14}$$

8.5 An Application

ABC Group is a vertically integrated full-service apparel manufacturing company that is proud of its international reputation as a manufacturer of top quality fabrics and garments. They produce more than 20 million fashion garments a year, destined for the markets of Western Europe and North America. They provide services ranging from product development through to logistics arrangements. The Group is today a major supplier for some of the most famous brands, and this success has been achieved by implementing key principles, which include:

- maintaining good quality;
- achieving reliable delivery;
- continuous innovation in research, design and development.

ABC Group has vast amount of experience and know-how in different manufacturing technologies as well as product development and design. The company works very closely with our customers to provide them developments of various fabric qualities, garment designs, embellishment or garment washing techniques. The ABC Group is famed for producing knitted fabrics of the highest quality, and all the stages from knitting through to finishing are conducted in-house.

For woven fabrics, they have strategic relationships with the most prominent mills both in Turkey and overseas. GIYSI Tekstil is one of the jersey wear garment manufacturing companies of the ABC Group with its liaison office in Istanbul and a modern premise in Malatya, 1120 km south-east of Istanbul. GIYSI Tekstil is one of the most developed and largest factories in the region with the flexible and efficient in-house production capability geared to serve internationally well-known brands that look for quality.

GIYSI Tekstil desires to determine the most appropriate supplier based on agile manufacturing principles. For this aim the criteria, explained in Section 8.3 and whose hierarchical structure is shown in Figure 8.3, are taken into consideration when the proposed AHP approach is used to evaluate the alternative suppliers. In this chapter, the most appropriate supplier alternative is selected by an FMCDM technique. Four experts, having different weights because of their experiences, evaluate the considered criteria to determine the most appropriate alternative. Each criterion of the hierarchy is evaluated by the experts under the defined criteria. A score system is shown in Figure 8.5. Each expert may provide a decision about his/her judgment as a precise numerical value, a possible range of numerical value, a linguistic term, or a fuzzy number. These evaluations are converted into STFNs.

Table 8.2 summarises the fuzzy weights of the criteria for supplier-I. The aggregations of the obtained scores are calculated by Equation 8.2. For instance, the aggregation of 'reliability' under 'delivery' is calculated as follows:

Table 8.2. Scores and converted STFNs for supplier-I

		E1		E2		E3		E4		Aggregated
		Score	STFN	Score	STFN	Score	STFN	Score	STFN	
Quality	C11	M	(2.5, 5, 5, 7.5)	5	(5, 5, 5, 5)	3,5	(3, 3, 5, 5)	A4	(3, 4, 4, 5)	(3.3, 4.45, 4.85, 6)
	C12	M	(2.5, 5, 5, 7.5)	5	(5, 5, 5, 5)	4,6	(4, 4, 6, 6)	4,5	(4, 4, 5, 5)	(3.65, 4.65, 5.2, 6.2)
	C13	H	(5, 7.5, 7.5, 10)	7	(7, 7, 7, 7)	6,7	(6, 6, 7, 7)	5	(5, 5, 5, 5)	(5.7, 6.7, 6.9, 7.9)
Delivery	C21	H	(5, 7.5, 7.5, 10)	H	(5, 7.5, 7.5, 10)	5,7	(5, 5, 7, 7)	A7	(6, 7, 7, 8)	(5.15, 6.925, 7.325, 9.1)
	C22	VG	(7.5, 10, 10, 10)	9	(9, 9, 9, 9)	3,4	(3, 3, 4, 4)	A4	(3, 4, 4, 5)	(6.3, 7.45, 7.65, 7.8)
	C23	G	(5, 7.5, 7.5, 10)	8	(8, 8, 8, 8)	G	(5, 7.5, 7.5, 10)	A4	(3, 4, 4, 5)	(5.45, 7.1, 7.1, 8.75)
	C24	H	(0, 2.5, 2.5, 5)	3	(3, 3, 3, 3)	H	(0, 2.5, 2.5, 5)	A2	(1, 2, 2, 3)	(0.9, 2.55, 2.55, 4.2)
Agility	C31	VH	(7.5, 10, 10, 10)	A5	(4, 5, 5, 6)	3,5	(3, 3, 5, 5)	A4	(3, 4, 4, 5)	(5.05, 6.45, 6.85, 7.25)
	C32	L	(0, 2.5, 2.5, 5)	3	(3, 3, 3, 3)	3,4	(3, 3, 4, 4)	A4	(3, 4, 4, 5)	(1.8, 2.95, 3.15, 4.3)
	C33	4	(4, 4, 4, 4)	4	(5, 5, 6, 6)	3,4	(3, 3, 4, 4)	A3	(2, 3, 3, 4)	(3.75, 3.9, 4.35, 4.5)
	C34	6.7	(6, 6, 7, 7)	A7	(6, 7, 7, 8)	6,7	(6, 6, 7, 7)	A5	(4, 5, 5, 6)	(5.7, 6.1, 6.7, 7.1)
Performance	C41	G	(5, 7.5, 7.5, 10)	6,7	(6, 6, 7, 7)	6,7	(6, 6, 7, 7)	A6	(5, 6, 6, 7)	(5.45, 6.6, 7.05, 8.2)
	C42	G	(5, 7.5, 7.5, 10)	6,8	(6, 6, 7, 7)	6,7	(6, 6, 7, 7)	A6	(5, 6, 6, 7)	(5.45, 6.6, 7.05, 8.2)
	C43	G	(5, 7.5, 7.5, 10)	7	(7, 7, 7, 7)	7	(7, 7, 7, 7)	G	(5, 7.5, 7.5, 10)	(5.9, 7.275, 7.275, 8.65)
Management	C51	7	(6, 7, 7, 8)	A7	(6, 7, 7, 8)	7	(6, 6, 6, 6)	A5	(4, 5, 5, 6)	(5.7, 6.5, 6.5, 7.3)
	C52	F	(2.5, 5, 5, 7.5)	3	(3, 3, 3, 3)	3,4	(3, 3, 4, 4)	F	(2.5, 5, 5, 7.5)	(2.725, 4.1, 4.3, 5.675)
	C53	P	(0, 2.5, 2.5, 5)	A2	(1, 2, 2, 3)	2,3	(2, 2, 3, 3)	A3	(2, 3, 3, 4)	(0.95, 2.35, 2.55, 3.95)
Service	C61	L	(0, 2.5, 2.5, 5)	2, 2	(2, 2, 2, 2)	2	(2, 2, 2, 2)	A4	(3, 4, 4, 5)	(1.35, 2.5, 2.5, 3.65)
	C62	H	(5, 7.5, 7.5, 10)	5	(5, 5, 5, 5)	3,5	(3, 3, 5, 5)	A3	(2, 3, 3, 4)	(4.15, 5.3, 5.7, 6.85)
	C63	H	(5, 7.5, 7.5, 10)	5,6	(5, 5, 6, 6)	3,5	(3, 3, 5, 5)	A5	(4, 5, 5, 6)	(4.45, 5.6, 6.25, 7.4)
	C64	G	(5, 7.5, 7.5, 10)	A7	(6, 7, 7, 8)	G	(5, 7.5, 7.5, 10)	5,6	(5, 5, 6, 6)	(5.25, 7, 7.15, 8.9)

Table 8.3. Fuzzy aggregation of 'agility' criteria for supplier-I

		Flexibility		Means of information		Electronic data interchange		Workforce	
	Experts	Scale	STFN	Scale	STFN	Scale	STFN	Scale	STFN
C31	E1			6.00 7.00	(6, 6, 7, 7)	1.00 2.00	(1, 1, 2, 2)	4.00 5.00	(4, 4, 5, 5)
	E2			5.00 6.00	(5, 5, 6, 6)	2.00 3.00	(2, 2, 3, 3)	4.00 5.00	(4, 4, 5, 5)
	E3			4.00 6.00	(4, 4, 6, 6)	1.00 1.00	(1, 1, 1, 1)	4.00 6.00	(4, 4, 6, 6)
	E4			6.00 7.00	(6, 6, 7, 7)	1.00 3.00	(1, 1, 3, 3)	3.00 5.00	(3, 3, 5, 5)
	Aggregation	1.00			(5.35, 5.35, 6.55, 6.55)		(1.25, 1.25, 2.2, 2.2)		(3.85, 3.85, 5.2, 5.2)
C32	E1					0.75 1.00	(0.75, 0.75, 1, 1)	3.00 5.00	(3, 3, 5, 5)
	E2					1.00 1.00	(1, 1, 1, 1)	4.00 5.00	(4, 4, 5, 5)
	E3					1.00 1.00	(1, 1, 1, 1)	4.00 4.00	(4, 4, 4, 4)
	E4					1.00 1.00	(1, 1, 1, 1)	2.00 4.00	(2, 2, 4, 4)
	Aggregation			1.00			(0.9, 0.9, 1, 1)		(3.3, 3.3, 4.65, 4.65)
C33	E1							3.00 5.00	(3, 3, 5, 5)
	E2							3.00 4.00	(3, 3, 4, 4)
	E3							3.00 3.00	(3, 3, 3, 3)
	E4							2.00 4.00	(2, 2, 4, 4)
	Aggregation					1.00			(2.85, 2.85, 4.2, 4.2)
C34	E1								
	E2								
	E3								
	E4								
	Aggregation							1.00	

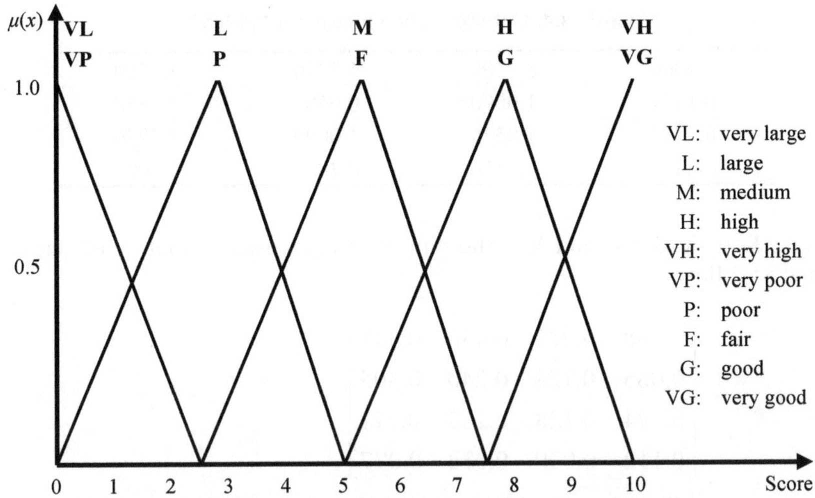

Figure 8.5. Membership functions for supplier evaluation

$$\tilde{S}_{reliability} = (5.0, 7.5, 7.5, 10.0) \otimes 0.40 \oplus (5.0, 7.5, 7.5, 10.0) \otimes 0.25$$
$$\oplus (5.0, 5.0, 7.0, 7.0) \otimes 0.20 \oplus (6.0, 7.0, 7.0, 8.0) \otimes 0.15$$

$$\tilde{S}_{reliability} = (5.15, 6.93, 7.33, 9.10)$$

The other values for aggregation are also shown in Table 8.2. The pair-wise comparisons of the 'agility' criterion and the corresponding STFNs are shown in Table 8.3. The aggregation of STFN scales are calculated from Equation 8.3. For example, the STFN scale of comparing 'flexibility' with 'means of information' is aggregated as follows:

$$\tilde{a}_{12} = (6.0, 6.0, 7.0, 7.0) \otimes 0.40 \oplus (5.0, 5.0, 6.0, 6.0) \otimes 0.25$$
$$\oplus (4.0, 4.0, 6.0, 6.0) \otimes 0.20 \oplus (6.0, 6.0, 7.0, 7.0) \otimes 0.20$$

$$\tilde{a}_{12} = (5.35, 5.35, 6.55, 6.55)$$

Then, the STFN scale of comparisons should be defuzzified. By using Equation 8.5, the STFN scale of comparing 'flexibility' with 'means of information' is defuzzified as

$$a_{24} = \frac{5.35 + 2(5.35 + 6.55) + 6.55}{6} = 5.95$$

The defuzzification matrix of criteria for 'agility' is determined as shown in Table 8.4.

Table 8.4. Defuzzification matrix for 'agility'

1.0000	5.9500	1.7250	4.5250
0.1681	1.0000	0.9500	3.9750
0.5797	1.0526	1.0000	3.5250
0.2210	0.2516	0.2837	1.0000

Using Equations 8.6 and 8.7, the pair-wise comparisons matrix of 'agility' is obtained as follows:

$$A_{\text{Agility}} = \begin{bmatrix} 0.508 & 0.721 & 0.436 & 0.347 \\ 0.085 & 0.121 & 0.240 & 0.305 \\ 0.294 & 0.128 & 0.253 & 0.271 \\ 0.112 & 0.030 & 0.072 & 0.077 \end{bmatrix}$$

By taking into account this matrix and using Equation 8.8, the weights of the sub-criteria of 'agility' are calculated using

$$w = \{0.5030, 0.1879, 0.2363, 0.0728\}$$

The final weights of the criteria are calculated by using Equation 8.8. Then, the $\tilde{FS}$ of supplier-I is calculated by using Equation 8.9 as follows:

$$\tilde{FS}_{SI} = (4.40, 5.60, 5.92, 7.07)$$

$\tilde{FS}$ values of supplier alternatives are also calculated. The results are summarised in Table 8.5. The membership functions of these fuzzy scores are shown in Figure 8.6.

In the final step of the proposed methodology, the fuzzy scores need to be ranked. To rank the fuzzy scores, the methods explained in step 8 are used. First, Yuan's method [8.72] is used and the ranking results are summarised in Table 8.6.

Table 8.5. Fuzzy scores of alternative projects

Suppliers	Fuzzy scores
I	(4.40, 5.60, 5.92, 7.07)
II	(4.82, 6.02, 6.37, 7.53)
III	(6.03, 7.23, 7.60, 8.48)

Table 8.6. The comparisons results for alternative suppliers

Supplier	Δ_{ij}	Δ_{ji}	Decision	Degree of largeness
I–II	1.075	1.976	II is better than I	0.647
II–III	0.478	2.773	III is better than II	0.853

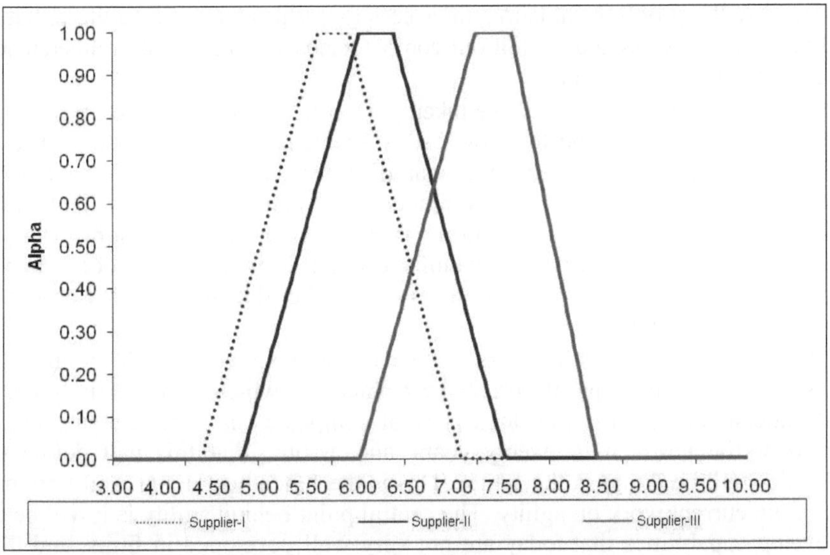

Figure 8.6. Membership functions of fuzzy scores for alternative suppliers

According to Table 8.6, II is better than I with a degree of 0.647 so that the second supplier alternative is more appropriate than the first one. The supplier alternatives are sorted as: {III, II, I}.

In addition, the fuzzy scores are ranking based on the second ranking method, *i.e.* Tran and Duckstein's method [8.73]. The ranking results for this method are summarised in Table 8.7.

Table 8.7. The comparisons results for alternative suppliers

Supplier	a_1	a_2	a_3	a_4	D_{max}	D_{min}
I	4.40	5.60	5.92	7.07	4.278	5.768
II	4.88	6.08	6.43	7.53	3.797	6.251
III	6.03	7.23	7.60	8.48	2.703	7.347

According to Table 8.7, supplier-III, whose D_{max} and D_{min} values are minimum and maximum, respectively, is determined as the best alternative. The ranking of supplier alternatives is determined as: {III, II, I}. The results are the same as those obtained by Yuan's ranking method.

8.6 Conclusions

Agile manufacturing enterprises will be capable of rapidly responding to changes in customer demand. They will be able to take advantage of the windows of opportunities that appear in the marketplace. With agile manufacturing, we will be able to develop new ways of interacting with our customers and suppliers. Our customers will not only be able to gain access to our products and services, but also be able to easily assess and exploit our competencies, and use these competencies to achieve what they are seeking.

Nowadays, globalisation has overtaken the world, and not a single country is left out to come up in front and to grow fast in manufacturing. Every moment, a new window of opportunity opens for manufacturers with the latest concept in manufacturing. Not only the latest equipments and machines, but new thoughts are equally important. Agile manufacturing is a new concept of manufacturing that helps with faster and better quality manufacturing. It is an adaptive process. In a few years, we will see agile manufacturing as a versatile globally accepted process for manufacturing products.

Most of the current literature, however, is concerned with defining and discussing agility in terms of current best practices, which is not entirely correct. Most understandings of agility look at what companies have been doing, in some cases over the past ten to twenty years, and assume that this will define what companies will be doing in the future. This is the fundamental and fatal flaw in the bulk of the current work on agility. The central point behind agility is it will be used to develop capabilities that today are not very well developed in firms, and this is why it is important to challenge taken-for-granted assumptions. Change, uncertainty and unpredictability in the current business environment are rendering invalid many of these assumptions as well as elements of current practice. A new and different sort of enterprise will be needed for agility, but such enterprises will not begin to emerge until people really understand what agility is actually about. In twenty years time, people will look back and see all this as obvious and will be puzzled why so many people today could not see it. Agility is a paradigm shift which implies that old ideas, including some lean production concepts, need to be re-evaluated, modified and in some cases abandoned.

Agility is not a new idea but it is essential for survival in the emerging global competitive environment. Possession of resources will matter far less in determining strategic advantage than the ability to configure and reconfigure resources rapidly. Agile manufacturing will be the future direction for the manufacturing industry in the twenty-first century. Partner selection is also an important decision problem in agile manufacturing environment.

References

[8.1] Ashall, D. and Parkinson, B., 2002, "Leaning towards agile," *Manufacturing Engineer*, February, pp. 27–32.

[8.2] http://www.technet.pnl.gov/dme/agile/index.stm.

[8.3] Sanchez, L.M. and Nagi, R., 2001, "A review of agile manufacturing systems," *International Journal of Production Research*, **39**(16), pp. 3561–3600.

[8.4] Gunasekaran, A., 1999, "Agile manufacturing: a framework for research and development," *International Journal of Production Economics*, **62**, pp. 87–105.

[8.5] Christopher, M. and Towill, D.R., 2000, "Supply chain migration from lean and functional to agile and customised," *Supply Chain Management: An International Journal*, **5**(4), pp. 206–213.

[8.6] http://www.navisioninfo.com/pdffiles/AgilityandManufacturing.pdf.

[8.7] Gunasekaran, A., 1998, "Agile manufacturing: enablers and an implementation framework," *International Journal of Production Research*, **36**(5), pp. 1223–1247.

[8.8] Ramesh, G. and Devadasan, S.R., 2007, "Literature review on the agile manufacturing criteria," *Journal of Manufacturing Technology Management*, **18**(2), pp. 182–201.

[8.9] Kahraman, C., Cebeci, U. and Ulukan, Z., 2003, "Multi-criteria supplier selection using fuzzy AHP," *Logistics Information Management*, **16**(6), pp. 382–394.

[8.10] Choy, K.L. and Lee, W.B., 2002, "On the development of a case-based supplier management tool for multinational manufacturers," *Measuring Business Excellence*, **16**(1), pp. 15–22.

[8.11] Cebeci, U. and Kahraman, C., 2002, "Measuring customer satisfaction of catering service companies using fuzzy AHP: the case of Turkey," In *Proceedings of International Conference on Fuzzy Systems and Soft Computational Intelligence in Management and Industrial Engineering*, pp. 315–325.

[8.12] Cebeci, U., 2001, "Customer satisfaction of catering service companies in Turkey," In *Proceedings of the 6th International Conference on ISO 9000 and TQM (6th ICIT)*, pp. 519–524.

[8.13] Ghodsypour, S.H. and O'Brien, C., 2001, "The total cost of logistics in supplier selection, under conditions of multiple sourcing, multiple criteria and capacity constraint," *International Journal of Production Economics*, **73**, pp. 15–27.

[8.14] Feng, C.-X., Wang, J. and Wang, J.-S., 2001, "An optimization model for concurrent selection of tolerances and suppliers," *Computers & Industrial Engineering*, **40**, pp. 15–33.

[8.15] Boer, L., Wegen, L. and Telgen, J., 1998, "Outranking methods in support of supplier selection," *European Journal of Purchasing & Supply Management*, **4**, pp. 109–118.

[8.16] Masella, C. and Rangone, A., 2000, "A contingent approach to the design of vendor selection systems for different types of co-operative customer/supplier relationships," *International Journal of Operations & Production Management*, **20**(1), pp. 70–84.

[8.17] Liu, J., Ding, F.-Y. and Lall, V., 2000, "Using data envelopment analysis to compare suppliers for supplier selection and performance improvement," *Supply Chain Management: An International Journal*, **5**(3), pp. 143–150.

[8.18] Braglia, M. and Petroni, A., 2000, "A quality assurance oriented methodology for handling trade-offs in supplier selection," *International Journal of Physical Distribution & Logistics Management*, **30**(2), pp. 96–112.

[8.19] Dowlatshahi, S., 2000, "Designer-buyer-supplier interface: theory versus practice," *International Journal of Production Economics*, **63**, pp. 111–130.

[8.20] Motwani, J., Youssef, M., Kathawala, Y. and Futch, E., 1999, "Supplier selection in developing countries: a model development," *Integrated Manufacturing Systems*, **10**(3), pp. 154–162.

[8.21] Ittner, C.D., Larcker, D.F., Nagar, V. and Rajan, M.V., 1999, "Supplier selection, monitoring practices, and firm performance," *Journal of Accounting and Public Policy*, **18**, pp. 253–281.

[8.22] Ganeshan, R., Tyworth, J.E. and Guo, Y., 1999, "Dual sourced supply chains: the discount supplier option," *Transportation Research Part E*, **35**, pp. 11–23.

[8.23] Verma, R. and Pullman, M.E., 1998, "An analysis of the supplier selection process," *Omega, International Journal of Management Science*, **26**(6), pp. 739–750.

[8.24] O'Brien, C. and Ghodsypour, S.H., 1998, "A decision support system for supplier selection using an integrated analytic hierarchy process and linear programming," *International Journal of Production Economics*, **56/57**, pp. 199–212.

[8.25] Noci, G., 1997, "Designing 'green' vendor rating systems for the assessment of a supplier's environmental performance," *European Journal of Purchasing & Supply Management*, **3**(2), pp. 103–114.

[8.26] Choi, T.Y. and Hartley, J.L., 1996, "An exploration of supplier selection practices across the supply chain," *Journal of Operations Management*, **14**, pp. 333–343.

[8.27] Mummalaneni, V., Dubas, K.M. and Chao, C.-N., 1996, "Chinese purchasing managers' preferences and trade-offs in supplier selection and performance evaluation," *Industrial Marketing Management*, **25**, pp. 115–124.

[8.28] Swift, C.O., 1995, "Preferences for single sourcing and supplier selection criteria," *Journal of Business Research*, **32**, pp. 105–111.

[8.29] Gupta, P. and Nagi, R., 1995, "Flexible optimization framework for partner selection in agile manufacturing," In *Proceedings of Industrial Engineering Research Conference*, pp. 691–700.

[8.30] Bocks, P., 1995, "Enterprise data management framework for agile manufacturing," *American Society of Mechanical Engineers*, **7**, pp. 41–46.

[8.31] Wang, Z.Y., Rajurkar, K.P. and Kapoor, A., 1996, "Architecture for agile manufacturing and its interface with computer integrated manufacturing," *Journal of Materials Processing Technology*, **61**(1–2), pp. 99–103.

[8.32] Yang, X., 1996, "Object-oriented model of agile manufacturing system," In *Proceedings of the IEEE International Conference on Industrial Technology*, pp. 171–175.

[8.33] McMullen, Jr., T.B., 1996, "Theory of constraints: the infrastructure of agile manufacturing," In *Proceedings of Annual International Conference – American Production and Inventory Control Society*, pp. 336–346.

[8.34] Merat, F.L., Barendt, N.A., Quinn, R.D., Causey, G.C., Newman, W.S., Velasco, V.B., Podgurski, A., Kim, Y., Ozsoyoglu, G. and Jo, J.Y., 1997, "Advances in agile manufacturing," In *Proceedings of IEEE International Conference on Robotics and Automation*, pp. 1216–1222.

[8.35] Aoyama, M., 1998, "Web-based agile software development," *IEEE Software*, **15**(6), 56–65.

[8.36] Monplaisir, L., 1997, "AI enhancements for a CSCW for manufacturing systems design," In *Proceedings of Industrial Engineering Research Conference*, pp. 158–163.

[8.37] Kusiak, A. and He, D.W., 1997, "Design for agile assembly: an operational perspective," *International Journal of Production Research*, **35**(1), 157–178.

[8.38] Lee, S., Kuo, H.C., Balkir, N.H. and Ozsoyoglu, G., 1997, "Database server architecture for agile manufacturing," In *Proceedings of IEEE International Conference on Robotics and Automation*, pp. 3048–3053.

[8.39] Song, L. and Nagi, R., 1997, "Design and implementation of a virtual information system for agile manufacturing," *IIE Transactions*, **29**(10), 839–857.

[8.40] Gunasekaran, A. and Yusuf, Y.Y., 2002, "Agile manufacturing: a taxonomy of strategic and technological imperatives," *International Journal of Production Research*, **40**(6), 1357–1385.

[8.41] Meade, L.M. and Sarkis, J., 1999, "Analyzing organizational project alternatives for agile manufacturing processes: an analytical network approach," *International Journal of Production Research*, **37**(2), pp. 241–261.

[8.42] Naylor, J.B., Naim, M.M. and Berry, D., 1999, "Leagility: Integrating the lean and agile manufacturing paradigms in the total supply chain," *International Journal of Production Economics*, **62**, pp. 107–118.

[8.43] Wu, N., Mao, N. and Qian, Y., 1999, "An approach to partner selection in agile manufacturing," *Journal of Intelligent Manufacturing*, **10**, 519–529.

[8.44] Mason-Jones, R., Naylor, B. and Towill, D.R., 2000, "Lean, agile or leagile? Matching your supply chain to the marketplace," *International Journal of Production Research*, **38**(17), 4061–4070.

[8.45] Chan, F.T.S. and Zhang, J., 2001, "Modelling for agile manufacturing systems," *International Journal of Production Research*, **39**(11), pp. 2313–2332.

[8.46] Aitken, J., Christopher, M. and Towill, D., 2001, "Understanding, implementing and exploiting agility and leanness," In *UK Symposium on Supply Chain Alignment*, Liverpool University.

[8.47] McCullen, P. and Towill, D., 2001, "Achieving lean supply through agile manufacturing," *Integrated Manufacturing Systems*, **12**(7), 524–533.

[8.48] Elkins, D.A., Huang, N. and Alden, J.M., 2004, "Agile manufacturing systems in the automotive industry," *International Journal of Production Economics*, **91**, pp. 201–214.

[8.49] Ip, W.H., Yung, K.L. and Wang, D., 2004, "A branch and bound algorithm for sub-contractor selection in agile manufacturing environment," *International Journal of Production Economics*, **87**, pp. 195–205.

[8.50] Lou, P., Zhou, Z., Chen, Y.P. and Ai, W., 2004, "Study on multi-agent-based agile supply chain management," *International Journal of Advanced Manufacturing Technology*, **23**, pp. 197–203.

[8.51] Yusuf, Y.Y., Gunasekaran, A., Adeleye, E.O. and Sivayoganathan, K., 2004, "Agile supply chain capabilities: determinants of competitive objectives," *European Journal of Operational Research*, **159**, pp. 379–392.

[8.52] Gunasekaran, A. and Ngai, E.W.T., 2005, "Build-to-order supply chain management: a literature review and framework for development," *Journal of Operational Management*, **23**, pp. 423–451.

[8.53] Agarwal, A., Shankar, R. and Tiwari, M.K., 2006, "Modelling the metrics of lean, agile and leagile supply chain: an ANP-based approach," *European Journal of Operational Research*, **173**, pp. 211–225.

[8.54] Cao, H. and Gao, Y., 2006, "Penalty guided genetic algorithm for partner selection problem in agile manufacturing environment," In *Proceedings of the 6th World Congress on Intelligent Control and Automation*, pp. 3276–3280.

[8.55] Dotoli, M., Fanti, M.P., Meloni, C. and Zhou, M, 2006, "Design and optimization of integrated e-supply chain for agile and environmentally conscious manufacturing," *IEEE Transactions on Systems, Man, and Cybernetics—Part A: Systems and Humans*, **36**(1), pp. 62–75.

[8.56] Ismail, H.S. and Sharifi, H., 2006, "A balanced approach to building agile supply chains," *International Journal of Physical Distribution and Logistics Management*, **36**(6), pp. 431–444.

[8.57] Krishnamurthy, R. and Yauch, C.A., 2007, "Leagile manufacturing: a proposed corporate infrastructure," *International Journal of Operations & Production Management*, **27**(6), pp. 588–604.

[8.58] Srinivasan, R., 2007, "Artificial intelligence methodologies for agile refining: an overview," *Knowledge Information Systems*, **12**(2), pp. 129–145.

[8.59] Forsberg, J.B. and Towers, N., 2007, "Creating agile supply networks in the fashion industry: a pilot study of the European textile and clothing industry," *Journal of the Textile Institute*, **98**(4), pp. 377–386.

[8.60] Demirli, K. and Yimer, A.D., 2008, "Fuzzy scheduling of a build-to-order supply chain," *International Journal of Production Research*, **46**(14), pp. 3931–3958.

[8.61] Hasan, M.A., Shankar, R. and Sarkis, J., 2008, "Supplier selection in an agile manufacturing environment using data envelopment analysis and analytical network process," *International Journal of Logistics Systems and Management*, **4**(5), pp. 523–550.

[8.62] Olsson, H., 1999, "A model for supplier selection & total cost of purchasing: a case study at Emhart Glass Sweden," Master's Thesis, Department of Business Administration and Social Sciences, Lulea University of Technology, Sweden.

[8.63] Gharehgozli, A.H., Rabbani, M., Zaerpour, N. and Razmi, J., 2008, "A comprehensive decision-making structure for acceptance/rejection of incoming orders in make-to-order environments," *International Journal of Advanced Manufacturing Technology*, **39**, pp. 1016–1032.
[8.64] Saaty, T.L., 1980, *The Analytic Hierarchy Process*, McGraw-Hill, New York.
[8.65] Kahraman, C. and Kaya, İ., 2009, "Realizing policies by projects using fuzzy multiple criteria decision making," In *Computational Intelligence in Complex Decision Systems*, Ruan, D. (ed.), Atlantis Press, Paris.
[8.66] Zadeh, L., 1965, "Fuzzy sets," *Information Control*, **8**, pp. 338–353.
[8.67] Zeng, J., An, M. and Smith, N.J., 2007, "Application of a fuzzy based decision making methodology to construction project risk assessment," *International Journal of Project Management*, **25**, pp. 589–600.
[8.68] Laarhoven, P.J.M. and Pedrycz, W., 1983, "A fuzzy extension of Saaty's priority theory," *Fuzzy Set Systems*, **11**, pp. 229–241.
[8.69] Buckley, J.J., 1985, "Fuzzy hierarchical analysis," *Fuzzy Sets and Systems*, **17**, pp. 233–247.
[8.70] Chang, D.Y., 1996, "Applications of the extent analysis method on fuzzy AHP," *European Journal of Operational Research*, **95**, pp. 649–655.
[8.71] Kahraman, C., Cebeci, U. and Ruan, D., 2004, "Multi-attribute comparison of catering service companies using fuzzy AHP: the case of Turkey," *International Journal of Production Economics*, **87**, pp. 171–184.
[8.72] Yuan, Y., 1991, "Criteria for evaluating fuzzy ranking methods," *Fuzzy Sets and Systems*, **43**, pp. 139–157.
[8.73] Tran, L. and Duckstein, L., 2002, "Comparison of fuzzy numbers using a fuzzy distance measure," *Fuzzy Sets and Systems*, **130**, pp. 331–341.
[8.74] Kaya, İ. and Kahraman, C., 2008, "Fuzzy process capability analyses: an application to teaching processes," *Journal of Intelligent & Fuzzy Systems*, **19**, pp. 259–272.

9

A Sustainable Green Supply Chain for Globally Integrated Networks

Balan Sundarakani[1], Robert de Souza[2], Mark Goh[2], David van Over[1], Sushmera Manikandan[2] and S.C. Lenny Koh[3]

[1] Faculty of Business and Management, University of Wollongong in Dubai Knowledge Village, Dubai, 20183, UAE
Emails: balansundarakani@uowdubai.ac.ae; davidvanover@uowdubai.ac.ae

[2] National University of Singapore, Singapore, 117574
Emails: tlihead@nus.edu.sg; mark_goh@nus.edu.sg; tlism@nus.edu.sg

[3] Logistics and Supply Chain Management Research Group, Management School The University of Sheffield, 9 Mappin Street, Sheffield S1 4DT, UK
Email: S.C.L.Koh@sheffield.ac.uk

Abstract

This study presents a sustainable supply chain platform in a globally integrated supply chain network. It asserts that this environmentally driven initiative has been launched in complex social environments and is inspired by the need for legitimacy, as well as social and economic fitness in a wider social structure. It proposes the importance of research-based improvements in the sustainable logistics field, and aims to bring about a better understanding of and provide a stronger scientific basis for the logistics industry in the sustainable supply chain platform, to allow them to be able to restructure their supply chain architecture. A systems-based approach allows emissions to be controlled across each stage of a global supply chain, by restructuring the existing complex globally integrated supply chain and performing lifecycle assessment in the drive for productivity. Following the preliminary analyses, this chapter offers some suggestions to help manufacturers and logistics service providers to restructure their supply chain strategies.

9.1 Introduction

The scientific evidence is overwhelming. Climate change presents a serious global risk, and demands an urgent global response. If companies act now, the mitigation of these effects will cost only 1% of global GDP (current or future) by 2050. However, the Stern review [9.1] on the economics of climate change iterates that the costs could be 20% of global GDP if left unchecked. Increasingly, organisations have realised that environmental management is an important strategic issue and the necessity to comply with mounting environmental regulations, address the environmental concerns of their customers, and enhance their competitiveness. There will be a need to modify existing business models to embrace a sustainable supply chain strategy with carbon emissions reduction at its core. In the near future, companies will have to be green to grow.

According to IBM [9.2], 'much of the opportunity to address CO_2 emissions rests on the supply chain, compelling companies to look for new approaches to managing carbon effectively – from sourcing and production, to distribution and product afterlife'. A report from PRTM [9.3] says that 'Environmental sustainability is a key consideration in the development of future globalisation strategies. Today, sustainability is mainly driven by the need for regulatory compliance and satisfaction of customer demand. It is yet considered a strategic differentiator'. The green supply chain strategy has become one of the most important initiatives for organisations trying to achieve a competitive advantage. Our detailed environmental scan about sustainable supply chain shows that the adoption of an environmental strategy is only driven by rational and clear orientation and that it is guided by economic and political goals. Success in today's business depends on superior supply chain planning and execution. Following the fast pace of trade liberation and globalisation since the 1990s, supply chain management has emerged as an important research field and has drawn much attention from both practitioners and academics. Simchi-Levi *et al.* [9.4] says that the competition in the twenty-first century is likely to be between supply chains, not individual firms. Historically, supply chain management concepts focused on managing upstream functionalities. However, in modern supply chain networks there has been a paradigm shift in the way companies operate and look to enhance their productivity.

The concept of a sustainable supply chain covers every stage in manufacturing, from the first to the last stage of the lifecycle. The definition of a green supply chain has ranged from green purchasing to an integrated green closed-loop supply chain. The adaptations of the supply chain can be at any of the following stages: product design, material sourcing and selection, manufacturing processes, delivery of the final product to the consumer, and end-of-life management of the product after its useful life. The literature contains many definitions on sustainable or green supply chains. However, few of them consider an end-to-end supply chain with green adoption. Hervani *et al.* [9.5] say that the green/sustainable supply chain comprises green purchasing, green manufacturing, green distribution/marketing and reverse logistics. We adapt this definition by incorporating green forward and reverse logistics, green consumption and green recycling, and express this as:

$$\text{Green Supply Chain} = \left\{ \begin{array}{l} \text{Green Supply} + \text{Green Forward and Reverse Logistics} \\ + \text{Green Manufacturing} + \text{Green Packaging and Distribution} \\ + \text{Green Consumption} + \text{Green Recycling} \end{array} \right\}$$

It is generally perceived that a green supply chain promotes efficiency and synergy among business partners, and helps to enhance environmental performance, minimise waste and achieve cost savings. This synergy is expected to enhance the corporate image, competitive advantage, quality of the product and marketing exposure. On the other hand, the use of environmentally sustainable products and production processes need to be balanced with external pressure from customers to achieve the requirements of reduced cost, higher quality, and faster delivery. As a result, there is usually a trade-off among cost, quality, carbon emissions, service and international trade.

In this chapter, we develop a systemic model, affecting systems, policies, procedures and processes, to measure the carbon footprint across the suite of supply chain services. The redefined supply chain network will focus on minimising raw material, product and process wastages; reducing carbon emissions and other environmental wastages; and bringing to light social awareness about these issues to supply chain players at each stage of their supply chain. Various economic and environmental policy scenarios will also be discussed for proactive green logistics policies. Several business implications are proposed for the business community, as well as governments, to work together on these initiatives.

9.2 The Importance of Going Green

Sustainable green supply chain practices have been given prime importance among supply chain leaders, brand manufacturers, third-party logistics service providers (3PLs) and information technology (IT)-enabled service providers. Many articles discuss the importance of green supply chains and green manufacturing. To mention a few, Zhang *et al.* [9.6] reviewed extensively on green manufacturing and green supply chain. Srivastara [9.7] reviewed extensively in an institutional perspective about sustainable supply chain across different industries. As most of the current research on the topic of 'green' employs qualitative, interview and case-study-based approaches which are largely interpretive in nature, more comprehensive methods, including quantitative models, are useful to focus on the details of green initiatives and provide more actionable solutions for green initiatives. Increasingly, organisations have realised that environmental management is an important strategic issue to comply with mounting environmental regulations around the world, to address the environmental concerns of their customers, and to enhance their competitiveness [9.8]. Existing business models are now being modified to include green and carbon emission reduction at their core.

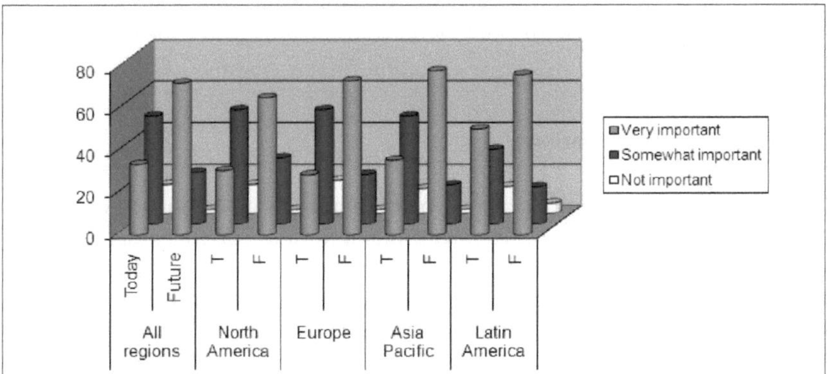

Figure 9.1. Importance of sustainability in supply chain [9.9]

A recent 3PL study [9.9] conducted by the Georgia Institute of Technology shows that 36% of the respondents from the Asia Pacific agree that a green supply chain is very important today and 72% of them believe it will be very important in the future (Figure 9.1). Interestingly in almost all regions across the world, 75% of the respondents agree that a sustainable supply chain becomes a future supply chain initiative among companies.

Another study conducted by *Supply Chain Management Review* [9.10] says that more than 50% of the respondents said they have a documented corporate sustainability plan, and about the same number said their company has a senior executive, often a vice president, dedicated to sustainability at the action level. Although the results were encouraging, if we look at the implementation level and return on investment view, many companies are just starting to get involved in sustainability.

9.2.1 Political Concern

The increasing interest in sustainable development in supply chains has drawn research interest globally. In Europe, Gonzalez-Benito [9.11] surveyed 186 medium- and large-sized Spanish companies and identified two dimensions of pressure, namely, governmental and non-governmental, to explain the implementation of environmental practices in logistics. A study by Hall [9.12], which investigated UK supermarket retailers and their suppliers over a four-year period, suggested that firms invest in environmental supply chain innovation because suppliers with poor environmental practices can expose the customer firm to high levels of environmental risk.

In Canada, using panel data across the oil and gas, mining and forestry industries, Bansal [9.13] reported that both resource-based and institutional factors influence corporate sustainable development. In Asia, researchers found that greening the different phases of the supply chain leads to an integrated green supply chain, and ultimately leads to competitiveness and economic performance [9.14]. Most recently, a survey study in China, with data collected from four typical manufacturing industrial sectors, suggested that different manufacturing industry types display different levels of green supply chain management implementation and outcomes [9.15]. All these studies iterate that sustainable green initiatives have been politically motivated.

9.2.2 Economic Considerations

While important contributions have been made in relation to environmental operations and policy, strategy, finance, product design, supplier relations and post-consumer product management, it is critical to move forward to the issues that exist at the intersection of sustainability, environmental management and supply chains. Previous research studies often look at an aspect of green supply chain, such as economic and environmental factors. However, due to the complexity of this issue, a holistic and systematic picture of the green supply chain is needed for both managers and policy makers.

9.2.3 Changing Business Model

The changing business structure from conventional channel distribution to direct marketing and the consumer's variety in product selection shifts an organisation toward a sustainable green paradigm. Shipment consolidation, direct delivery and inventory replenishment policy changes the business model and reduces the impact of frequency and carbon. IBM [9.2] states that the heightened service level frequency reduces the inventory pipeline while increasing the transportation cost and carbon emissions. Changing this to fewer and larger shipments increases the inventory and warehouse cost but reduces the transportation cost and carbon emissions.

9.2.4 Public Image

Consumers are concerned about brand equity, product quality and price. Oracle [9.16] says that consumers are interested in looking at the product's SECH (social, ethical, cultural and health) rating. Consumers are setting an agenda for organic food purchases and anti-sweatshop labour practices (but what has this got to do with green?). There is a trend for buying local products that can demonstrate their good citizen credentials. A PRTM survey [9.3] shows that an organisational focus on environmental sustainability is more likely to be due to government regulations and customer requirements than a desire to improve a company's image or achieve competitive advantage.

9.2.5 Innovation and Technology Adaption

Technology provides a crucial link to creating a centralised supply chain network and enhances supply chain visibility, traffic scheduling, re-routing the vehicle and thereby reducing the cost, inventory and carbon emissions. Information and communications technology (ICT), global positioning systems (GPS), radio frequency identification (RFID), bar code and routing optimisation packages are some of the latest technologies platform that 3PLs can use to minimise the footprint, to efficiently plan the schedule and to maximise the profit. It is predicted that tomorrow's supply chain will be characterised by free-flowing information, data sharing and collaborative networking [9.10].

9.3 Examining the Sustainable Green Supply Chain

In today's global supply chain network, organisations are looking at moving closer to their market, particularly towards emerging markets, so as to increase their profits. They try to relocate either their manufacturing facility or their distribution centres. In a globally integrated supply chain environment, manufacturing and logistics account for major emissions. In particular, industrial manufacturing, which accounts for about 80% of the industrial energy consumption, contributes about 80% of industrial energy-related carbon emissions. Of these, the petroleum, chemicals, and iron and steel industries produce nearly 60% of the total energy-related carbon emissions from manufacturing (US DoE, 2006). The logistics sector accounts for 14% of global greenhouse gas (GHG) emissions, in which the majority of the emissions are from road transport (76%) and aviation (12%).

Stern [9.1] indicates that the total emissions by source will reach 9.3 Gt CO_2 by 2030. However, data from International Energy Agency (IEA) suggest that transport emissions are expected to reach 8.7 Gt CO_2 by 2030 under BaU (business as usual) conditions. The logistics industry is always working on a sustainable platform using the latest technologies. Table 9.1 compares the results of two major surveys conducted across the world and Europe in 2008 to differentiate some of the major initiatives adopted in this field. The result shows that almost half of the respondents are actively measuring carbon emissions and/or reducing their footprint. More than three-quarters of respondents rate consolidation, routing and mode selection as top services that could contribute to green strategies; the rest are trialling and looking for alternate fuels whilst other inventory policies are less important to their roadmap.

Table 9.1. Green initiatives in 2008

	3PL Logistics Study 2008	European 3PL Market Report 2008
Top current green plan	• Improve transportation efficiency and reducing carbon emissions, through effective shipment consolidation, routing and mode selection • Reduce the use of non-recyclable packaging materials • Manage energy efficient distribution centres • Improve transportation scheduling to reduce carbon emissions • Green implementation advice	• Improve energy efficiency • Measure/reduce emissions/carbon footprint • Strategic location of warehouse/ distribution facilities • Switch to more fuel-efficient modes off transport • Vehicle re-routing to reduce miles
Less interested sustainable action plan	• Use of alternate fuels to reduce greenhouse gas emissions • Facilitate reverse logistics process to recover wastages • Provide effective inventory management plan • Use hybrid electric vehicles	• Switch to more fuel efficient road vehicles • Emission advice/request from suppliers or carriers • Scoping for alternate fuels • Any other green initiatives

Source: *Third-Party Logistics* [9.9], *European 3PL Market* [9.17]

9.4 Critical Drivers that Stimulate Companies to Adopt a Green Supply Chain

Although the green supply chain strategy involves a large investment and uncertain economic paid-offs in the short term, organisations should be willing to adopt the green supply chain strategy for development in the long term. Today's rising energy costs, and global concern about greenhouse gases and climate change are driving the world's largest companies to pressure suppliers to go green and to focus on environmental and social issues. The recent North American Environmental keynote [9.18] provides a compelling blueprint for how forward-thinking companies can address critical environmental issues. Addressing topics from climate change to energy and water conservation, the keynote recommended how companies can improve their manufacturing and supply chain performance, gain competitive advantage and increase profits.

The September 11th attacks on the World Trade Center and the Pentagon in the USA, free trade agreements, and the phenomenon of globalisation have created enormous dimensional changes across the supply chain boundary. While many regulatory issues have served to streamline supply chain activities, innovation and competition among supply chain players creates a redefined supply chain channel (Figure 9.2). To cope with these pressures, an organisation operating in one domain may look for alternative ways to green their business, including outsourcing their activities, becoming more environmentally conscientious by using reusable or recyclable packaging, and adopting reverse manufacturing practices.

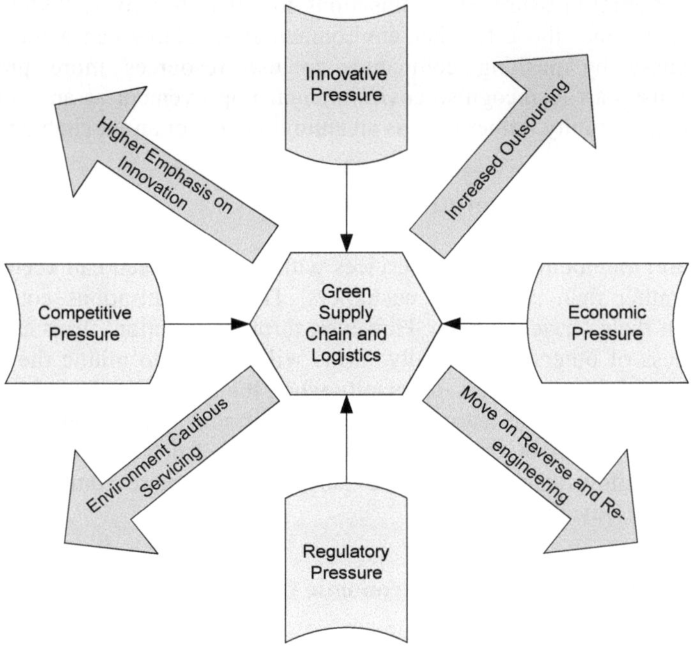

Figure 9.2. Organisational strategies in the sustainable supply chain

9.4.1 Regulatory Issues, Mandates and Standards

Regulatory forces, external standards and mandates are considered as the powerful driving forces for moving towards a green supply chain paradigm. Organisations such as the Coalition for Environmentally Responsible Economies (CARES), the European Parliament Commission (EuP), the Global Reporting Initiative (GRI), the European Community Regulation on Registration, Evaluation, Authorisation and Restriction of Chemicals (REACH), the Restrictive of Hazardous Substances Directive (RoHS), the Directive on Waste Electrical and Electronic Equipment (WEEE), Spring Singapore, and the International Standard for Organisation for Environmental Management System (ISO-EMS) are some of the regulatory organisations working with companies to address the sustainability changes with the aim of making companies endorse green principles. Failing to comply with these forces, particularly those imposed by powerful stakeholders (such as regulations), can result in loss of earnings, a damaged reputation, or even the loss of the licence to operate.

9.4.2 Market Competitiveness

The competitive pressure existing in a market often acts as a driver to become cleaner in production and to opt for a sustainable supply chain. Consumers now prefer eco-friendly and toxic-free recyclable products as their option. This has created an external pressure on organisations for such initiatives. Porter and Linde [9.19] point out that those tougher environmental standards can actually enhance competitiveness by pushing companies to use resources more productively. Managers must start to recognise environmental improvement as an economic and competitive opportunity, rather than as an annoying cost or an inevitable threat.

9.4.3 Differentiation by Innovative Strategies

Providing and maintaining unique services with quality indeed can keep and create customers rather than search for customers. These organisations could be yield innovators in their service domain. However, through imitation, firms can capitalise on the success of others. Specifically, firms will be able to mimic the visible and well-defined activities of others, especially when their activities have been regarded as success stories, and can learn how to avoid certain organisational practices that have failed for others in the past [9.20]. Imitating fruitful practices from early adaptors may allow an organisation to unwittingly acquire some unexpected or unsought unique advantages.

9.4.4 Supplier Consolidation and Economic Gain

Working with business partners (suppliers and customers) provides significant sustainable green practice. It changes the conventional way of a supplier's services. Suppliers provide not only raw materials and finished products, but also transportation, energy, packaging and waste management services. However, business consolidation reinforces greater concern on adopting green supply chain practices with brand manufacturers. The greater the extent of adoption of a practice in an industry, the more likely the potential adopters in that industry would adopt the innovation to avoid being perceived as being less environmentally aware.

9.5 Important Things to Consider while Designing a Network

9.5.1 Controlling Emissions Across the Supply Chain

Many progressive companies have realised the importance of measuring and controlling carbon emissions across the supply chain and have invested heavily in creating carbon-neutral supply chains. We will now attempt to understand the sources of carbon emissions at various stages in a closed-loop supply chain (Figure 9.3).

In an automotive supply chain, carbon emissions come from the processing of raw materials to the dispatching of finished goods. At the supplier side, processing of ore/raw materials and preparing the semi-finished parts emits hydrocarbons, oxides of sulphur and waste in the form of gaseous and acidic compounds. At this stage, the proper use of technologies and the latest equipment could reduce the carbon footprint considerably. In logistics, the levels and type of carbon emissions depend upon the mode of transportation and the distance travelled. At this stage, the total logistics emissions are calculated from the emissions by the various modes of transportation, total sea or air port link emissions, and total warehouse emissions.

The total carbon emissions at the manufacturing stage can be measured from direct and indirect emissions at different manufacturing points. Finally, the total carbon emissions at the distribution and consumer side depend upon the type of packaging used, trade policy, consumer density and the level of reuse. In general, the heat flux influencing the drivers of a supply chain controls the emissions from upstream to downstream in a supply chain. As the product enters each node of the supply chain, its heat flux increases. Figure 9.4 explains this systemic approach to capturing the emissions from the various stages of the supply chain.

Controlling this flux and carbon emission requires a company to monitor the entire supply chain and redesign this based on the scientific approach presented here. However, in today's globally integrated supply chain environment, implementation from end to end can be a challenging task, requiring huge investment and the active participation of all supply chain members.

9.5.2 Restructuring the Network

With the complexity of the automotive supply chain, its many suppliers and its substantial impact on the environment, it is an excellent example to help explain the present landscape. In a typical passenger automotive supply chain in the Asia Pacific region, a manufacturer (such as Toyota, General Motors or Ford) needs to procure thousands of auto-parts from various suppliers with the help of inbound logistics providers (ILPs). Subsequently, the company manufactures some essential parts such as engines by themselves, and assembles all these parts into a complete car. Some cars are sold to retail dealers directly through outbound logistics providers (OLPs). Others are first transported through OLPs to multiple distribution centres (DCs) in various countries, reach retail dealers and finally become accessible to customers. This supply chain thus includes multiple suppliers, ILPs, the manufacturer, OLPs, DCs, retailers and customers.

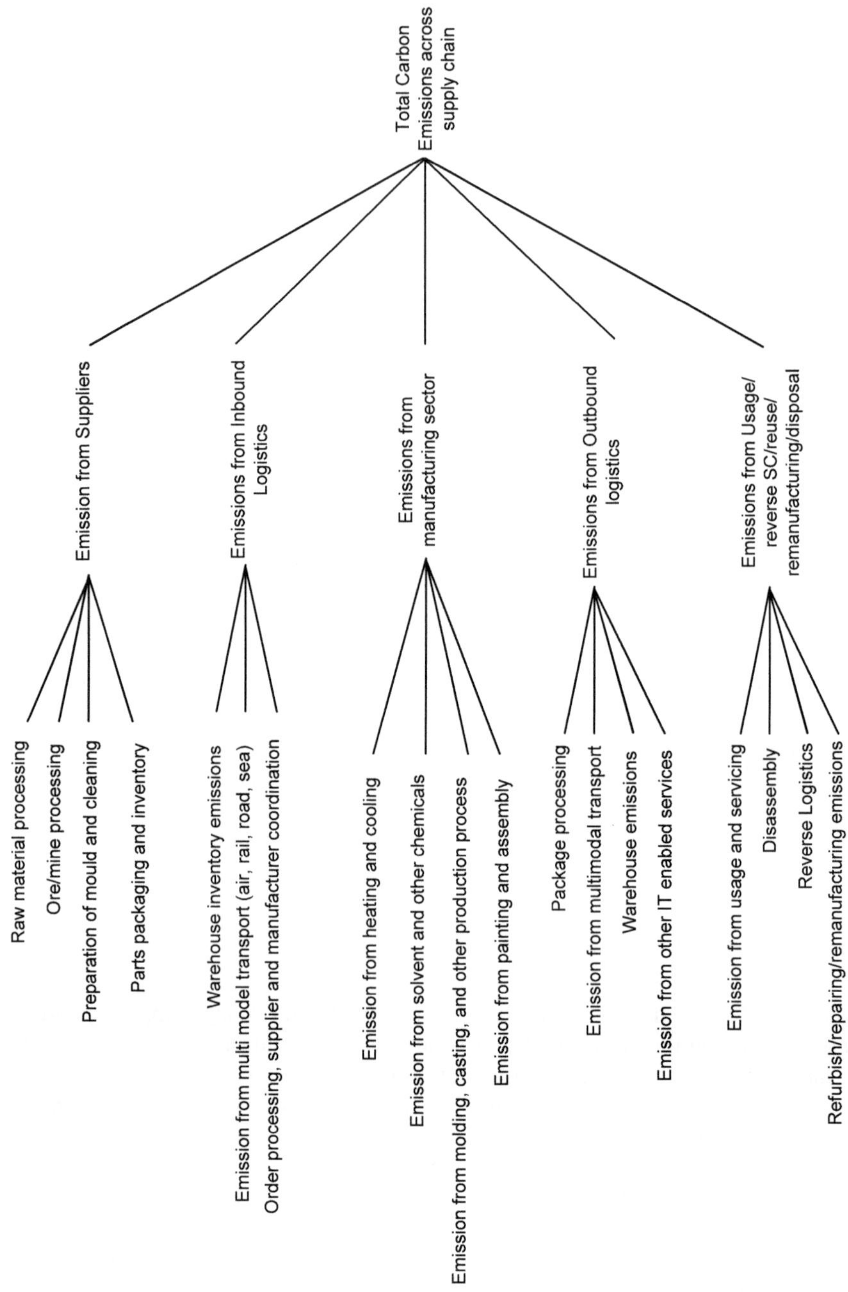

Figure 9.3. Carbon emissions at different stages of the supply chain

A complete supply chain does not end with the customer. As more and more products are returned to the company for a variety of reasons, reverse supply chain design is important for both value creation and environment protection [9.21]. In the automotive industry, there are two main routes in the reverse supply chain. The first is from the retailers to local DCs for reuse. The car is still fully workable and moved back to the local retail system for re-selling as a new car or as a used car. The second is from the retailers to the manufacturer for refurbishment or re-manufacturing. In the case of refurbishment, the car is essentially workable but there are small defects needing some manufacturing refurbishment by the manufacturer.

In the case of remanufacturing, there are serious defects in the car and it needs to be disassembled, and various parts need to be recovered and then remanufactured. Some auto parts may even be moved back to suppliers, sometimes called reverse logistics. Here, OLPs are engaged for the transportation from retailers to the manufacturer. Given the huge transportation cost for cars, the reverse supply chain in automotive industry tends to take the approach of decentralisation. The redesigned network considering closed-loop supply chain architecture is shown in Figure 9.4. Every stage of this supply chain contains carbon emissions, wastage elimination, energy consumption and optimal usage of parts. This restructured supply chain essentially needs the re-engineering, the re-manufacturing, the refurbishing and the re-usage at each echelon in the network.

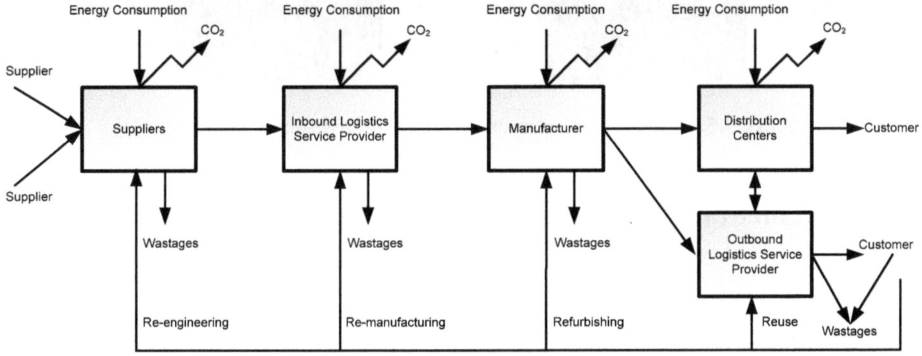

Figure 9.4. Closed-loop supply chain, reconfigured with a green focus

9.5.3 Performing Life-cycle Assessments

According to the European Commission [9.22], life-cycle assessment (LCA) is an internationally standardised method to evaluate the environmental burdens and resources consumed along the life-cycle of products; from the extraction of raw materials, the manufacture of goods and their use by final consumers, or for the provision of a service, recycling, energy recovery and ultimate disposal.

Performing LCA benefits a company by securing cost savings, improving network efficiency and reducing the carbon footprint, thus resulting in an overall improvement in brand image and product image. Continuous assessments in a proactive way shifts the decision cycle towards an organisation's green productivity

(Figure 9.5). However, performing LCA in a globally integrated network is time consuming and requires much investment. Another challenge associated with performing LCA in a network is the exchange of environmental data between supply chain players, because of sensitivity in the information being exchanged.

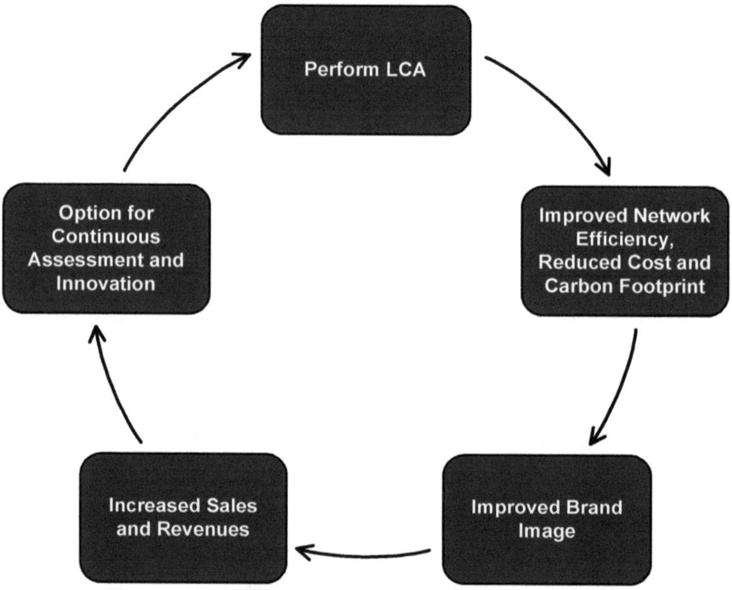

Figure 9.5. Life-cycle assessment in an integrated supply chain network

9.6 Implementation Challenges of a Sustainable Supply Chain

There are several actors involved in a complete end-to-end sustainable supply chain implementation, namely, private sector organisations pushing for environmentally friendly practices; government agencies that enforce environmental controls; industry thought leaders; academic thought leaders; and supply chain thought leaders. Green supply chain implementation challenges are influenced by different dimensional pressures such as innovative, regulatory, economic and competitive pressures. Challenges also abound in terms of looking for alternate modes of transportation, carbon absorption across the supply chain and technology adaption.

The outcome of these pressures could be increased outsourcing to companies with 'green' strategic initiatives, a move towards re-engineering of the supply chain, innovation at the design level and environmentally sustainable design. Every player involved in the supply chain must be able to understand the difficulties faced in the implementation of green supply chain management. Adequate resource commitments are required for green supply chain management and continuous learning is important at every stage of their implementation cycle. Some companies have already taken several steps towards green supply, green packaging, green transporting and optimal energy usage, but in Asia the journey is long.

9.6.1 Green Logistics Initiatives in the UAE

The following three examples show some of our observations of different directions for green initiatives in the United Arab Emirates (UAE). Two of them come out of the UAE itself whereas the third one is affected by a global initiative of international logistics providers.

9.6.1.1 Green Buildings for Maxx 3PL Logistics in Dubai

In 2009, Maxx 3PL logistics inaugurated its $13 million multi-purpose logistics facility in Jebel Ali Free Zone to consolidate its operation in the Middle East. The construction will be a 'green building' having an infrastructure certified under Dubai government guidelines.

9.6.1.2 Masdar City in Abu Dhabi

In Abu Dhabi, the UAE government is establishing Masdar City as the world's first carbon-neutral, zero-carbon, zero-waste, car-free city. It has been selected in 2009 as the headquarters of the International Renewable Energy Agency (IRENA) [9.23]. Masdar also signed an agreement with Abu Dhabi Ports Company (ADPC) to explore carbon emissions capture and greenhouse gases reduction at Khalifa Port and Industrial Zone (KPIZ) in Taweelah. KPIZ is a multi-billion dollar project involving the construction of a world-scale container and industrial port, and developing zones of industrial, logistics, commercial, educational, as well as residential special economic and free zones.

9.6.1.3 Emission Reduction Emissions by UPS/DHL/FedEx

In July 2009, UPS announced a plan to reduce its carbon emissions by 20% by the year 2020, which will be a cumulative 42% since 1990. Its first goal will be aircraft engines because they contribute up to 53% of UPS' carbon output. FedEx rolled out a plan in May 2009 to convert 30% of jet fuel to biofuels by 2030. In 2008, DHL introduced its 'Go-Green Program' which will reduce DHL's carbon footprint by 30% by 2020 [9.24]. Since UAE is one of the major logistics hubs in Asia, we expect these positive measures to drive UPS/DHL's suppliers in the UAE to follow the same path as well.

9.6.2 Implementation Challenges Perceived in UAE

In the Middle East, Rettab and Brik [9.25] highlighted some critical challenges of green supply chain (GSC) during implementation. In conclusion, only 38% of the companies factored GSC in their strategic decision up to last year. This shows that GSC is not a current priority item on the agenda of most companies in the UAE. The three highest barriers for implementation were:

1. insufficient GSC knowledge (41%);
2. lack of supplier awareness (30%);
3. high costs (11%).

Their recommendations to drive GSC forward focus on three main areas:

1. corporate social responsibility is the key factor in engaging suppliers to implement green supply chain;
2. incorporating best practices in supply chain functions from end to end with special emphasis on environmentally friendly purchase for raw materials and semi-finished goods;
3. implementation of green supply chain training programs and implementation of environmental management systems such as ISO 14000.

It is perceived that everyone and every entity involved in the supply chain network should adopt green practices and work together to reduce the overall carbon footprint. It is conceived as bottom-up strategy associated with the decision-making triangle.

9.7 Managerial Implications and Concluding Remarks

To be a leader in sustainable logistics, it is essential that a company can overcome the carbon conflicts noted above. In particular, understanding the role of a carbon-conscious supply chain and related innovation practices is essential for industries to maintain a competitive position. This chapter outlines some strategic implications for companies, focusing on minimising raw material, product and process waste, the dollars spent on each stage and other environmental waste to give competitive advantage at each stage in its operation. We suggest the following key strategies for organisations to follow to maintain carbon neutrality across their supply chain:

- innovate at the design level to reduce redundant rework and restructure the company's products, thus reducing carbon emissions;
- have a green supplier selection policy, taking into consideration their green strategic initiatives and their location to drastically reduce emissions;
- have green supply and purchasing policies;
- observe environmental regulations on transhipment;
- enforce acceptable carbon regulation at the manufacturing level;
- leverage on green innovation in logistics services;
- reduce inventory and increase visibility at the distribution level;
- have green packaging and distribution strategies;
- have a reduce, reuse, recycle policy at the consumption stage;
- collaborate with other partners;
- create awareness among consumers about carbon; and
- sustain in green practices to improve agility and adaptability, and to promote alignment.

Invoking these strategies can help manufacturers and 3PLs to position themselves proactively. Corporate awareness on social responsibility has created immense change across the manufacturing and business communities. However, in Asia, a sustainable supply chain environment is far from being achieved. This chapter looked at the sustainable supply chain from the tactical and strategic levels, and studied the carbon issues at every stage across the supply chain. We attempted to define some of the apparent trends, the importance of the area to supply chains, and to define the key research questions. The implementation challenges associated with the sustainable supply chain were also discussed. The strategies and managerial implications proposed above will aid supply chain practitioners, industrialists, and governments to develop their own strategies for effective green supply chain policies. Moreover, companies can use this as a guide to gain significant competitive advantage over their competitors.

References

[9.1] Stern, N., 2007, *Stern Review: The Economics of Climate Change*, Cambridge University Press, Cambridge.

[9.2] IBM Global Business Services, 2008, *Mastering Carbon Management: Balancing Trade-Offs to Optimize Supply Chain Efficiencies.*

[9.3] PRTM, 2008, *Global Supply Chain Trends 2008–2010.*

[9.4] Simchi-Levi, D., Kaminsky, P. and Simchi-Levi, E., 2002, *Designing and Managing the Supply Chain: Concepts, Strategies & Case Studies*, 2nd edition, Irwin/McGraw-Hill, Boston, MA.

[9.5] Hervani, A.A., Helms, M.M. and Sarkis, J., 2005, "Performance measurement for green supply chain management," *Benchmarking: An International Journal*, **12**(4), pp. 330–353.

[9.6] Zhang, H.C., Kuo, T.C. and Lu, J., 1997, "Environmentally conscious design and manufacturing: a state-of-the-art survey," *Journal of Manufacturing Systems*, **16**(5), pp. 352–371.

[9.7] Srivastara, S.K., 2007, "Green supply-chain management: a state-of-the-art literature review," *International Journal of Management Reviews*, **9**(1), pp. 53–80.

[9.8] Bacallan, J.J., 2000, "Greening the supply chain," *Business and Environment*, **6**(5), pp. 11–12.

[9.9] Langley, C., Morton, J., Wereldsma, D., Swaminathan, S., Murphy, J., Deakins, T.A., Hoemmken, S. and Baier, J.M., 2008, *Third-Party Logistics: The State of Logistics Outsourcing.*

[9.10] SCMR, 2008, "Green supply chain study and survey results," *Supply Chain Management Review*, August.

[9.11] Gonzalez-Benito, J. and Gonzalez-Benito, O., 2006, "The role of stakeholder pressure and managerial values in the implementation of environmental logistics practices," *International Journal of Production Research*, **44**(7), pp. 1353–1373.

[9.12] Hall, J., 2006, "Environmental supply chain innovation," In *Greening the Supply Chain*, Sarkis, J. (ed.), Springer, London.

[9.13] Bansal, P., 2005, "Evolving sustainably: a longitudinal study of corporate sustainable development," *Strategic Management Journal*, **26**(3), pp. 197–218.

[9.14] Rao, P. and Holt, D., 2005, "Do green supply chains lead to competitiveness and economic performance?" *International Journal of Operations & Production Management*, **25**(9), pp. 898–916.

[9.15] Zhu, Q. and Sarkis, J., 2007, "The moderating effects of institutional pressures on emergent green supply chain practices and performance," *International Journal of Production Research*, **45**(18–19), pp. 4333–4355.

[9.16] Oracle Report, 2008, *The Shape of Tomorrow's Supply Chain, the Science of Sustainability*, Oracle Corporation.

[9.17] Eyefortransport, 2008, *The European 3PL Market: A Brief Analysis of Eyefortransport's Recent Survey*.

[9.18] Winston, A., 2008, *Green to Gold: How Smart Companies Use Environmental Strategy to Innovate, Create Value, and Build High Performance Supply Chains*, http://www.na2008.org/attendees/conf_forum.aspx.

[9.19] Porter, M.E. and Linde, C., 1995, "Green and competitive: ending the stalemate," *Harvard Business Review*, September–October.

[9.20] Simon, H., 1979, "Rational decision making in business organizations," *American Economic Review*, **69**(4), pp. 493–513.

[9.21] Hui, K.H., Spedding, T.A., Bainbridge, I. and Taplin, M.R., 2007, "Creating a green supply chain: simulation and modelling approach," In *Greening the Supply Chain*, Sarkis, J. (ed.), Springer, Heidelberg.

[9.22] EPA Report, 2000, *The Lean and Green Supply Chain: A Practical Guide for Materials Managers and Supply Chain Managers to Reduce Costs and Improve Environmental Performance*, EPA 742-R-00-001.

[9.23] http://www.masdarcity.ae/en/index.aspx.

[9.24] Logistics Management, 2009, "Green logistics: UPS lays out CO_2 emissions reduction goals," In *New Sustainability Report*, http://www.logisticsmgmt.com/article/CA6669461.html.

[9.25] Rettab, B. and Brik, B.A., 2008, *Green Supply Chain in Dubai*, Chamber, Dubai.

10

A Multi-agent Framework for Agile Outsourced Supply Chains

N. Mishra[1], V. Kumar[2] and F.T.S. Chan[3]

[1] School of Computer Science and Information Technology
University of Nottingham, Nottingham, NG8 1BB, UK
Email: nxm@cs.nott.ac.uk

[2] Department of Management, Exeter Business School
University of Exeter, Exeter, EX4 4PU, UK
Email: vk211@ex.ac.uk

[3] Department of Industrial and Systems Engineering
The Hong Kong Polytechnic University, Hung Hom, Hong Kong, China
Email: f.chan@inet.polyu.edu.hk

Abstract

The primary goal of an agile supply chain is to meet the varying demand of customers. Therefore, the supply chain nowadays involves coordination among partners, and this has raised issues of effective networking and logistics. The present chapter proposes a reconfigurable multi-agent architecture framework that can assist in selecting outsourcing partners and develop effective coordination among the partners and between manufacturing units. The proposed multi-agent architecture is inspired by the human self-healing mechanism and is capable of managing disruptions that occur during manufacturing operations. When a new production order is introduced, or during the disruptions, this agent framework uses a string matching algorithm to generate a better plan. The proposed agent architecture also learns continuously from its past experiences. This framework will also help to manufacture better quality products at minimum cost and within the due date.

10.1 Introduction

Over the last two decades, time-based competition is gradually emerging as a new standard in supply chain environment. The uncertain demand pattern has pressurised manufacturing firms to incorporate agility in their supply chain to efficiently tackle problems relating to demand management [10.1]. Therefore, the agile supply chain network is gaining the attention of researchers worldwide [10.2–10.6]. The key characteristic of an agile supply chain network is to allow effective collaboration between partners in a manner that enables them to quickly respond to customers. Supply chain partners can be classified as primary and secondary partners [10.7,

10.8]. The primary partners are the business units that are part of the already established supply chain network. Nevertheless, when the supply chain network fails to respond in time to unprecedented demand, then the agile supply chain network looks to the secondary partners, which are actually the outsourced units. Previously, the supply chain network used to respond to the demand following the decision maker's judgement [10.9, 10.10]. However, this sometimes causes bias in the decision making. Nevertheless, the newly emerging field of agile supply chain partnership is characterised by certain key features, such as automated quick response, effective coordination between the partners, ability to adjust to demand, and ability to explore their characteristics in an effective manner [10.11–10.13]. In order to meet the objectives of effective coordination, computational intelligence has been contemplated at every stage of the agile supply chain to make it capable of analysing, predicting, and optimising the performance during the demand-handling phase as well as properly utilising the resources available.

However, the control and management of such a large-scale network of diverse operations presents a challenging research problem, due to the immense complexity of the current supply chain network [10.14, 10.15]. Under such a complex situation, the existing management tools have shown to be insufficient. Therefore, current research is focused towards a novel technique inherited from autonomous systems that has already been successful in controlling and managing complex, interactive and constrained systems. This research aims to design a system that can automatically coordinate with both primary and secondary partners, identify the available resources, and allocate demand to available resources, with less human intervention using a string matching algorithm.

Recently, there have been several research papers on building a self-healed system that can manage itself by self-configuration, self-healing, self-optimisation and self-protection [10.16–10.18]. A generic framework for building a self-adaptive system is to model them as a collection of frequently similar coordinating agents that can take decisions regarding their behaviour and communicate among themselves. The implementation of such an approach can lead to robustness. Therefore, self-healed systems can play a vital role in managing complex supply chain networks. In order to deal with the former intricacies, the proposed research focuses on the development of a multi-agent system guided by an artificial immune system (AIS) inspired control framework [10.19–10.21]. The motive behind such an approach is the incorporation of self-learning and self-healing concepts through the agglomeration of artificial intelligence techniques, with computational potential that makes it able to handle fluctuating and unpredictable customer demand.

This chapter is organised as follows. Section 10.2 briefly explains the agile manufacturing concept. The problem scenario for which the multi-agent framework has been proposed is described in Section 10.3. The agent framework is discussed in Section 10.4. Section 10.4.1 explains the proposed agent architecture for the agile outsourced supply chain network. This section also discusses the communication ontology and string matching algorithm used in this research. Section 10.4.2 illustrates the communication channel and sections 10.4.2.1 and 10.4.2.2 describe the attributes of the communication module and encoding format, respectively. Section 10.5 summaries the chapter and concludes with some suggestions for future research.

10.2 Agile Manufacturing

The global market is increasingly demanding better goods and services at lower cost and with a shorter delivery cycle. To meet these challenges, manufacturing industry is striving to find an appropriate way to deal with the dynamic competitive market. Certainly, the agile philosophy is one of the more viable and competent ways to tackle the various challenges posed by uncertain customer demand [10.22, 10.23]. Agile manufacturing is an emerging concept, which is adopted to improve the competitiveness of firms. It is more pragmatically defined and closely associated with *quick response* [10.24]. Manufacturing enterprises adopting the agile concept are characterised by customer–supplier integrated processes for product design, manufacturing, marketing and support services. This requires stable unit cost, flexible manufacturing, easily accessible integrated data, modular production facilities and decision making at functional levels. Agility connects the interface between the company and the market. Essentially, it is a set of abilities for meeting widely varied customer requirements in terms of price, specification, quality, quantity and delivery. Agility has been expressed as having four underlying principles [10.25]:

1. delivering value to the customer;
2. being ready for change;
3. valuing human knowledge and skills; and
4. forming virtual partnerships.

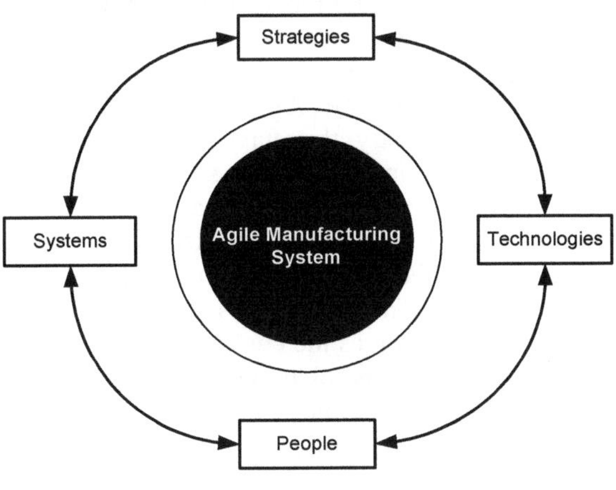

Figure 10.1. Agile manufacturing system

The successful execution of an agile manufacturing system in an organisation seeks the integration of design processes, process planning and scheduling at enterprise levels. Figure 10.1 shows a conceptualised model for an agile manufacturing system adapted from the work of Gunasekaran [10.26]; the figure illustrates an integrated agile manufacturing system developed using appropriate strategies and techniques in order to establish rapid partnership formation, virtual enterprise and re-configurability for mass customisation. Hence, it can be inferred that the increased range of product varieties, specialised and fragmented customers, and markets compels the enterprise to adopt agile strategies. The next section describes the problem scenario for which the multi-agent framework is proposed in an agile outsourced supply chain environment.

10.3 Problem Scenario

Supply chain coordination in manufacturing industry depends upon the relationship among suppliers, manufacturers and customers. Nowadays, there are many manufacturing units that produce similar kinds of products using different available technological alternatives. Therefore, today's supply chain has become more customer-oriented, and manufactures products at minimum cost as per customer choice. In this scenario, it is difficult for any manufacturing unit to produce a complete product of good quality on their own while keeping the cost low. However, a manufacturing unit can overcome this problem of producing a good quality product at minimum cost through collaboration with small companies nearby, known as outsourcing units, who are specialists in manufacturing certain products in minimum time. Therefore, the fierce competition in the market and the continuously changing demand pattern has forced manufacturing units to focus on their core competencies and leverage the specialised expertise of their partners. This helps the manufacturing units to eliminate their investments in non-core activities [10.27].

Additionally, this shifts the traditional manufacturing pattern towards the collaborative manufacturing pattern, also known as outsourcing. Although the outsourcing strategy helps to reduce the cost and increase the quality of a product, it increases its complexity at the same time. As collaborators are based outside and are not part of the manufacturing industry, they have their own policies and decisions to make, which makes the communication process more complex. However, this complexity can be resolved by effective coordination among different units. The coordination should not just be information oriented (*i.e.* resource availability and expertise), but rather should also be feasible in terms of logistics, *i.e.* in-time response and cost-effectiveness. This is only possible through effective networking and coordination. Therefore, as soon as the customer order arrives, the entire supply chain network needs to explore its own available resources, *i.e.* selection of available suppliers/partners and allocation of the task to the respective units.

The supply chain also needs to be capable of managing itself if new orders are introduced in between the manufacturing processes. Additionally, if there is a fault on the machines or a lack of material and human resources or power, then the supply chain also needs to be flexible and capable of adjusting itself. Generally, the faults are of two types, *i.e.* machine or product. Hence, if there is a machine fault, the product needs to be transferred to a new machine within the plant or to the outsourced partners, whichever is the best available option. If the parts produced are of poor quality, it is re-manufactured using the best available resources at a given point in time. Also if the part runs out of resources, then new alternative resources are made available. Moreover, during the selection of the partners, if more than one alternative is available then their past records must also be taken into account. This arrangement increases the reliability of the decision. To solve the given problem scenario, this research proposes a multi-agent architecture. The next section will discuss the proposed agent architecture in detail.

10.4 Agent Framework

This research framework proposes an intelligent reconfigurable multi-agent self-healing architecture for carrying out manufacturing operations while taking into account new orders and minimising losses due to sudden breakdowns. Recent research on intelligent agent system architectures have proven that problems that are inherently distributed can be efficiently implemented in a multi-agent framework [10.28], and thereby different distributed resources in the agile supply chain network are assumed to constitute a multi-agent architecture. Here, the autonomous agents are able to self-organise or manipulate their activities and patterns, and thereby obtain maximum benefit from a dynamic environment to achieve goals that exceed their individual skills. In order to confer self-organisation properties on the system, the supply chain activities are considered to be performed whereby the agents form a network of collaborative, yet autonomous, units modelled as interacting agents that proliferate, monitor, control and organise all activities involved in a distributed, dynamic and observable environment. In the following subsections, the agent model is explained in detail.

10.4.1 Agent Architecture

The proposed architecture framework is shown in Figure 10.2. This framework consists of ordering agent, planning agent, inventory agent, data-mining agent, corporate memory agent, distribution agent and learning agent. Every agent poses certain skills and that can be represented in the form of symbolic code schemes. A symbolic coding scheme will be utilised for the representation of the achieved or inherited skills of an agent. Simple alphabetical symbols such as *a*, *b*, *c*, *etc*., will be utilised to represent the fundamental skills, while combinations such as *aa*, *ab*, *abc*, *etc*., will represent the compound skills. For example, different outsourcing units perform different types of operations or have expertise in manufacturing certain types of products. Based on the symbol chosen to represent a particular property, a skill stream is formulated. The skill stream thus represents the knowledge status of the agent, *i.e.* the knowledge regarding the task (represented through a sequence of symbols) that it can perform or previously performed. The utilised symbolic coding stream is generic to different types of operations and can be manipulated (acquired learning) to suit different types of application through the use of an AIS-based controller.

The agents also have the ability to perceive their neighbourhood. While perceiving their neighbourhood, they pass on the status among themselves. The status of an agent that is busy on a job or helping another agent is represented by the status quo. The status quo represents the current state of the agent depending upon which agents respond to a particular situation. The different status quos are listed in

Table 10.1 along with their denotation and significance. The status quo decides the ability of an agent to perform a task and forms the basis for achieving a cooperative task utilising inter-agent communication. For example, when a task is found, only those agents that are in the 'wander' state are suited to perform the tasks. As soon as a new task arrives, the centralised system will contact the appropriate agents to acquire their status, and then it will coordinate the activities according to the updated status of the agents. The detailed descriptions of the tasks of each agent and their communication methods are described below.

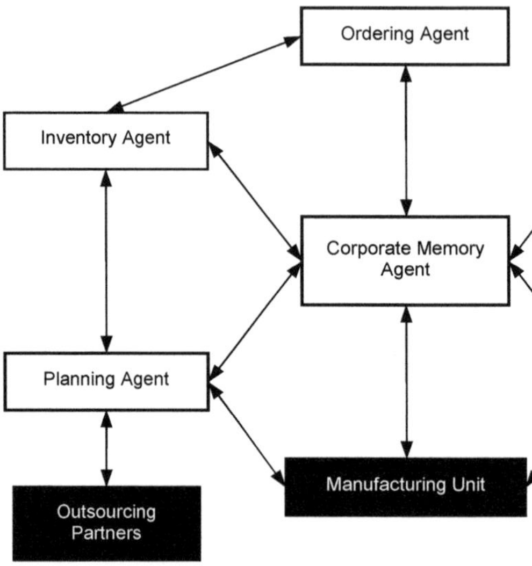

Figure 10.2. Reconfigurable multi-agent architecture framework

Table 10.1. Status quo of agents

Status quo	Denotation	Significance
Engaged	En	The agent is busy in performing some task and cannot render help instantly
Detected	Dt	The agent has detected a particular task and approaching it to perform it
Collaborate	Co	Offer help to perform a cooperative task
Wander	Wa	To randomly search the space for tasks
Idle	Id	Waiting for help from other agents to perform a task
Summon	Sn	To summon or call other agents for providing 'help'

10.4.1.1 Ordering Agent

The job of the *ordering agent* is to take orders from customers. Afterwards, this agent gathers all the information related to complete the order, such as the required part types and the assembly sequence. Since many parts are common to products of different types, in agile manufacturing a postponement strategy is preferred and common parts are manufactured and only assembled at the end, depending on the demand. Therefore, the ordering agent decides the production of common parts based on the demand pattern. According to the order, this agent communicates with the *planning* and *inventory agents* (see below) to estimate the type and size of the order. The planning agent provides further information on the part of the order being manufactured, and the inventory agent provides an estimate of the material available at the manufacturing plant. Once the order is finalised, the ordering agent contacts the *corporate memory agent* to find the possible available alternatives (in-plant or outsourced units) where the order can be placed at the minimum cost. If more than one alternative is available, then it will check the past records and assign the orders accordingly. This agent also takes into account the distance between the alternatives, the logistics medium used, and the available warehouse. If an outsourced unit is selected, then products are transferred by the logistics medium through the available warehouse. Therefore, every time an outsourced unit is selected, the agent not only evaluates the capacity and capability of the outsourced unit but also takes into account the logistics medium and the warehouse. These considerations are also taken into account while transferring the products within the in-plant manufacturing units. Also, whenever the raw material is ordered, the logistics cost is considered by the ordering agent.

10.4.1.2 Inventory Agent

The main task of the *inventory agent* is to check and keep the records of the available materials within the manufacturing unit. This agent also continuously exchanges information with machines and the ordering agent. Furthermore, this agent also forecasts the required inventory level based on past experience. After forecasting, this agent communicates with the ordering agent and selects appropriate suppliers to place the orders. Additionally, the inventory agent decides where to store the raw materials while simultaneously minimising the travelling and storage costs. Moreover, it makes raw materials easily available to appropriate machines as required.

10.4.1.3 Planning Agent

The main task of the *planning agent* is to decide where and how many parts need to be manufactured. Taking into account the capacity of the manufacturing plant as well as the available manufacturing resources and the due dates, this agent decides whether the products need to be manufactured within the plant or in the outsourced units using a string matching algorithm. The planning agent receives information regarding the availability of the manufacturing resources as well as information on the status of the machines, such as whether they are idle, in process, or in a breakdown condition or maintenance state from the *corporate memory agent*. As soon as a new order arrives, this agent helps in re-planning. If during the processing of a part a fault occurs, this agent executes the string matching algorithm again and automatically relocates the part to the appropriate machine. If a new part is introduced, the string matching algorithm will be executed to generate a schedule so that it is manufactured within the due date. The string matching algorithm mimics the self-healing mechanism of the human body to adjust the system automatically when any disturbances occur. As soon as any problem is detected, the agents communicate among themselves and seek help. The help ontology used by the agents to communicate among themselves and the string matching algorithm are described below in more detail.

Communication Ontology

As mentioned earlier, an agent can communicate only with those agents that come into its perceptual area. A communication channel ensures agent collaboration and knowledge diffusion between the agents involved. Conceptually, it standardises the interaction between computational agents and defines a communication language. An additional aspect is the use of standard ontologies that define the vocabulary used in the communication between agents and will be detailed below. The major application of communication ontology is during the approach to a cooperative task, *i.e.* whenever an agent need help from another agent, it establishes a communication channel with the agent in its perception range by sending a 'help' signal. Mathematically, the help signal H_i^t send by the *i*th agent for collaborative help in task *t* is defined as

$$H_i^t = (a_i, t_t) \ \ \forall A^{PR_i} \tag{10.1}$$

where A^{PR_i} represents the set of agents within the perception range (*PR*) of agent *i* and is defined as

$$A^{PR_i} = \{a_j \in a / a_j^{PR_i}\} \tag{10.2}$$

$$\exists A^{PR_i} \in a, \quad A^{PR_i} \wedge \neg(a_i) \tag{10.3}$$

and $a_j^{PR_i}$ represents the *j*th agent lying in the perception range of agent *i* and is defined as

$$a_j^{PR} < |d_{ij}| < PR_i \tag{10.4}$$

$|d_{ij}|$ represents the distance between agents *i* and *j*. In this architecture, Manhattan distance has been taken into account because of its effectiveness compared to Euclidean distance in parallel computing scenario [10.21, 10.29]. Upon receiving a 'help' signal from agent *i*, agent *j* sends a 'reply' signal r_j^i defined as

$$r_j^i = (a_j, a_i, H_i^t) \tag{10.5}$$

String Matching and AIS-based Control Framework

The aim to develop an intelligent agent architecture control system based on AIS is due to its efficacy in solving complex problems. AIS agents show characteristics such as specificity, inducibility, diversity, distinguishing self from non-self, and self-regulation that are similar to the human immune mechanism [10.30]. For the ease of understanding of AIS, a brief overview of immune system followed by its role in the development of a multi-agent control framework as well as the string matching algorithm is discussed in the following subsections. Thereafter, the advantages procured by utilising such a control framework are listed.

Overview of Human Immune System

The human immune system (IS) [10.31–10.33] is an extremely effective and complex system that can identify abnormal activities, solve the problem using existing knowledge, and generate new solutions for unseen events. In short, the immune system can be viewed as a network of players who mutually cooperate to get things done. It consists of diverse organs, tissues, innate cells and acquired cells acting in a highly coordinated and specific manner to recognise, eliminate and remember foreign macromolecules and cells. The immune system is basically divided into two major parts, the *innate* immune system (also known as natural immunity) and the *acquired* (or *adaptive*) immune system.

Innate immunity is inborn, unchanging and provides limited protection against infections, while acquired immunity is developed during the lifetime of a person and acts as a powerful supplement to innate immunity. Acquired immunity is antigen specific and is activated as a result of the interaction of the immune system with antigens in which antibody and immune cells eliminate the antigens. After the elimination process, immune cells become memory cells and are then used to eliminate the same antigen at a faster rate on subsequent encounters. Lymphocytes, the main antigen killer immune cell, have special binding areas known as receptors that can structurally determine and react with specific foreign antigens. The two important types of lymphocytes are B-cells and T-cells. B-cells have direct interactions with the antigens during the elimination process while T-cells act as mediators in the control of immune responses by providing specific cells capable of helping or suppressing these responses. Whenever an antigen is recognised by immune cell surface receptors, this interaction activates the proliferation and differentiation of the population of immune cells specific for that individual antigen. After the elimination of the antigen, some of the immune cells become memory cells. Due to this immunologic memory, the next time the body encounters the same antigen a much faster and stronger immune response results, known as secondary immune response. The concept of the immune system is employed for the development of a robust meta-heuristic known as an *artificial immune system* (AIS).

AIS-inspired Multi-agent Controller

This research aims to develop an AIS-based control framework to organise a fleet of agents with different skills and knowledge in a dynamic environment. In AIS, artificial intelligence techniques are utilised to mimic the behaviour shown by the human immune system and thereby obtain enhanced cooperation among distributed agents. In order to implement the framework, skill streams of agents are treated as antibodies with a unique set of functionalities and intelligence. This intelligence can be increased by performing newer explored tasks or cooperative tasks with other agents. For manufacturing operations, we assume that certain operations are predefined for each agent (equivalent to innate immunity). However, dynamic manipulation is also possible to adapt to the corresponding working environment (equivalent to acquired immunity). This control framework provides a set of rules to guide the behaviour of AIS agents within dynamically changing environment. The knowledge base of agents is bifurcated into long-term and short-term memory (Figure 10.3). Long-term memory stores the knowledge required by the agent for long-term usage as AIS rules, specific intelligence and acquired intelligence, whereas short-term memory only stores data pertinent to the current operation.

The basic attributes of the AIS-based control framework include: a set of agents that operate in the system; a set of tasks located in the workplace; and the perception range of an agent, which enables it to gain information about its surroundings and communicate and exchange information with other agents nearby.

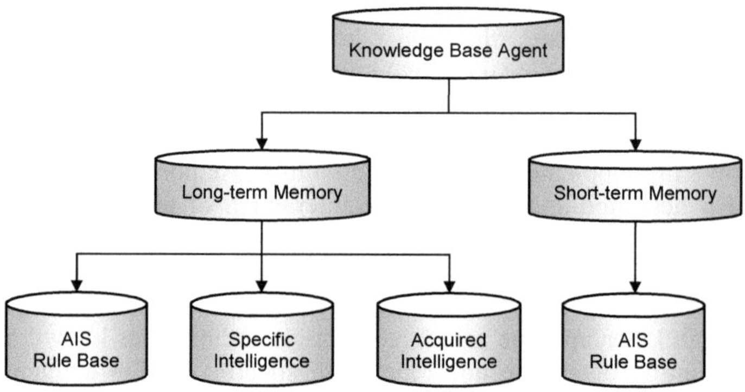

Figure 10.3. Knowledge base of agents

Affinity Function

The rule base of an agent stores an affinity function that measures its suitability to recognise and approach a particular task. The affinity for an agent to perform a particular task is measured in terms of the distance of the agent from that task and its specificity with respect to that task, *i.e.*

$$\rho = f(d_{ij}, \sigma_{ij}) \tag{10.6}$$

where ρ is the agent's affinity to perform a particular task, d_{ij} is the distance between agent and task (Manhattan distance, in this case) and σ_{ij} is the agent's specificity to perform a particular task.

Specificity in AIS refers to the extent of the similarity between an antibody and an antigen, and its evaluation involves several processes, such as pattern recognition, hydrogen binding, and non-covalent and Van der Waals interactions. This research tends to utilise the pattern recognition as the criterion for evaluating the specificity of an agent for a particular task. Since the recognition of an antigen by an immune cell is performed structurally, in a similar manner, skill streams of agents are matched with task strings and the extent to which these match determines the magnitude of the affinity between the task and the agent. Mathematically, the specificity matching function is described as

$$\sigma_{ij} = \begin{cases} \dfrac{1}{(R_L)^S} & \text{if } S \geq 1 \\ \dfrac{1}{(R_L)^S + p} & \text{if } S < 1 \end{cases} \tag{10.7}$$

where R_L is a measure of the match between the string and a particular task, and S is the relative strength of the agent to that required for the completion of a task. When the strength of agent is greater than required for the completion of a job, its affinity for that job is proportionately decreased; while if the agent is incapable of performing a task (its strength is less than that required for the completion of a job), its affinity is reduced by the inclusion of a big penalty term p. The motive behind this is to encourage those agents that can most efficiently perform the incumbent task.

The mathematical expressions for R_L and S are given in Equations 10.8 and 10.9, and the designation of the parameters utilised in it is explained through the help of Figure 10.4.

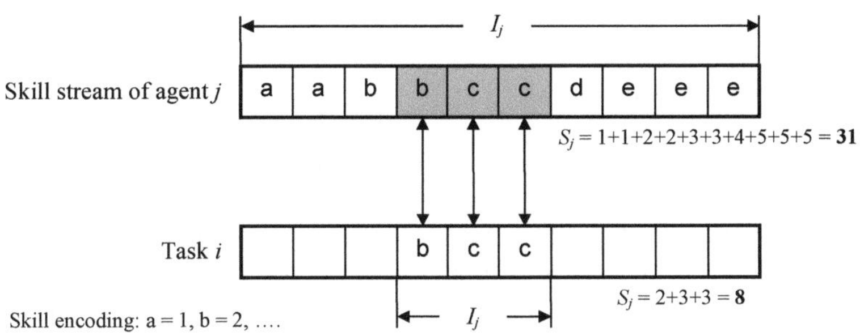

Figure 10.4. Specificity matching

$$R_L = \frac{l_j}{l_i} \tag{10.8}$$

$$S = \frac{S_j}{S_i} \tag{10.9}$$

String Matching Algorithm

The core of the proposed framework is the manipulation of the different skills of an agent towards the various problems encountered. In this process, when a task first appears, the specific skills (predefined) of an agent are matched with the task string utilising the string matching algorithm [10.34]. This specificity determines the ability of an agent to perform a task. In the natural immune system, antibody cells recognise antigens structurally from their antigen receptors. Utilising this pattern recognition concept, tasks are matched with the agent's skill streams. In case that a mismatch results, the acquired skills are matched with the task strings. If a mismatch results again, the string matching algorithm recombines or rearranges the elementary skills (process described later) in the specific skill stream to generate a new set of acquired skill streams specifically suited to the task. This acquired ability is stored in the long-term memory database for further utilisation, and thereby enhances the potential and knowledge base of the agent through self-learning. The process of skill manipulation is carried out using two separate algorithms – string matching set development and append string algorithm. String matching involves breaking down the task string into separate skill requirements, which are then matched with skill streams to identify a new skill set to be added to the skill stream of an agent. The generic steps of string manipulation are given below:

Step 1: Dismantle the task complexity chain into separate task requirements.
Step 2: Check if the task requirement matches the skill stream.
Step 3: Generate all possible combinations of task requirements.
Step 4: Compare the generated combinations with the skill stream.
Step 5: Find the shortest path to generate the new skill stream using Dijkstra's algorithm [10.35].

The newly generated sets are appended to the skill stream using the append string algorithm [10.30], which avoids any repetition during addition.

10.4.1.4 Corporate Memory Agent

This agent is the main hub of the useful information. Therefore, it stores all the information relating to the manufacturing plant, such as available resources and outsourcing partners. Further, this agent keeps information on the status of the products, such as on which machine they are being processed, their processing times and the order in which they are being processed. It also tracks information on the machines, such as their maintenance condition, idle or in-process stage, breakdown status, *etc.* It continuously communicates and learns with the *learning agent*. It also coordinates with all the other agents, such as the planning, inventory and ordering agents. This agent inherits the property of updating itself both online and offline.

10.4.1.5 Data-mining Agent

This agent analyses the manufacturing processes and order data. It further ranks the outsourcing partners, suppliers and manufacturing machines according to their past performances. All the information collected by this agent is continuously shared with the corporate memory agent. The rankings assigned by this agent are also used to resolve any allocation conflict in the future if more than one alternative is available.

10.4.1.6 Distribution Agent

This agent (not shown in Figure 10.2) gathers information from the planning agent on the manufacturing units and accordingly assigns the particular operation to an appropriate machine, which can be either in-plant or outsourced. If the outsourcing machine is selected, then this agent passes all the information related to the parts such as due date, type of material and CAD (computer aided design) diagram to the outsourcing partner. This agent also remains in continuous touch with the outsourcing and/or in-plant machines. If the outsourced order or in-plant machine fails to meet the due dates, then it instantly informs the planning agent. The planning agent then finds an alternative to complete the order within the due date. This agent also remains in continuous contact with the learning agent.

10.4.1.7 Learning Agent

The learning process in agents can be viewed as an alternative way of acquiring knowledge to increase the adaptiveness of the agents. With the presence of continuous noise and variation in the system, it is almost impossible to detect and take preventive action passively without updating the knowledge base. Thus, the knowledge base is exposed to the dynamics of the stage discrepancy as well as the impact of the variation in market determiners over time. For this purpose, a specific agent architecture for symptom recognition at each stage as well as for the whole architecture is employed. The learning agent learns through both online and offline learning. In online learning, data gathered through in-process stage is used, such as the quality of the product, tardiness and any fault in the machines. This data will be used either in new planning or in rescheduling. In offline learning, the information is collected from the data-mining agent. Online and offline learning are explained below in detail.

Offline Learning

The overall multi-agent architecture for offline learning consists of three agents (Figure 10.5), *viz.* the *collector agent*, *data-mining agent* and *corporate memory agent*. The process starts by collecting the response from the agent, *i.e.* considering time and quality aspects from the data-mining agent. Thereafter, the collector agent collects data from the partners based on their expertise and their distance from the different units. This data consists of valuable information for operation and control strategies as well as data on normal and abnormal operational patterns. This information is passed to the corporate memory agent. The information gathered by the corporate memory agent is exploited for extracting useful and understandable knowledge by finding patterns or fitting models to the observed data. This information is further used for the offline generation of a model bank.

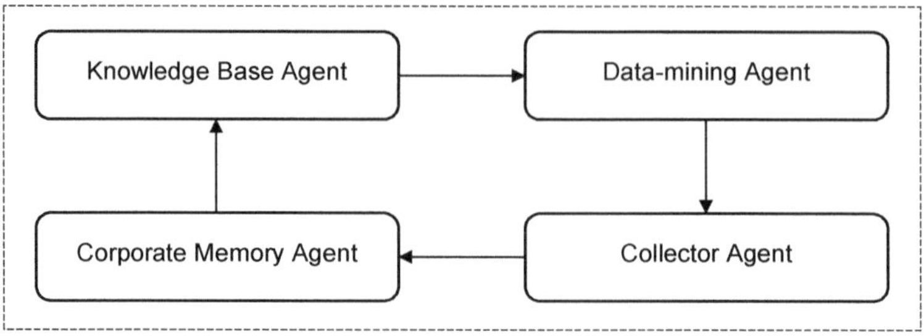

Figure 10.5. Multi-agent architecture for offline learning

Online Learning

Online learning of the dynamic model bank is performed to autonomously generate control solutions for unexpected faults while explicitly generating input–output maps through the use of a neural network. In the proposed research work (Figure 10.6), online learning of the neural network is performed through an artificial immune algorithm (AIA). AIA emulates the human immune system in general and the clonal selection in particular; therefore, researchers have adopted biological terminology to describe their structural elements and algorithmic operators [10.21, 10.36]. While implementing AIA, the antibodies are the solutions present in the bank, while antigens are the faults diagnosed by the diagnosis agent. The sequential procedure of this algorithm is as follows [10.37]:

i. Initially, build a database as per the offline learning and store them in the form of an antibody set.
ii. Receive the information according to the response corresponding to the order.

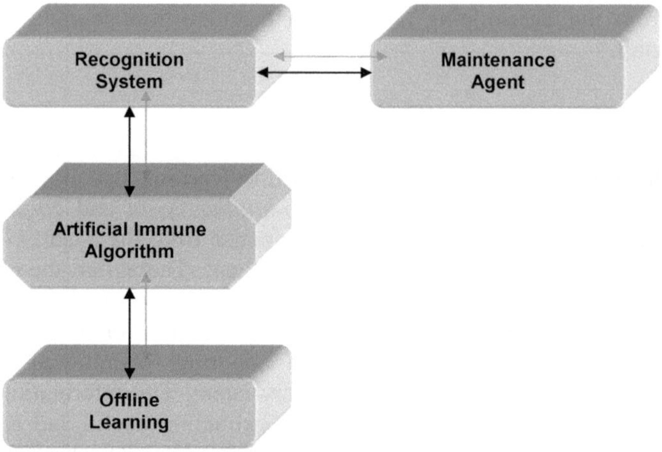

Figure 10.6. Multi-agent architecture for online learning

iii. A reconfigurable controller (further detailed below) is used to maintain the performance by continuously modifying the system, unless the offline diagnoser provides any corrective measures. In the present work, the abovementioned action agent works with a reconfigurable controller. The action agent, the core of the adaptive controller, is responsible for mapping of the output of the plant to the control input. The action agent is trained with the goal of producing the control sequence of feedback agents to minimise the quality problems in the system. In case if one agent is inefficient in solving the problem, another agent completely accordant with the first one is used to minimise the effect of the quality problems.

iv. Subsequently, the dynamic model bank is updated with the control solution generated from the offline diagnoser or the temporary solution generated by the neural network.

v. Repeat steps ii–iv until the system is under monitoring and diagnosis processes.

10.4.2 Communication Channel

The success of a multi-agent architecture depends on the effective communication among agents. These agents communicate by sending signals, which are used by the agents to easily interpret and manipulate any unexpected changes and the courses of action. Over years, researchers have developed many languages through which the agents can communicate, such as knowledge query and manipulation language (KQML) and agent communication language (ACL) [10.38–10.41]. Recently, a multi-agent logic language for encoding teamwork (MALLET) was developed by [10.42] to encourage team-oriented programming in the first sense. This agent language framework facilitates and manages the activities of agents through a proactive information exchange and based on their information need. MALLET facilitates knowledge encoding (*i.e.* declarative and procedural) and the information flow in the system. In our multi-agent framework, MALLET is used to set up the communication flow among the agents using a sequential and iterative process, and CAST (collaborative agents for simulating teamwork) [10.42] is adopted as the interpreter of MALLET. The main attributes of the communication module are described below.

10.4.2.1 Attributes of Communication Module

The three essential attributes and the requirements for launching communication in a multi-agent system are expressivity, understandability and reusability.

1. *Expressivity* – the attribute ensures that the communication language should be expressive, clear and precise.
2. *Understandability* – this attribute ensures that the information transferred should be encoded in an easily understandable manner.
3. *Reusability* – the communication module also needs the attribute of reusing information when needed, in order to reduce the cost of developing and maintaining agent systems.

10.4.2.2 Encoding Format

Before agents start communicating with each other, they pre-define their tasks, plans and capabilities. According to the requirements, one agent starts synchronising the shared task with other agents. The agent then transfers its own pre-requirements (needed information and knowledge) to the next agent, and the communication can be sequential (SEQ), parallel (PAR), iterative (WHILE, FORALL), conditional (IF) and choice (CHOICE). The agent does not execute any operation until its pre-requirements are met. If one agent fails to meet all the specified requirements, it will seek assistance of other agents and thus work collaboratively with each other. CAST as an interpreter of MALLET is used in the team-oriented agent architecture. Figure 10.7 shows the workflow of the CAST framework. A CAST agent consists of six components: reasoning engine (RE), shared mental model (SMM), individual mental model (IMM), team process tracking (TPT), proactive behaviour (PB) and goal management (GM). This agent helps to extract information from the knowledge base using Java-based reasoning engine, known as JARE [10.42].

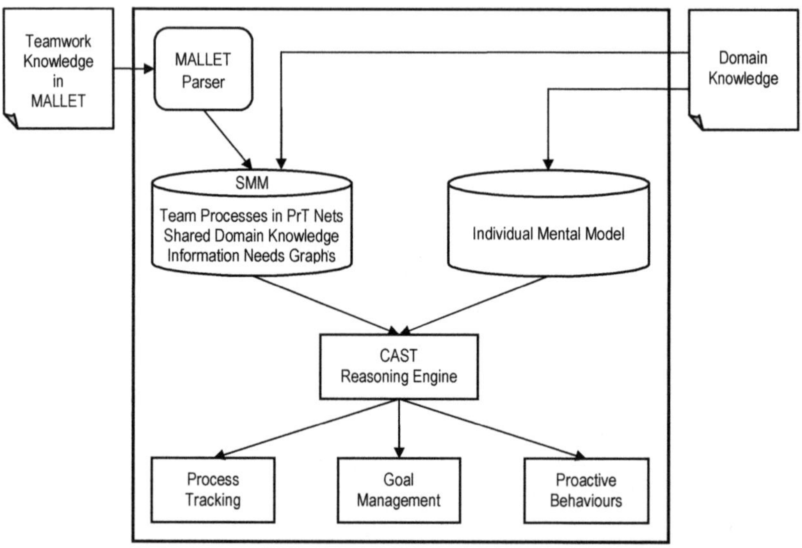

Figure 10.7. The CAST architecture [10.42]

10.5 Conclusions

The supply chain network has drawn the attention of the research community over the last couple of decades. In addition, as noted at the start of the chapter, the concept of agility in the supply chain has been widely discussed in the literature. The inclusion of the agile concept increases the complexity; however, the scenario becomes more complex with the introduction of collaboration among partners, commonly known as outsourcing. The complexity associated with the supply chain demands a strong communication network to effectively handle the problem. Within manufacturing units today, an effective communication system can be developed. However, in the outsourcing supply chain where the collaborative partners work closely with each other, it is difficult to develop an efficient communication channel. An agent-based architecture has evolved over time to tackle such problems and has been widely used and discussed in manufacturing scenarios. Inspired by this, a reconfigurable multi-agent-based architecture is proposed in this chapter for the agile outsourcing supply chain environment.

The reconfigurable multi-agent-based architecture is inspired by the self-healing mechanism of the human immune system. As the human self-healing mechanism automatically responds to repair any damage caused to the body, in a similar fashion this reconfigurable multi-agent-based architecture resolves the problems occurred during manufacturing operations. In this research, the proposed multi-agent-based architecture consists of ordering agent, planning agent, inventory agent, data-mining agent, corporate memory agent, distribution agent and learning agent. As soon as an order arrives, the agent architecture uses a string matching algorithm to allocate tasks to appropriate machines and outsourced partners. If there are any disruptions during the manufacturing process, this agent architecture automatically recovers and reassigns the necessary tasks. This chapter also briefly explains the communication channel used within the agent framework.

This agent architecture will assist the agile outsourced supply chain network to effectively communicate within the plant and among the outsourced partners. This will further aid in manufacturing good quality products at minimum cost while simultaneously meeting the due dates. Depending upon the demand pattern, the agent can identify common parts and the quantity of those parts that need to be manufactured in advance. If any disruption occurs during the manufacturing process, this agent framework is capable of making automated decisions to resolve any problems through effective communication among themselves. This agent also properly utilises the available resources, *i.e.* in-plant and outsourced resources.

Future research needs to focus on testing this agent framework under different manufacturing scenarios. Since manufacturing scenarios are quite complex and vary significantly from industry to industry, the viability of this agent framework will reveal its robustness. Therefore, according to the industry requirements, this agent framework needs to be modified to be able to effectively respond to any problems, which may include inclusion of more agents for specific tasks, assigning more tasks to existing agents, using different communication methods, *etc.* Nowadays, product recycling issues are grabbing the attention of the research community; therefore, this agent framework can play a crucial role in deciding the kinds of products to be manufactured that can be easily recycled, as well as in deciding when, where and how to recycle them. Therefore, in the future, this multi-agent framework needs to be modified, tested and explored under diverse complex manufacturing environment and multiple supply chain scenarios.

References

[10.1] Chan, F.T.S. and Kumar, V., 2009, "Performance optimization of a leagility inspired supply chain model: a CFGTSA algorithm based approach," *International Journal of Production Research*, **47**(3), pp. 777–799.

[10.2] Richards, C.W., 1996, "Agile manufacturing: beyond lean?" *Production and Inventory Management Journal*, 2nd Quarter, pp. 60–64.

[10.3] Jones, R.M., Naylor, J.B. and Towill, D.R., 2000, "Lean, agile or leagile? Matching your supply chain to the market place," *International Journal of Production Research*, **38**(17), pp. 4061–4070.

[10.4] Christopher, M., 2000, "The agile supply chain: competing in volatile markets," *Industrial Marketing Management*, **29**(1), pp. 37–44.

[10.5] Christopher, M. and Towill, D.R., 2000, "Supply chain migration from lean and functional to agile and customised," *Supply Chain Management: an International Journal*, **5**(4), pp. 206–213.

[10.6] Power, D.J., Sohal, A.S. and Rahman, S.U., 2001, "Critical success factors in agile supply chain management: an empirical study," *International Journal of Physical Distribution & Logistics Management*, **31**(4), pp. 247–265.

[10.7] Lambert, D.M., Cooper, M.C. and Pagh, J.D., 1998, "Supply chain management: implementation issues and research opportunities," *International Journal of Logistics Management*, **9**(2), pp. 1–20.

[10.8] Min, H. and Zhou, G., 2002, "Supply chain modelling: past, present and future," *Computers and Industrial Engineering*, **43**, pp. 231–249.

[10.9] Cakravastia, A., Toha, I.S. and Nakamura, N., 2002, "A two-stage model for the design of supply chain networks," *International Journal of Production Economics*, **80**(3), pp. 231–248.

[10.10] Chan, F.T.S., Chung, S.H. and Wadhwa, S., 2004, "A heuristic methodology for order distribution in a demand driven collaborative supply chain," *International Journal of Production Research*, **42**(1), pp. 1–19.

[10.11] Schonsleben, P., 2000, "With agility and adequate partnership strategies towards effective logistics networks," *Computers in Industry*, **42**, pp. 33–42.

[10.12] Perry, M. and Sohal, A.S., 2001, "Effective quick response practices in a supply chain partnership – an Australian case study," *International Journal of Operations & Production Management*, **21**(5–6), pp. 840–854.

[10.13] Yusuf, Y.Y., Gunasekaran, A., Adeleye, E.O. and Sivayoganathan, K, 2004, "Agile supply chain capabilities: determinants of competitive objectives," *European Journal of Operational Research*, **159**(2), pp. 379–392.

[10.14] Wilding, R., 1998, "The supply chain complexity triangle: uncertainty generation in the supply chain," *International Journal of Physical Distribution & Logistics Management*, **28**(8), pp. 599–616.

[10.15] Lambert, D.M. and Cooper, M.C, 2000, "Issues in supply chain management," *Industrial Marketing Management*, **29**(1), pp. 65–83.

[10.16] Sterritt, R. and Bustard, D., 2003, "Towards an autonomic computing environment," In *Proceedings of IEEE 14th International Workshop on Database and Expert Systems Applications*, pp. 694–698.

[10.17] White, S.R., Hanson, J.E., Whalley, I., Chess, D.M. and Kephart, J.O., 2004, "An architectural approach to autonomic computing," In *Proceedings of the International Conference on Autonomic Computing*, pp. 2–9.

[10.18] Parashar, M. and Hariri, S., 2005, "Autonomic computing: an overview," *Unconventional Programming Paradigms*, Springer, Berlin/Heidelberg, pp. 257–269.

[10.19] De Castro, L.N. and Von Zuben, F.J., 1999, *Artificial Immune Systems: Part I – Basic Theory and Applications*, Technical Report – RT DCA 01/99, Campinas, SP: State University of Campinas, Brazil.

[10.20] De Castro, L.N. and Von Zuben, F.J., 2001, "aiNet: an artificial immune network for data analysis," In *Data Mining: a Heuristic Approach*, Abbass, H.A., Sarker, R.A. and Newton, C.S. (eds.), Idea Group, Hershey, PA, pp. 1–37.

[10.21] De Castro, L.N. and Timmis, J., 2002, *Artificial Immune Systems: a New Computational Intelligence Approach*, Springer-Verlag, London.

[10.22] Kidd, P., 1994, *Agile Manufacturing: Forging New Frontiers*, Addison-Wesley, Wokingham.

[10.23] Chan, F.T.S., Kumar, V. and Tiwari, M.K., 2009, "The relevance of outsourcing and leagile strategies in performance optimization of an integrated process planning and scheduling model," *International Journal of Production Research*, **47**(1), pp. 119–142.

[10.24] Agarwal, A., Shankar, R. and Tiwari, M.K., 2006, "Modeling the metrics of lean, agile and leaglie supply chain: an ANP-based approach," *European Journal of Operational Research*, **173**(1), pp. 211–225.

[10.25] Goldman, S., Nagel, R. and Preiss, K., 1995, *Agile Competitors and Virtual Organizations*, Van Nostrand Reinhold, New York.

[10.26] Gunasekaran, A., 1999, "Agile manufacturing: a framework for research and development," *International Journal of Production Economics*, **62**, 87–105.

[10.27] Mishra, N., Choudhary, A.K. and Tiwari, M.K., 2008, "Modeling the planning and scheduling across the outsourcing supply chain: a Chaos-based fast Tabu–SA approach," *International Journal of Production Research*, **46**(13), pp. 3683–3715.

[10.28] Ferber, J., 1999, *Multi-Agent System: An Introduction to Distributed Artificial Intelligence*, Addison-Wesley Longman, Boston, MA.

[10.29] Freitas, A.A. and Timmis, J., 2003, "Revisiting the foundations of artificial immune systems: a problem-oriented perspective," *Artificial Immune Systems*, 2787, Springer, Berlin/Heidelberg, pp. 229–241.

[10.30] Lau, H.Y.K. and Wong, V.W.K., 2006, "An immunity-based distributed multi agent-control framework," *IEEE Transactions on Systems, Man and Cybernetics, Part A: Systems and Humans*, **36**(1), pp. 91–108.

[10.31] Segerstrom, S.C. and Miller, G.E., 2004, "Psychological stress and the human immune system: a meta-analytic study of 30 years of inquiry," *Psychological Bulletin*, **130**(4), pp. 601–630.

[10.32] Somayaji, A., Hofmeyr, S. and Forrest, S., 1998, "Principles of a computer immune system," In *Proceedings of the 1997 Workshop on New Security Paradigms*, Langdale, Cumbria, pp. 75–82.

[10.33] Kim, J. and Bentley, P.J., 2002, "Towards an artificial immune system for network intrusion detection: an investigation of dynamic clonal selection," In *Proceedings of the 2002 Congress on Evolutionary Computation*, Honolulu, pp. 1015–1020.

[10.34] Cormen, T.H., Leiserson, C.E., Rivest, R.L. and Stein, C., 2001, *Introduction to Algorithms*, 2nd Edition, MIT Press, Boston.

[10.35] Dijkstra, E.W. 1959, "A note on two problems in connexion with graphs," *Numerische Mathematik*, **1**, 269–271.

[10.36] Watkins, A., Timmis, J. and Boggess, L., 2004, "Artificial Immune Recognition System (AIRS): an immune-inspired supervised learning algorithm," *Genetic Programming and Evolvable Machines*, **5**(3), pp. 291–317.

[10.37] Fan, X., Yen, J., Miller, M. and Volz, R., 2004, "The semantics of MALLET – an agent teamwork encoding language," In *Declarative Agent Languages and Technologies II: Proceedings of the Second International Workshop*.

[10.38] Barbuceanu, M. and Fox, M.S., 1995, "COOL: a language for describing coordination in multi-agent systems," In *Proceedings of the First International Conference on Multi-Agent Systems*, Lesser, V. (Ed.), AAAI Press/MIT Press, San Francisco, CA, pp. 17–24.

[10.39] Barbuceanu, M. and Fox, M.S., 2006, "The design of a coordination language for multi-agent systems," *Intelligent Agents III Agent Theories, Architectures, and Languages*, Springer, Berlin/Heidelberg, pp. 341–355.

[10.40] Pitt, J. and Mamdani, A., 1999, "Designing agent communication languages for multi-agent systems," *Multi-Agent System Engineering*, Springer, Berlin/ Heidelberg, pp. 102–114.
[10.41] Da Silva, V.T. and De Lucena, C.J.P., 2004, "From a conceptual framework for agents and objects to a multi-agent system modeling language," *Autonomous Agents and Multi-Agent Systems*, **9**(1–2), pp. 145–189.
[10.42] Fan, X., Yen, J., Miller, M., Ioerger, T.R. and Volz, R., 2006, "MALLET – a multi-agent logic language for encoding teamwork," *IEEE Transactions on Knowledge and Data Engineering*, **18**(1), pp. 123–138.

11

Agent-based Simulation and Simulation-based Optimisation for Supply Chain Management

Tehseen Aslam and Amos Ng

Virtual Systems Research Centre, University of Skövde
Högskolevägen, Skövde 541 28, Sweden
Emails: tehseen.aslam@his.se; amos.ng@his.se

Abstract
Agent-based simulation (ABS) represents a paradigm in the modelling and simulation of complex and dynamic systems distributed in time and space. Since manufacturing and logistics operations are characterised by distributed activities as well as decision making – in both time and in space – and can be regarded as complex, the ABS approach is highly appropriate for these types of systems. The aim of this chapter is to present a new framework of applying ABS and simulation-based optimisation techniques to supply chain management, which considers the entities (supplier, manufacturer, distributor and retailer) in the supply chain as intelligent agents in a simulation. This chapter also gives an outline on how these agents pursue their local objectives/goals as well as how they react and interact with each other to achieve a more holistic objective(s)/goal(s).

11.1 Introduction

Today, as the globalisation of product markets continues and the competition between original equipment manufacturers (OEMs) increases, the majority of OEMs are emphasising integration, optimisation and management of their entire supply chain from component through manufacturing, inventory management and distribution to end customers [11.1, 11.2].

According to Archibald *et al.* [11.3], the majority of the operating expenses of most companies are related to supply chain management (SCM) costs, which can be as high as 75% of overall operating expenses. When compared with some years ago, the challenge that these companies are facing has shifted away from being forced to achieve internal efficiency to achieving overall supply chain (SC) efficiency, due to the global competition. SC can be defined as a network of autonomous organisations (*i.e.* suppliers, manufacturers, distributors and retailers), through which raw materials and components are acquired, transformed and delivered to the customers (see, *e.g.* [11.4–11.8]). The aim of SC is to create agile and independent, but cooperative groups of companies, which are able to reduce overall costs and increase their competitiveness in the market through shorter time-to-market, higher agility and flexibility to meet customer demands while incurring minimum costs [11.1, 11.9].

According to Christopher [11.4], the agility concept has only been considered through/as ‘agile manufacturing’ in the same sense that the lean concept was earlier only coupled to ‘lean production’. However, limiting the agility concept within the four walls of a manufacturing entity is not preferred. The manufacturer might be an important entity, but it is not the only entity involved in satisfying customer demand. Hence, making the manufacturer agile does not make the other entities in the SC agile. The same applies to the interacting logistics operations.

The next question that arises is what the mapping/connection between ‘agile manufacturing’ and ‘agile supply chains’ is. Swafford *et al.* [11.10] explained it by outlining that a manufacturing entity on its own consists of an internal supply chain that comprises product development, manufacturing, procurement and distribution functions, where the flexibility of the supply chain represents various abilities in these internal functions. For instance, a manufacturing entity’s ability to vary its production mix has a direct effect on procurement and its ability to supply materials to support a new production schedule. Reducing supply chain lead-time, ensuring production capacity and providing product variety while fulfilling customer expectations are some abilities that can be gained from having flexible supply chains. All these abilities work in synergy and affect each other in one way or another regardless of whether you are optimising your process on a local level (*e.g.* manufacture entity) or a global level (*e.g.* entire supply chain).

Hence, it is in the best interest of all managers to start thinking in terms of *global optimisation*, instead of achieving *local optimisations*, which are obtained when SC entities optimise their processes without taking into account their impact on other entities [11.11–11.14]. The impacts of such local optimisations often result in great variations in inventories and demands, which then result in insufficient material flows, creating longer lead times [11.1, 11.15].

One of the major research areas within the SC domain is the bullwhip effect, which refers to a phenomenon in SC where the demand variability of incoming orders are amplified as they move up the supply chain [11.16–11.18].

In principle, OEMs in the manufacturing industry have always had to address fluctuations in demand, with perhaps some incidental exceptions when demand exceeded world-wide production capacity (which is rarely the case today). Those OEMs, which are often striving after lean operations whilst at the same time needing to be agile, have traditionally passed the agility problem on to their suppliers. Many manufacturing companies have recognised that the issue of demand fluctuations as well as other fluctuations such as delayed deliveries not only is causing disturbances in their own production, but has an even more widespread effect on their suppliers. Typically, a supply network is subject to knock-on effects; small variations in demand (output from the market) result in somewhat larger variations for manufacturers, which in turn result in larger variations for the suppliers.

The bullwhip effect is one of the major contributors to excess costs within SC; it contributes to significant piles of stocks, inefficient utilisation and overtime, and frequent stock-outs, as well as added transportation costs due to inefficient scheduling [11.2, 11.3].

The existence of the bullwhip effect has repeatedly been exposed in industrial businesses, such as Procter & Gamble and Hewlett Packard [11.18–11.20], and in macro-economics [11.21–11.24]. Forrester [11.19, 11.20] initiated the analysis of demand variability back in 1961; his studies showed that SC is distorted by large demand swings when companies within the SC tend to solve the issues from their own perspective. Later on, other researchers [11.22, 11.25, 11.26] have continued the research into the bullwhip effect and its impact on SC. Lee *et al.* [11.18, 11.27] identified five causes of the bullwhip effect: demand forecast updating, order batching, price fluctuation, rationing and shortage gaming, and non-zero lead time.

Researchers have over the years presented different proposals on how to solve the bullwhip effect issue; some have investigated order batching and studied its impact on the bullwhip and the total supply chain inventory levels [11.21, 11.28–11.30]. Another research direction is on investigating demand research updating and information sharing issues [11.16, 11.27, 11.31–11.34]. The impact of lead time on the bullwhip effect has been examined by [11.35–11.37], where they emphasise the priority of reducing the lead time. They show that the bullwhip effect could dramatically increase the lead time.

Most of the above-mentioned solutions have been compiled empirically; it is only in recent years that the idea of using agent-based simulation (ABS) for addressing the bullwhip issue has gained interest in the research community. Liang *et al.* [11.38] developed an ABS system to control inventory and minimise total costs for an SC by sharing forecast and information knowledge. Fu *et al.* [11.39] presented a collaborative inventory management framework in SC using ABS. Zarandi *et al.* [11.17] addressed the bullwhip effect by developing an ABS model to minimise total costs and reduce the bullwhip effect by implementing fuzzy logic, genetic algorithms and neural networks.

ABS represents a paradigm in the modelling and simulation of complex and dynamic systems distributed in time and space [11.40, 11.41]. Since manufacturing and logistics operations are characterised by distributed activities as well as decision making – both in time and in space – and can be regarded as complex, the ABS approach is highly appropriate for these types of systems [11.41–11.45]. The aim of this chapter is to present a new framework of applying ABS and simulation optimisation techniques to SCM problems, which considers the entities (supplier, manufacturer, distributor and retailer) in the supply chain as intelligent agents in a simulation, not only how these agents pursue their local objectives/goals but also how they react and interact with each other to achieve a more holistic objective(s)/goal(s). The contents of this chapter include a literature review of related work to justify the argument that ABS is an appropriate tool for solving demand amplification issues; an introduction of simulation-based optimisation and multi-objective optimisation; and a presentation of the ABS framework.

11.2 Literature Review: Agent-based Simulation

The agent technology originates from distributed artificial intelligence (DAI) where agents are used to bridge the gap between humans and machines by means of interaction and intelligence [11.46]. Nowadays, agent technology is used in many different domains [11.47–11.49]. A definition proposed by Wooldridge and Jennings [11.50] classifies agents as hardware- or software-based computer systems with the following characteristics:

- *Autonomy* – agents are autonomous in the sense that they operate by themselves without any direct human intervention and they have some kind of control over their internal state and behaviour in an environment.
- *Social ability* – their social ability is basically an agent's capability of interacting with other agents and with its environment.
- *Reactivity* and *proactiveness* – agents have also the capability to perceive their environment and in respond react to the changes in the environment. Simultaneously, they are also capable of pursuing their own goals by controlling their future in a proactive manner [11.51].

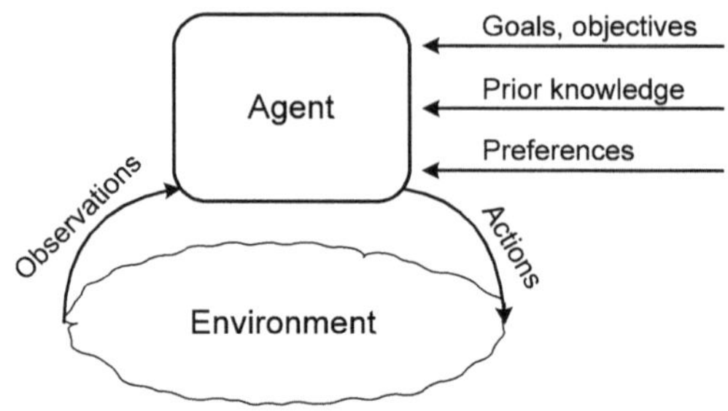

Figure 11.1. Theoretical concept of an agent

An agent (Figure 11.1) is basically a computational system that is situated in a dynamic environment and is capable of exhibiting autonomous and intelligent behaviour. It has some stated goals/objectives, prior knowledge and preferences that govern its internal decision making, and based on this the agent performs some actions to influence its surroundings (*e.g.* environment, other agents, *etc.*) so that the agent's stated objectives can be reached. The agent also observes changes in its surroundings, and based on the changes and the above-mentioned mechanisms, it executes actions.

Multi-agent systems (MAS) are formed when more than one agent interact and communicate with each other (see Figure 11.2) in order to achieve some shared goal(s). The agent's ability to collaborate, coordinate and interact with other agents is the most important feature of MAS. By sharing information, knowledge and tasks among the agents in MAS, a collective intelligence may emerge that cannot be derived from the internal mechanism of an individual agent. The ability to coordinate within an agent community makes it possible for agents to coordinate their actions among themselves, *i.e.* taking the effect of another agent's actions into account when making a decision about what to do [11.51].

ABS is an approach to model systems of interacting autonomous agents, hence ABS is used to design and understand MAS [11.52]. Currently, there is no single definition of ABS; however, Sanchez and Lucas [11.53] proposed that ABS is a simulation made up of agents, objects or entities that behave in an autonomous way.

Where the agents are aware of and interact with their environment through internal rules for decision making, movement and action, and as result of this interaction of the relatively simple behaviours of individual simulated agents, we obtain the collective behaviour of the simulated system.

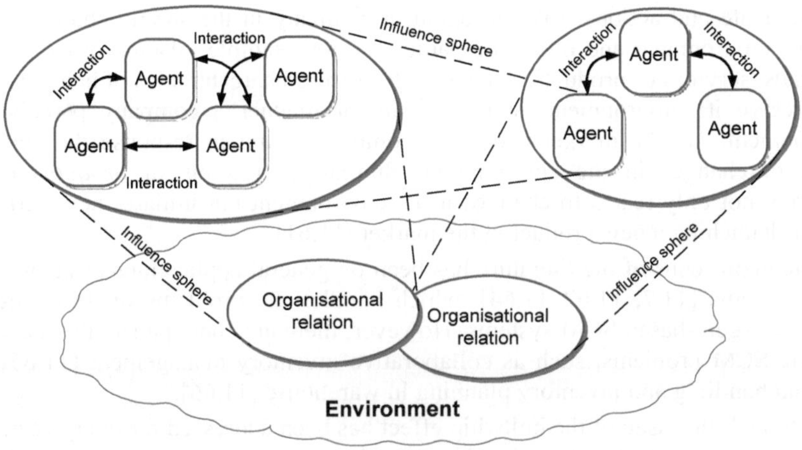

Figure 11.2. A multi-agent system

The concept of MAS and ABS has gained a great deal of momentum since the early 1990s. Agent-based applications have in recent years started to appear in many different fields [11.52, 11.54]. The manufacturing domain has started embrace the idea of agents due to their capabilities of *autonomy, responsiveness, redundancy* and *distributedness* [11.51]. MAS and ABS applications are successfully being used in different manufacturing areas. In terms of production planning and resource allocation, Bruccoleri *et al.* [11.55, 11.56] have addressed a framework where five different levels of production planning in a reconfigurable enterprise are distinguished and a multi-agent production planning system is built. Within such a framework, the traditional production planning activity is executed by agents that make their own specific planning decisions, while the global planning decisions are achieved through coordination and negotiation among the agents. Koussis *et al.* [11.57] presented an agent-based application for production scheduling and control applications where agents, who are dedicated to work centres, dynamically select the most suitable dispatching rules in an agent-based scheduling system. An agent-based collaborative production framework with the ability to carry out scheduling and dispatching functions among production entities is presented in [11.58]. There are also different agent-based applications for manufacturing process monitoring, control and diagnostics. For example, an agent-based diagnostic system was developed for an automotive manufacturer to be used in a PLC-controlled assembly line [11.59]. Another agent-based application for production monitoring has also been developed to control the body shop, paint shop and assembly line for a German car manufacturer [11.60].

ABS has gained a great deal of interest in the supply chain community because of the similarity between supply chain participants (*e.g.* factories, customers, *etc.*)

and agents in ABS models. Another motivating aspect for implementing ABS for supply chain issues is the resemblance of characteristics between a company in a supply chain and characteristics of an agent (autonomy, social ability, reactivity and proactiveness). A company in a supply chain carries out some tasks *autonomously* without any intervention from external entities and has some kind of control over its internal state and actions. Like an agent, a company in the supply chain *interacts* with its environment and other companies by, for example, placing orders for raw materials, services or products. An agent also shows characteristics of being *reactive* to perceive its environment, and in a similar manner a company perceives its environment, *i.e.* the market and other companies, and in response the company reacts to changes in the environment. Similarly, in terms of *proactiveness*, a company not only reacts to changes in its environment but initiates new activities, such as launching a new product in the market [11.61].

The main focus of the literature has been on general applications of agent-based SCM systems [11.7, 11.62–11.64], which handles the problems of designing and operating agent-based SCM systems. However, there are some papers that deal with specific SCM problems, such as collaborative inventory management [11.65], and material handling and inventory planning in warehouse [11.66].

Although the issue of the bullwhip effect has been addressed for many years, it is only in recent years that ABS has been used to evaluate this issue. The *MIT beer game* [11.67] and the *wood supply game* [11.68] have been the basis for many studies of the bullwhip effect using ABS. The beer game is used by Kimbrough *et al.* [11.69] to investigate the concept of an intelligent supply chain run by software agents to see whether the artificial agents are able to cope with the bullwhip issue. The results from the study have shown that the agents are able to reduce the bullwhip effect, discover optimal ordering policies and outperform humans playing the beer game. The study also showed that the supply chain managed by agents was adaptable to its changing environment. Moyaux *et al.* [11.70–11.73] have repeatedly shown the advantages of using ABS when managing the bullwhip effect. In all their studies, they have implemented ABS on the *Québec wood supply game* (which is an adaptation of the original wood supply game) [11.72], to evaluate different strategies to reduce the bullwhip effect. In [11.73], they simulated the effect of collaboration and information sharing between the supply chain entities that are represented by agents; and in [11.71], they investigated various coordination techniques and proposed a new technique based on tokens to coordinate the agents in the supply chain. Other researchers such as Yung and Yang [11.74] have also investigated the approach of agents coordinating the supply chain in which each company is represented as an agent with the goal to minimise its costs in relation to some constraints. In a similar approach, Zarandi *et al.* [11.17] addressed the bullwhip effect by developing an ABS model to control the order quantity for every supply chain entity, minimise total costs, and reduce the bullwhip effect by implementing fuzzy logic, genetic algorithms and neural networks. All of the research studies presented clearly show the applicability and advantages of ABS when dealing with the bullwhip effect.

Based on this literature review, we outline an agent-based simulation framework for the multi-level and multi-objective optimisation (MLO and MOO) of SCM design. The speciality and advantages of such a framework are shown using an example that includes typical entities in a supply chain, *i.e.* supplier, manufacturer, distributor and retailer in a supply chain in the next section.

11.3 An ABS Framework for Multi-objective and Multi-level Optimisation

Most supply chain design problems require the simultaneous optimisation of more than one conflicting objective. For example, while cost and delivery service level can be the most common indicators to determine the performance of a supply chain, there are other important metrics used in supply chain analysis, for example, lead time, final goods inventory and work-in-process (WIP). A short average lead time means that the total time a product is stored in the system is short, which means that customer orders can be fulfilled within a shorter time and thus leverages the overall performance of the supply chain. A low WIP means that the cost spent on transportation and inventory is lowered and thus is also highly desired. Therefore, to a decision maker, an ideal configuration is the one that maximises delivery service level while simultaneously minimising lead time and WIP. Unfortunately, this is never an easy task because in most real-world complex systems, these objectives are in conflict with each other – in many cases, delivery service level increases proportionally with inventory level and cost. In a general MOO problem, there exists no single best solution with respect to all objectives, as improving the performance on one objective would reduce the performance of one or more other objectives [11.75]. A simple method to handle an MOO problem is to form a composite objective function as the weighted sum of the conflicting objectives. Because a weight for an objective is proportional to the preference factor assigned to that specific objective, this method is also called preference-based strategy. Apparently, preference-based MOO is simple to apply, because by scalarising an objective vector into a single composite objective function (*e.g.* combining all performance measures into a weighted average objective function to represent the system investment cost), an MOO problem can be converted into a single-objective optimisation problem and thus a single trade-off optimal solution can be sought, effectively. However, the major drawback is that the trade-off solution obtained by using this procedure is very sensitive to the relative preference vector. Therefore, the choice of the preference weights and thus the obtained trade-off solution is highly subjective to the particular decision maker.

At the same time, it is also argued that using preference-based MOO to obtain a single 'global' optimal solution for multi-tier systems, like supply chains, is not desirable if the 'global' optimum suggests a set of decision variable values that may sacrifice the performance of the sub-system level. For example, the optimal solution found by the simulation optimisation may be optimal when considering the overall supply chain but not at all acceptable to the company that plays the role of the manufacturer. Therefore, for a decision maker, it would be useful if the posterior Pareto front can be generated quickly by using an MOO algorithm, as shown in Figure 11.3, so that he/she can choose the most suitable configuration among the trade-off solutions generated. The meaning of 'trade-off' is two-fold: (1) trade-off between the conflicting objectives, and (2) trade-off between sub-systems and the overall system. While the concept of applying MOO to find trade-off solutions for a multi-tier system design is sound, in practice it is very challenging because the search space, constituted by the possible values of the multi-level decision variables, is often huge.

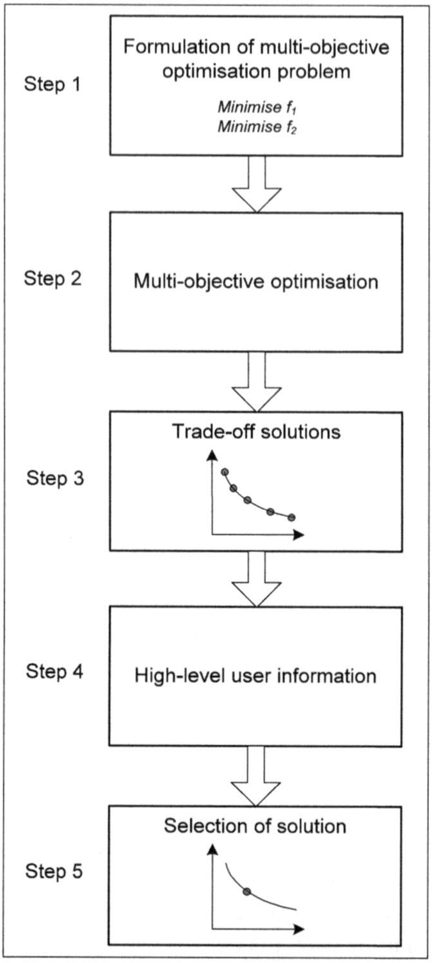

Figure 11.3. General multi-objective optimisation procedure

Therefore, in this chapter, we propose a novel ABS environment that supports the MOO and MLO for SCM that can significantly reduce the search space. As mentioned above, an ABS architecture allows the characteristics of the different entities in a supply chain to be modelled as autonomous agents. Using MOO, these agents would be able to perform local optimisations for the represented entity in the supply chain and global optimisation for the entire supply chain, and MLO makes it possible to run high-fidelity process models that feed the overall system process (supply chain) with Pareto-optimal solutions. With the multi-level architecture, simulation optimisation can be carried out to different levels using models with varying fidelity. Considerable reduction of computing time can therefore be achieved by avoiding a high-fidelity 'global simulation'. At the same time, a significant reduction in the search space of the overall system level can be achieved by using the MOO algorithms with the intelligent agents so that only solutions lying on the Pareto fronts of the sub-system level are transferred to the optimisation process in the system level. The agent-based environment presented in Figure 11.4 illustrates the different levels and the agents within those levels and how they interact within the framework.

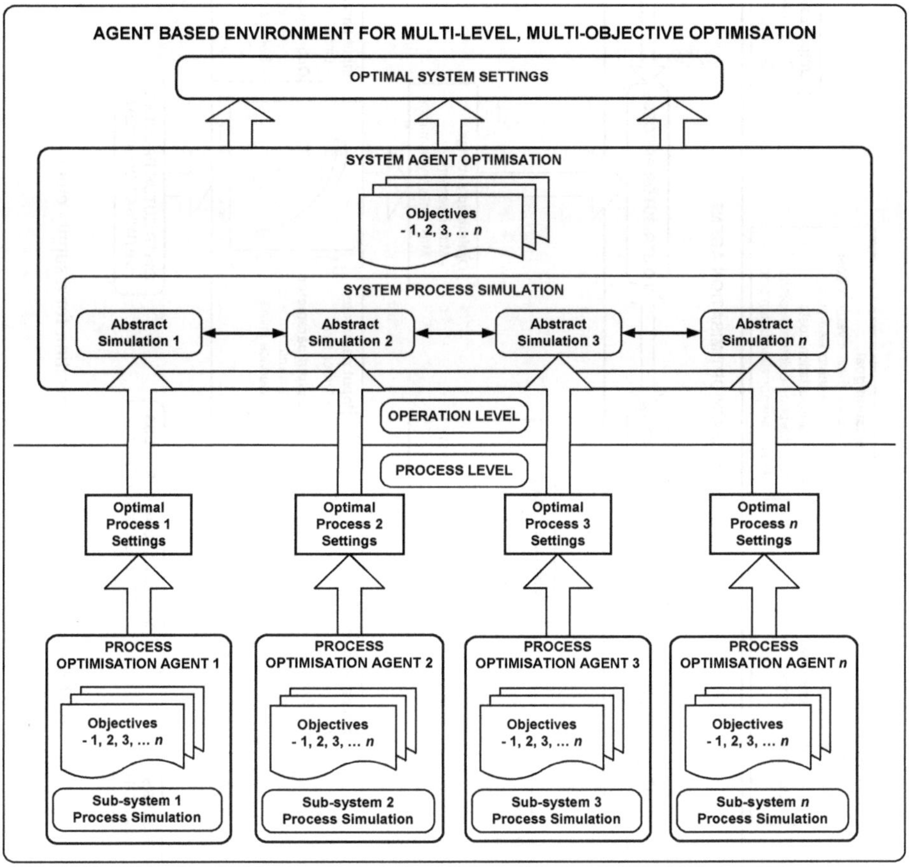

Figure 11.4. Agent-based framework for multi-level, multi-objective optimisation

Within this framework, a system is divided into two levels, namely, the *process level* and the *operation level*, which represent the overall system and its sub-systems, respectively. In the process level, all optimisations occur within each process optimisation agent that has its own stated goals to peruse. The results from a process optimisation agent simulation run are the optimised process settings that are sent to the operation level. In the operation level, the overall system optimisation is

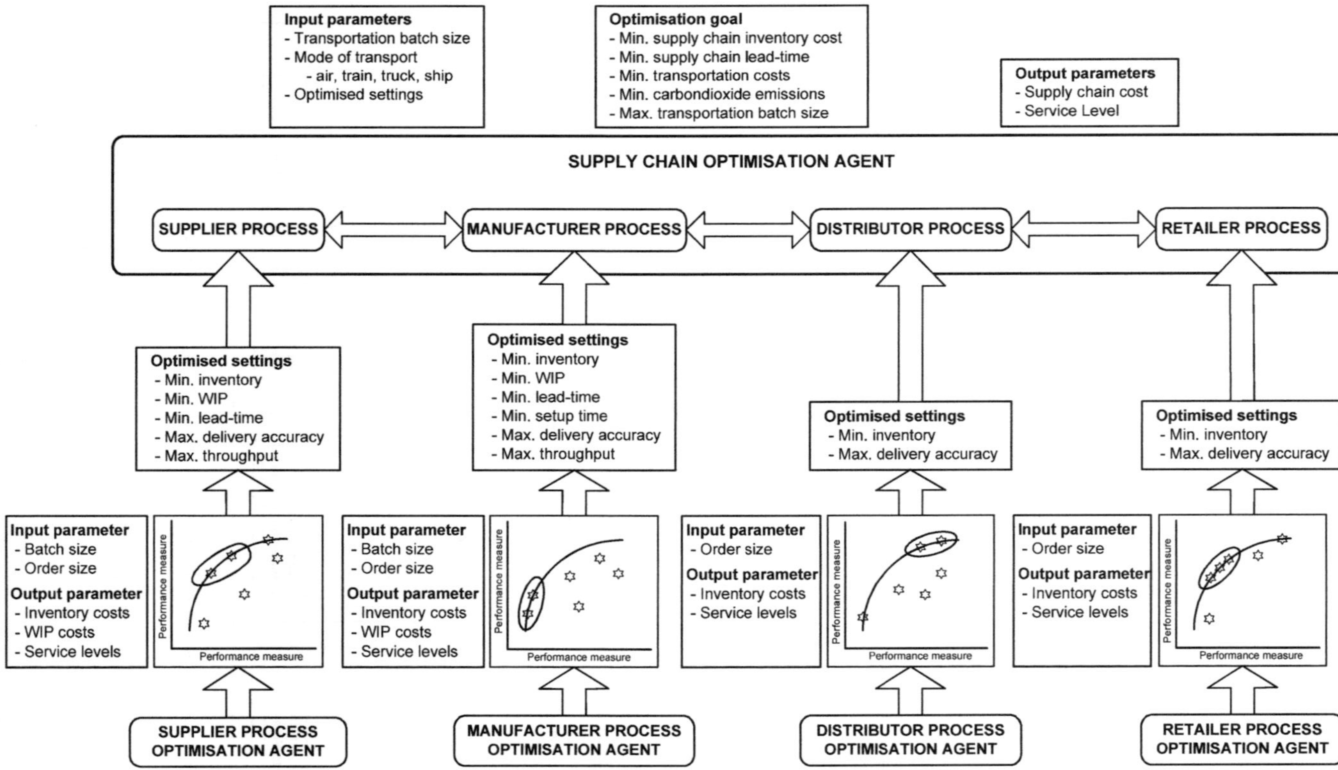

Figure 11.5. Multi-level optimisation for a supply chain

performed. The optimised process settings from the process optimisation agents are used as input parameters to the system agent optimisation. In the framework each process optimisation agent consists of a process that the agent needs to optimise and a set of objectives that the agent has to reach. Other process optimisation agents in the framework are configured similarly. The agents may have similar or different objectives in optimising their own processes, but the objectives are tied to the internal processes of individual agents. Nevertheless, all sub-system processes must have some kind of relationship with their downstream and upstream processes; this is a necessity to be able to perform system-level optimisation. Thus, in the process level that includes sub-system processes, it simulates all the sub-system processes simultaneously but in an individual manner; whereas in the operation level, the relationships between the sub-systems are simulated to achieve holistic optimisation of the overall system.

The basic rationale behind this multi-level architecture is that the simulations in the process level usually consist of models with very high fidelity while models with higher abstraction are commonly found in the operation level. In this case, the multi-level architecture can minimise the number of evaluations with the high-fidelity process models without reducing the accuracy of the simulation optimisation on the overall operation level.

To clarify the application of the proposed framework, one can look at the concept of a supply chain network. Based on the agent framework presented in Figure 11.4, one can see the process optimisation agents as entities of a supply chain, *i.e. supplier*, *manufacturer*, *distributor* and *retailer* (Figure 11.5), where each entity is defined by its internal process. As mentioned earlier, each entity has its own objectives to pursue. In the process level, for instance, the supplier and manufacturer have to consider their internal production, inventory and service levels. The optimisation objectives of the supplier agent might be to gain high delivery accuracy while minimising inventory levels and lead-time, and to maximise the throughput while minimising the work in process. The manufacturer would have the same internal objectives to peruse, but might also have to consider maximising the batch size to minimise the set-up time. While in the operation level, the overall system optimisation objectives are to minimise the overall supply chain costs by minimising inventory levels, lead-time and simultaneously achieving high delivery accuracy; the system agent would also consider minimising the transportation costs and minimising carbon dioxide emissions from transportation between the entities by maximising transportation batch sizes and minimising the amount of deliveries. The system agent is provided with optimal process settings from the process-optimisation agents. The input parameters that are sent to the system agent are the finished goods inventory levels, WIP and incoming goods inventory levels for each process-optimisation agent. Other input parameters for the system agent would be data regarding different transportation possibilities and their related vehicle size, cost and emissions.

As shown in Figure 11.6 and explained above, only the feasible solutions from the process-level optimisation will be considered and sent to the operation-level optimisation. In this way, all the process-level entities will only send their internal optimised solutions, containing the process settings that will be incorporated in the operation-level optimisation.

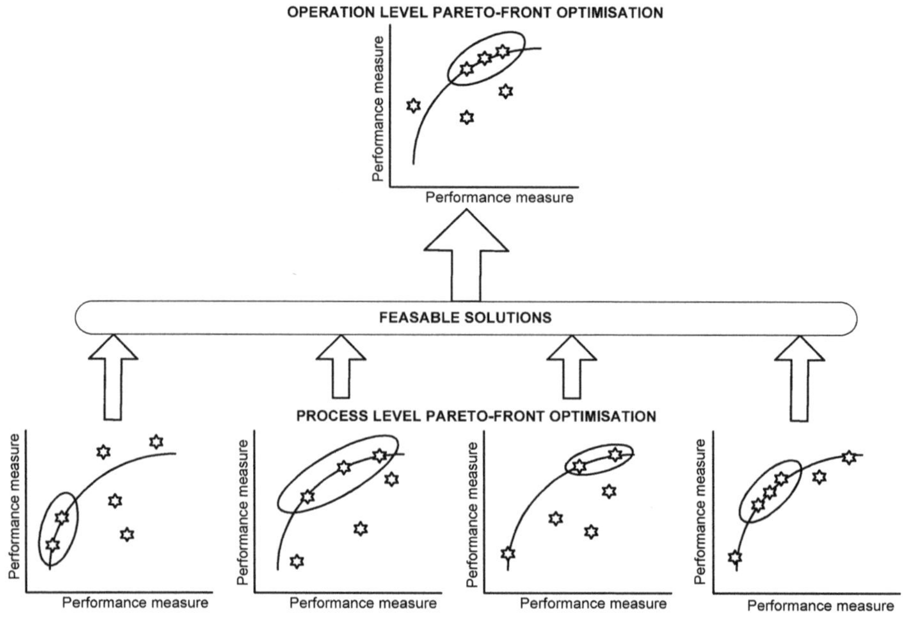

Figure 11.6. Multi-level optimisation solution

11.4 A Simple Case Study

In order to verify the agent-based framework, a simple supply chain simulation model containing three entities, *i.e.* supplier, manufacturer and distributor, has been developed. MOO has been run on the model using the previous outlined framework, in the sense that each entity has its own detailed simulation model on which the process-level optimisations are performed individually. The Pareto-optimal solutions generated from these process-level optimisation results are then transferred to the operation-level optimisation.

Figure 11.7 shows the manufacturing process at the supplier. There are two raw-materials inventories (RMIs), SRMI1 and SRMI2, which contain product *A* and product *B*, respectively. The raw material inventories are followed by two operations, SOP1 and SOP2, which are single machines. At the end of the line is SOP3, which is a transfer machine that needs setup between product changes. Before the setup for a new product starts, *i.e.* product *A* or product *B*, SOP3 needs to empty itself of the current processed product. After SOP3, the products continue to the finished-goods inventories (FGIs), SFGI1 and SFGI2, from where the customer demand is satisfied. The decision of what to manufacture in the production line or which product to release to SOP1 is based on the total WIP level – the total amount of product *A* and product *B*, maintained in the line from SOP1 to the FGIs. For instance, if there is more of product *A* in the system, then the model will start to release product *B* from the RMI until the batch size is reached and then the model checks the WIP level again. If there is the same amount of product *A* and product *B*, the model will continue to manufacture the current product to avoid extra setups on

SOP3. The supplier works only one 8-hour shift a day, 5 days a week, and the shift starts at 06:00 and ends at 14:00, including one short break and one lunch break. At the end of the shift, the products in the FGIs are sent to the customer to fulfil the demand. The supplier has a defined production schedule for 20 days based on the customer demand. However, there is variability in the customer demand. At the end of each shift, the customer might require more or less products from the supplier.

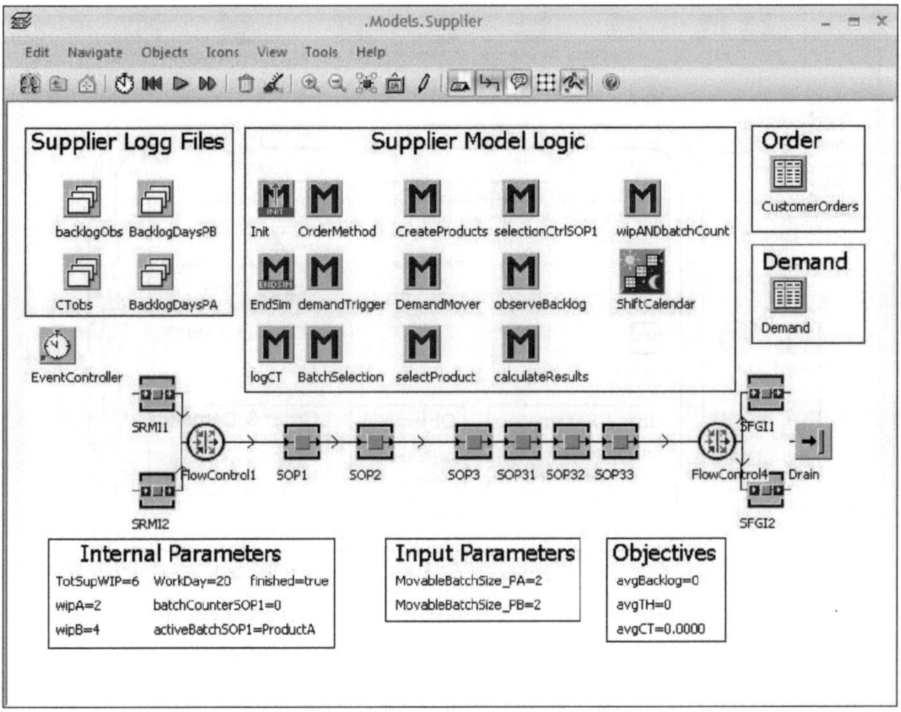

Figure 11.7. Process-level optimisation – supplier

The manufacturer (Figure 11.8) has a similar production management as the supplier regarding shifts, customer demand, scheduling and order release. The main difference between the supplier entity and the manufacturer entity lies in the manufacturing process. Within the manufacturer, there are nine single-machine operations: MOPIN, MOP1, MOP2, MOP3, MOP5, MOP6, MOP7, MOPOUT1 and MOPOUT 2. MOP4 is a transfer machine similar to SOP3 in the supplier process and needs setup when switching product type. There is a closed pallet loop in the manufacturer's process: the products are placed on pallets before MOP1 and are taken off after MOP7.

Figure 11.9 shows the last simulation entity, *i.e.* the distributor. The distributor has a sequential process consisting of five single operations: DOPIN1, DOPIN2, DOP, DOPOUT1 and DOPOUT2. There is no operation that needs setup, but operation DOP has different cycle time, depending on whether it is product *A* or product *B* that is being processed.

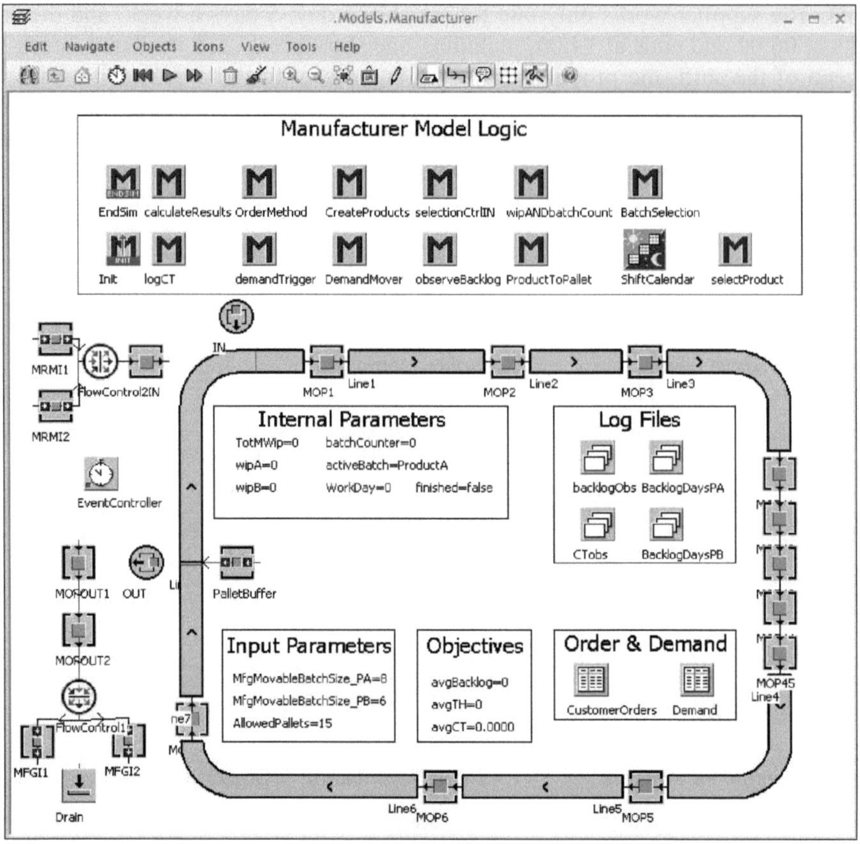

Figure 11.8. Process-level optimisation – manufacturer

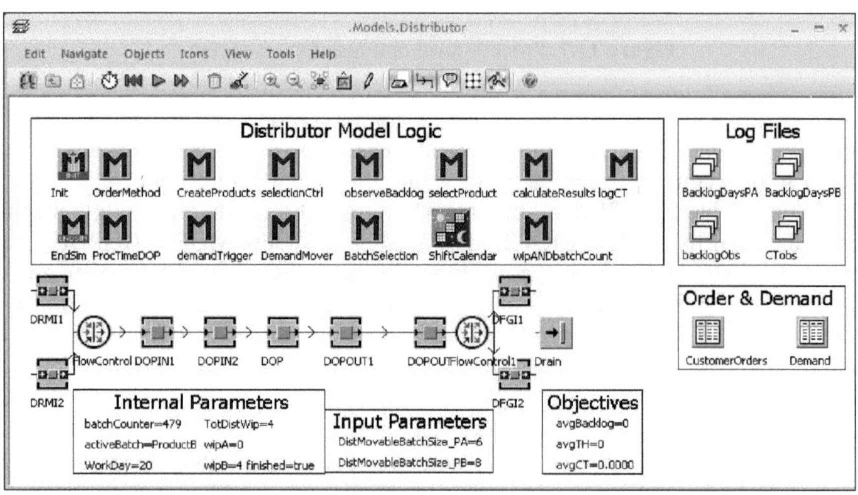

Figure 11.9. Process-level optimisation – distributor

Process-level optimisation has been run for each entity by an optimisation agent. There are three objectives in each agent-based MOO: (1) to maximise the average throughput, (2) to minimise the average customer backlog, and (3) to minimise the average cycle time for the products in the manufacturing processes. As mentioned earlier, each entity manufactures two products: product *A* and product *B*. The batch size for each of these products is an input (decision) variable for the optimisations. The manufacturer has a third input parameter that is the amount of pallets in the system. The simulation horizon for one simulation evaluation is 20 days, of which five days are warm-up time.

On the operation-level optimisation, the supplier, manufacturer and distributor entities are simulated simultaneously to perform an SC optimisation. Figure 11.10 shows the SC model that includes the above-mentioned entities with their internal manufacturing processes and internal system logic. In this model, the optimisation objective is to minimise the overall average customer backlog and average lead time for the SC and to maximise the average throughput for the whole SC. The input parameters for the SC optimisation are the batch sizes for product *A* and product *B* for each entity and the amount of pallets that can be used in the manufacturer's internal process. In this simple case study, no consideration has been made of the transportation between the entities – all products in the FGIs at each entity are moved to the RMIs of the subsequent entity at the end of each shift.

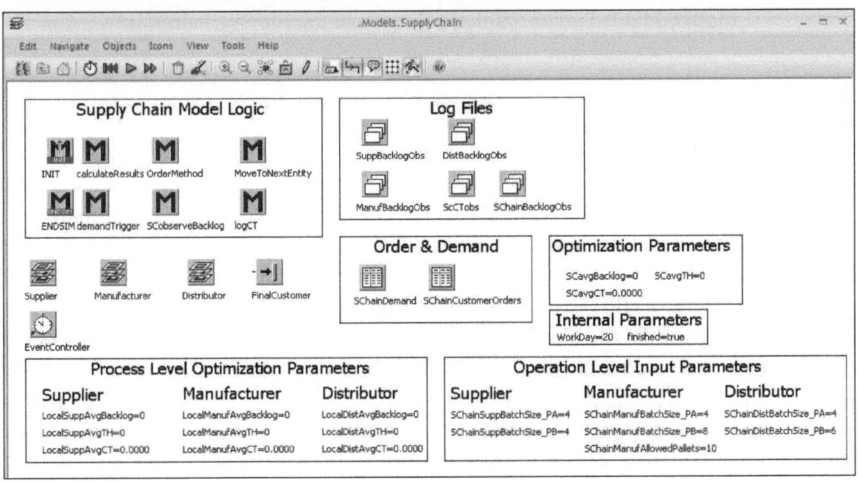

Figure 11.10. Operation-level optimisation – supply chain

Figure 11.11 shows the Pareto-optimal solutions generated at the operation level by the SC agent and plotted on the Backlog–Throughput space, based on the optimisation results gathered from the process-level optimisation. As a matter of fact, since the search space of this case study is not huge, the problem can be easily solved without an agent-based framework. Nevertheless, through such a simple supply chain case study, we have illustrated how the multi-level and multi-objective ABS architecture can be used to find the optimal batch sizes that can satisfy the objectives (throughput, cycle time and backlog as the optimisation objectives) at two levels. In the next step of our future work, the same framework and implementation will be tested with more complex models and/or real-world problems that involve more decision variables.

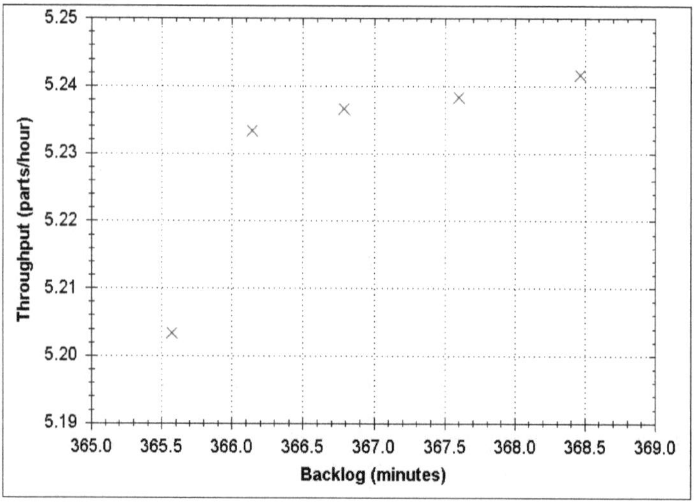

Figure 11.11. Pareto-optimal solutions in the supply chain level

11.5 Conclusions

In recent years, the concept of ABS has gained interest in the supply chain research community because of the similarity between supply chain participants (*e.g.* factories, customers, *etc.*) and agents in ABS models. Agents, as supply chain participants, act autonomously, pursuing their own goals and objectives, but they also display a social ability by communicating, coordinating and collaborating with each other and their environment to fulfil their stated goals and objectives. They also display reactivity and proactiveness by perceiving their environment and respond to changes in the environment, but at the same time they are also capable of pursuing their own goals by controlling their future in a proactive manner. All of the research studies reviewed in this chapter clearly show the applicability and advantages of ABS within the supply chain domain.

In this chapter, we have presented an agent-based environment for multi-level and multi-objective optimisation. With the multi-level architecture, simulation optimisation can be carried out at different levels using models with varying fidelity. Considerable reduction of computation time can, therefore, be achieved by avoiding a high fidelity 'holistic simulation'. Significant reduction in the search space of the overall system level can also be achieved by using the multi-objective optimisation algorithms with the intelligent agents so that only solutions lying on the Pareto fronts of the sub-system level are transferred to the optimisation process in the system level. The simple supply chain case study has illustrated how this multi-level and multi-objective ABS architecture can be used to find the optimal batch sizes at different levels, using throughput, cycle time and backlog as the optimisation objectives.

We believe that combining the characteristics of these techniques (Figure 11.12) for supply chain management is most beneficial. For the ABS, we have the resemblance of characteristics between an entity in a supply chain and those of an agent. Using MOO, we would be able to perform local optimisations for each entity in the supply chain and global optimisation for the entire supply chain, and MLO makes it possible to run high-fidelity process models that feed the overall system process (supply chain) with Pareto-optimal solutions.

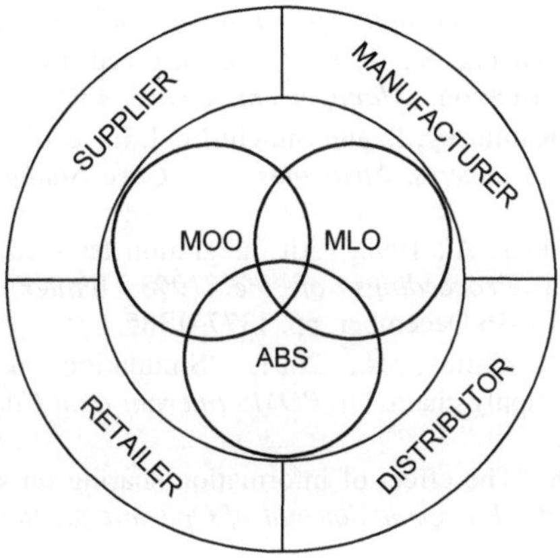

Figure 11.12. ABS, MOO and MLO for a supply chain

References

[11.1] Jung, H., Song, I. and Jeong, B., 2007, "Genetic algorithm-based integrated production planning considering manufacturing partners," *The International Journal of Advanced Manufacturing Technology*, **32**(5–6), pp. 547–556.

[11.2] Kim, C.-S., Tannock, J., Byrne, M., Farr, R., Cao, B. and Er, M., 2004, "State of the art review of techniques to model the supply chain in an extended enterprise," available at: http://www.vivaceproject.com/content/engine/tmscee_full.pdf (accessed December 8, 2009).

[11.3] Archibald, G., Karbakal, N. and Karlsson, P., 1999, "Supply chain vs. supply chain: using simulation to compete beyond the four walls," In *Proceedings of the 1999 Winter Simulation Conference*, Phoenix, AZ, USA, available at: http://www.informs-sim.org/wsc99papers/175.PDF (accessed December 9, 2009).

[11.4] Christopher, M.C., 2000, "The agile supply chain: competing in volatile markets," *Industrial Marketing Management*, **29**(1), pp. 37–44.

[11.5] Fox, M.S., Barbuceanu, M. and Teigen, R., 2000, "Agent-oriented supply-chain management," *International Journal of Flexible Manufacturing Systems*, **12**(2–3), pp. 165–188.

[11.6] Fung, R.Y.K. and Chen, T., 2005, "A multiagent supply chain planning and coordination architecture," *International Journal of Advanced Manufacturing Technology*, **25**(7–8), pp. 811–819.

[11.7] Kaihara, T., 2003, "Multi-agent based supply chain modelling with dynamic environment," *International Journal of Production Economics*, **85**(2), pp. 263–269.

[11.8] Swaminathan, J.M., Smith, S.F. and Sadeh, N.M., 1998, "Modeling supply chain dynamics: a multiagent approach," *Decision Sciences*, **29**(3), pp. 607–632.

[11.9] Iannone, R., Miranda, S. and Riemma, S., 2007, "Supply chain distributed simulation: an efficient architecture for multi-model synchronization," *Simulation Modelling Practice and Theory*, **15**(3), pp. 221–236.

[11.10] Swafford, P.M., Ghosh, S. and Muthy, N., 2008, "Achieving supply chain agility though IT integration and flexibility," *International Journal of Production Economics*, **116**(2), pp. 288–297.

[11.11] Liang, W.-Y. and Huang, C.-C., 2006, "Agent-based demand forecast in multi-echelon supply chain," *Decision Support Systems*, **42**(1), pp. 390–407.

[11.12] Lee, H.L. and Whang, S., 1999, "Decentralized multi-echelon supply chains: incentives and information," *Management Science*, **45**(5), pp. 633–640.

[11.13] Simchi-Levi, D., Kaminsky, P. and Simchi-Levi, E., 2000, *Designing and Managing the Supply Chain Concepts, Strategies, and Case Studies*, McGraw-Hill, Boston, MA.

[11.14] Umeda, S. and Jones, A., 1998, "An integration test-bed system for supply chain management," In *Proceedings of the 1998 Winter Simulation Conference*, Washington DC, 13–16 December, pp. 1377–1385.

[11.15] Chandra, C. and Chilov, N., 2001, "Simulation modeling for information management in a supply chain," In *POMS International Conference*, Orlando, FL, 30 March–2 April.

[11.16] Ouyang, Y., 2006, "The effect of information sharing on supply chain stability and the bullwhip effect," *European Journal of Operational Research*, **182**(3), pp. 1107–112.

[11.17] Zarandi, M.H., Pourkbar, M. and Turksen, I.B., 2007, "A fuzzy agent-based model for reduction of bullwhip effect in supply chain systems," *Expert Systems with Applications*, **34**(3), pp. 1680–1691.

[11.18] Lee, H.L., Padmanabhan, V. and Whang, S., 1997, "The bullwhip effect in supply chains," *Sloan Management Review*, available at: http://www.uta.edu/faculty/swafford/opma3308/BullWhipEffect.pdf (accessed December 8, 2009).

[11.19] Forrester, J.W., 1958, "Industrial dynamics – a major breakthrough for decision makers," *Harvard Business Review*, **36**(4), pp. 37–66.

[11.20] Forrester, J.W., 1961, *Industrial Dynamics*, MIT Press, Cambridge, MA.

[11.21] Baganha, M.P. and Cohen, M.A., 1998, "The stabilizing effect of inventory in supply chains," *Operations Research*, **46**(3), pp. 572–583.

[11.22] Kahn, J., 1987, "Inventories and the volatility of production," *American Economic Review*, **77**, pp. 667–678.

[11.23] Naish, H.F., 1994, "Production smoothing in the linear quadratic inventory model," *Economic Journal*, **104**(425), pp. 864–875.

[11.24] Ramey, V.A., 1991, "Nonconvex costs and the behaviour of inventories," *Journal of Political Economy*, **99**, pp. 306–334.

[11.25] Blanchard, O.J., 1983, "The production and inventory behaviour of the American automobile industry," *Journal of Political Economy*, **91**(3), pp. 365–400.

[11.26] Blinder, A.S., 1982, "Inventories and sticky prices," *American Economic Review*, **72**, pp. 334–349.

[11.27] Lee, H.L., Padmanabhan, V. and Whang, S., 2004, "Information distortion in a supply chain: the bullwhip effect," *Management Science*, **50**(12), pp. 1875–1886.

[11.28] Disney, S.M. and Towill, D.R., 2003, "On the bullwhip and inventory variance produced by an ordering policy," *Omega*, **31**(3), pp. 157–167.

[11.29] Disney, S.M., Farasyn, I., Lambrecht, M., Towill, D.R. and Van de Velde, W., 2006, "Taming the bullwhip effect whilst watching customer service in a single supply chain echelon," *European Journal of Operational Research*, **173**(1), pp. 151–172.

[11.30] Potter, A. and Disney, S.M., 2006, "Bullwhip and batching: an exploration," *International Journal of Production Economics*, **104**(2), pp. 408–418.

[11.31] Zhang, C. and Zhang, C., 2007, "Design and simulation of demand information sharing in a supply chain," *Simulation Modelling Practice and Theory*, **15**(1), pp. 32–46.

[11.32] Chen, F., Drezner, Z., Ryan, J.K. and Simichi-Levi, D., 2000, "Quantifying the bullwhip effect in a simple supply chain: the impact of forecasting, lead times and information," *Management Science*, **46**(3), pp. 436–443.

[11.33] Li, G., Whang, S. and Yu, H., 2005, "Information transformation in a supply chain," *Computers and Operations Research*, **35**(3), pp. 707–725.

[11.34] Dejonckheere, J., Disney, S.M., Lambrecht, M. and Towill, D.R. 2004, "The impact of information enrichment on the bullwhip effect in supply chains: a control engineering perspective," *European Journal of Operational Research*, **153**(3), pp. 727–750.

[11.35] Zhang, X., 2004, "The impact of forecasting methods on the bullwhip effect," *International Journal of Production Economics*, **88**(1), pp. 15–27.

[11.36] Van Ackere, A., Larsen, E.R. and Morecroft, J.D.W., 1993, "Systems thinking and business process redesign: an application to the beer game," *European Management Journal*, **11**(4), pp. 412–423.

[11.37] De Treville, S., Shapiro, R.D. and Hameri, A.-P., 2004, "From supply chain to demand chain: the role of lead time reduction in improving demand chain performance," *Journal of Operations Management*, **21**(6), pp. 613–627.

[11.38] Liang, W.-Y. and Huang, C.-C., 2006, "Agent-based demand forecast in multi-echelon supply chain," *Decision Support Systems*, **42**(1), pp. 390–407.

[11.39] Fu, Y., Piplani, R., de Souza, R. and Wu, J., 2000, "Multi-agent enabled modeling and simulation towards collaborative inventory management in supply chains," In *Proceedings of the 2000 Winter Simulation Conference*, http://www.informs-sim.org/wsc00papers/242.PDF.

[11.40] Jennings, N.R., Sycara, K. and Wooldridge, M., 1998, "A roadmap of agent research and development," *Autonomous Agents and Multi-Agent Systems*, **1**(1), pp. 7–38.

[11.41] Lim, M.K. and Zhang, Z., 2003, "A multi-agent based manufacturing control strategy for responsive manufacturing," *Journal of Materials Processing Technology*, **193**(1–3), pp. 379–384.

[11.42] Choi, T.Y., Dooley, K.J. and Rungtusanatham, M., 2001, "Supply networks and complex adaptive systems: control versus emergence," *Journal of Operations Management*, **19**(3), pp. 351–366.

[11.43] McCarthy, I.P., 2004, "Manufacturing strategy: understanding the fitness landscape," *International Journal of Operations & Production Management*, **24**(2), pp. 124–150.

[11.44] Nilsson, F. and Darley, V., 2006, "On complex adaptive systems and agent-based modeling for improving decision-making in manufacturing and logistics settings," *International Journal of Operations & Production Management*, **26**(12), pp. 1351–1373.

[11.45] Nilsson, F., 2005, "Adaptive logistics – using complexity theory to facilitate increased effectiveness in logistics," *PhD Thesis*, Lund University, Lund.

[11.46] Tweedale, J., Ichalkaranje, N., Sioutis, C., Jarvis, B., Consoli, A. and Phillips-Wren, G., 2007, "Innovations in multi-agent systems," *Journal of Network and Computer Applications*, **30**(3), pp. 1089–1115.

[11.47] Chaib-Draa, B., Moulin, B., Mandiau, R. and Millot, P., 1992, "Trends in distributed artificial intelligence," *Artificial Intelligence Review*, **6**, pp. 35–66.

[11.48] Chaib-Draa, B., 1995, "Industrial applications of distributed artificial intelligence," *Communications of the ACM*, **38**(11), pp. 49–53.

[11.49] Hilletofth, P., Aslam, T. and Hilmola, O.-P., 2009, "Multi-agent based supply chain management: case study of requisites," *International Journal of Networking and Virtual Organisations*, (accepted).

[11.50] Wooldridge, M. and Jennings, N.R., 1995, "Intelligent agents: theory and practice," *The Knowledge Engineering Review*, **10**(2), pp. 115–152.

[11.51] Monostori, L., Váncza, J. and Kumara, S.R.T., 2006, "Agent-based systems for manufacturing," *CIRP Annals – Manufacturing Technology*, **55**(2), pp. 697–720.

[11.52] Macal, C.M. and North, M.J., 2005, "Tutorial on agent-based modeling and simulation," In *Proceedings of the 2005 Winter Simulation Conference*, http://www.informs-sim.org/wsc05papers/002.pdf.

[11.53] Sanchez, S.M. and Lucas, T.W., 2002, "Exploring the world of agent-based simulations: simple models, complex analyses," In *Proceedings of the 2002 Winter Simulation Conference*, San Diego, CA, 8–11 December, http://www.informs-sim.org/wsc02papers/015.pdf.

[11.54] Moyaux, T., Chaib-Draa, B. and D'Amours, S., 2004, "Multi-agent simulation of collaborative strategies in a supply chain," In *Proceedings of 3rd International Joint Conference on Autonomous Agents and Multi Agent Systems*, ACM Press, New York.

[11.55] Bruccoleri, M., Lo Nigro, G., Federico, F., Noto La Diega, S. and Perrone, G., 2003, "Negotiation mechanisms for capacity allocation in distributed enterprises," *Annals of the CIRP*, **52**(1), pp. 397–400.

[11.56] Bruccoleri, M., Lo Nigro, G., Perrone, G. and Noto La Diega, S., 2005, "Production planning in reconfigurable enterprises and reconfigurable production systems," *Annals of the CIRP*, **54**(1), pp. 433–436.

[11.57] Koussis, K., Pierreval, H. and Mebarkim, N., 1997, "Using multi-agent architecture in FMS for dynamic scheduling," *Journal of Intelligent Manufacturing*, **8**(1), pp. 41–47.

[11.58] Lu, T. and Yih, Y., 2001, "An agent-based production control framework for multiple-line collaborative manufacturing," *International Journal of Production Research*, **39**(10), pp. 2155–2176.

[11.59] Jarvis, D.H. and Jarvis, J.H., 2003, "Holonic diagnosis for an automotive assembly line," In *Agent-Based Manufacturing: Advances in the Holonic Approach*, Deen, S.M. (ed.), Springer, Berlin, pp. 193–206.

[11.60] Sauer, O., 2004, "Modern production monitoring in automotive plants." In *Proceedings of the FISITA World Automotive Congress*, Barcelona, 23–27 May.

[11.61] Chen, Y., Peng, Y., Finin, T., Labrou, Y. and Cost, S.R., 1999, "Negotiating agent for supply chain management," http://www.csee.umbc.edu/~ypeng/Publications/1999/Aaai99-YChen.pdf,.

[11.62] Julka, N., Karimi, I. and Srinivasan, R., 2002, "Agent-based supply chain management: framework and a refinery application," *Computers and Chemical Engineering*, **26**, pp. 1755–1769.

[11.63] Lou, P., Zhou, Z.D., Chen, Y.P. and Ai, W., 2004, "Study on multi-agent-based agile supply chain management," *International Journal of Advanced Manufacturing Technology*, **23**, pp. 197–203.

[11.64] Mondal, S. and Tiwari, M., 2003, "Formulation of mobile agents for integration of supply chain using the KLAIM concept," *International Journal of Production Research*, **41**(1), pp. 97–119.

[11.65] Kádár, B. and Monostori, L., 1998, "Agent-based control of novel and traditional production systems," In *Proceedings of ICME98, CIRP International Seminar on*

Intelligent Computation in Manufacturing Engineering, Capri, Italy, 1–3 July, pp. 31–38.

[11.66] Ito, T. and Abadi, S., 2002, "Agent-based material handling and inventory planning in warehouse," *Journal of Intelligent Manufacturing*, **13**(3), pp. 201–210.

[11.67] Sterman, J., 1989, "Modelling managerial behaviour: misperceptions of feedback in a dynamic decision making experiment," *Management Science*, **35**, pp. 321–339.

[11.68] Fjeld, D.E., 2001, "The wood supply game as an educational application for simulating dynamics in the forest sector," In *Supply Chain Management for Paper and Timber Industries*, Växjö, Sweden, pp. 241–251.

[11.69] Kimbrough, S., Wu, D.J. and Zhong, F., 2001, "Computers play the Beer Game: can artificial agents manage supply chains," In *Proceedings of the 34th Hawaii International Conference on System Sciences*, Hawaii.

[11.70] Moyaux, T., Chaib-Draa, B. and D'Amours, S., 2003, "Agent-based simulation of the amplification of demand variability in a supply chain," In *Agent Based Simulation 4*, Montpellier, France, 28–30 April.

[11.71] Moyaux, T., Chaib-Draa, B. and D'Amours, S., 2003, "Multi-agent coordination based on tokens: reduction of the bullwhip effect in a forest supply chain," In *Proceedings of the Second International Joint Conference on Autonomous Agents and Multiagent Systems*, Melbourne, Australia, 14–18 July, pp. 670–677.

[11.72] Moyaux, T., 2004, *Design, Simulation and Analysis of Collaborative Strategies in Multi-Agent Systems: The Case of Supply Chain Management*, Department of Computer Science and Software Engineering, Laval University.

[11.73] Moyaux, T., Chaib-Draa, B. and D'Amours, S., 2007, "Information sharing as a coordination mechanism for reducing the bullwhip effect in a supply chain," *IEEE Transactions on Systems, Man, and Cybernetics, Part C*, **37**(3), pp. 396–409.

[11.74] Yung, S.K. and Yang, C.C., 1999, "A new approach to solve supply chain management problem by integrating multi-agent technology and constraint network," In *Proceedings of the 32nd Hawaii International Conference on System Sciences*, University of Hong Kong.

[11.75] K. Deb, 2001, *Multi-Objective Optimization Using Evolutionary Algorithms*, Wiley, Chichester.

12

Analysing Interactions among Battery Recycling Barriers in the Reverse Supply Chain

P. Sasikumar and A. Noorul Haq

Department of Production Engineering, National Institute of Technology
Tiruchirappalli, 620 015, India
Emails: sasi_me75@yahoo.com; anhaq@nitt.edu

Abstract

Because of growing environmental concerns and possible cost reductions in the total supply chain, original equipment manufacturers are under pressure to take back their used or end-of-life (EOL) products through reverse supply chain systems. Recycling is widely accepted as a sustainable supply chain management method because of its potential to reduce disposal costs and waste transport costs, and to prolong the lifespan of sanitary landfill sites. Individuals recycle for various reasons, but the basic principle is that of environmental concerns. For increasing participation in recycling, it is necessary to understand what motivates people to recycle and what discourages them. It involves a complex chain of behaviours that involves government legislation, financial support, local governmental support through policy decisions, education, and distribution of information and services that encourage recycling. The main objective of this research work is to identify the major barriers facing a battery recycling system and to analyse the interaction among these barriers. For this purpose, an interpretive structural modelling (ISM) approach is used to understand the mutual influences among the barriers so that driving barriers, which can aggravate other barriers, and independent barriers, which are most influenced by driving barriers, can be identified. By analysing the barriers using this model, we may find the crucial barriers that hinder the recycling activities.

12.1 Introduction

The reverse supply chain (RSC) involves the movement of used products from customers to manufacturers or suppliers, for possible recycling and reuse. The existence, effectiveness and efficiency of service management activities such as repair services and value recovery depend heavily on effective reverse logistics operations. Reverse logistics is defined as 'the process of planning, implementing and controlling the efficient, cost-effective flow of materials and related information from the point of consumption to the point of origin for the purpose of recapturing value or proper disposal' [12.1]. Reverse logistics encompasses activities of processing and transporting EOL products from the end user to the manufacturer with the goals of maximising value from the returned item or minimising the total reverse logistics cost.

Carter and Ellram [12.2] emphasised the environmental aspect of reverse logistics, which they defined as the 'process whereby companies could become more environmentally efficient through recycling, reusing, and reducing the amount of materials used'. Beullens [12.3] described some important frameworks, models and insights of reverse logistics that had been developed in recent years. Prahinski and Kocabasoglu [12.4] reviewed the literature in RSC and developed ten research propositions to be studied using empirical research methods.

Due to the revolution in green manufacturing for the global market, reverse logistics concepts have become an important issue that can play a pivotal role in a company's competitive advantage and help strategic decision making. Srivastava [12.5] classified the green supply chain management (GSCM) literature into three broad categories: literature highlighting the importance of GSCM; literature on green design; and literature on green operations. Rubio *et al.* [12.6] analysed the main characteristics of articles on reverse logistics in the production and operations management field. Sasikumar and Kannan [12.7] presented two classification schemes and a simple analysis for the RSC. The first classification scheme is based on the content related issues on RSC and the second is based on the solution methodology.

Reverse logistics is practiced in many industries, such as steel, aircraft, computers, cellular phones, photocopiers, single-use cameras, automobiles, plastics, refillable containers, carpets, paper, chemicals, appliances and pharmaceuticals. Now, companies realised that RSC should be integrated with the forward supply chain (*i.e.* closed-loop supply chain or CLSC) to reduce the overall supply chain costs and also to meet the environmental regulations. Thierry *et al.* [12.8] presented a CLSC framework for product recovery activities such as repair, reuse, refurbish, remanufacture and recycle. Dekker *et al.* [12.9] reviewed multi-echelon reverse logistics network models for CLSC.

Due to government regulations, market requirements and the hidden economic value of solid waste, recovery of used products has become a field of rapidly growing importance in RSC management. Moyer and Gupta [12.10] conducted a comprehensive survey of works related to environmentally conscious manufacturing practices, recycling, and the complexities of disassembly in the electronics industry. Gungor and Gupta [12.11] also presented the development of research in environmentally conscious manufacturing and product recovery and provided a state-of-the-art survey of the published work in that area. Products are returned from the consumer to the original supply point, for various reasons, which include warranty returns, end-of-use returns, commercial returns, bad delivery, over-supply, damage, expiry, failing inspection tests at the customer point, products unsold, *etc.*

Johnson and Wang [12.12] defined product recovery as a combination of remanufacturing, reuse and recycling. Product recovery aims to minimise the amount of waste sent to landfills by recovering the materials and parts from old or outdated products by means of reuse, recycling and remanufacturing [12.13]. Among these, recycling is a very important field of product recovery because it not only protects the environment, but also saves natural resources, energy, landfill space and money on raw materials. Recycling can take place during the production process itself or after the product's life. Recycling seeks to recover the material content of returned products by performing the necessary disassembly, sorting and reprocessing operations. Examples of recycling include: plastics [12.14], paper [12.15], glass [12.16], metal from scrap [12.17–12.20], fibre optic cables [12.21], sand [12.22], electronic waste [12.23], carpets [12.24–12.27] and batteries [12.28–12.35].

The present chapter focuses on the recycling of lead-acid batteries, since used lead-acid batteries may be considered as hazardous waste because of their corrosivity, reactivity or toxicity and also because of the presence of heavy metals such as lead, mercury and cadmium. Battery manufacturers are responsible for the implementation of fiscal policies and for any infrastructure development for the collection, storage, transportation and processing of used batteries. Since the birth of the motor car, lead-acid batteries are used for starting, lighting and ignition (SLI) purposes in automobiles and trucks, as well as providing power for automobiles, forklifts and submarines [12.35].

The increased use of lead-acid batteries will further increase the demand for lead, and to meet this increasing demand for lead for new battery manufacturing, used batteries have been identified as an important source of lead through recycling. Since recycled lead is a costly commodity, the market potential for reclaiming the secondary lead from the used batteries has been growing, whilst at the same time these lead-acid batteries are generally having a shorter service life. Espinosa *et al.* [12.36] analysed the environmental laws for battery recycling in Brazil and presented some suggestions for other countries in order to manage this kind of dangerous waste. Andrews *et al.* [12.37] described the latest technology in the recycling of secondary lead to be used as raw material for lead industries, and Bernardes *et al.* [12.38] presented the status of the technologies involved in the collection, sorting and processing of portable batteries.

The main components of a lead-acid battery are [12.28]:

1. active mass:
   - anode (negative electrode) consisting of PbO_2;
   - cathode (positive electrode) consisting of Pb;
2. metallic grids, metallic connections;
3. electrolyte (aqueous solution of H_2SO_4);
4. polypropylene casing (box);
5. other components (wood, paper, PVC).

To realise the potential benefits of battery recycling, management needs to consider appropriate options for recycling programmes with regard to financial constraints, the existing situations, environmental regulations, and socio-cultural and technical issues. However, the success of recycling will depend not only on participation levels in recycling programmes or the effectiveness of the programmes, but also on the efficiency of such a programme. In order to achieve sustainable supply chain management, it is essential to identify and understand the barriers in battery recycling.

The main objectives of this chapter can be stated as:

(i) to identify the major barriers of battery recycling program;
(ii) to establish a contextual relationship between the variables;
(iii) to develop an ISM model to analyse the interactions among these barriers; and
(iv) to test the model with a case study.

This chapter is organised as follows. Section 12.2 presents a survey of previous work. Section 12.3 provides a description of battery recycling barriers. Next, Section 12.4 discusses the solution methodology for developing the ISM model. Application of the model to the case study and the ISM model is provided in Sections 12.5 and 12.6, respectively. MICMAC analysis of the recycling barriers is presented in Section 12.7. Finally, Section 12.8 summarises the work presented in this chapter.

12.2 Survey of Previous Work

Bloemhof-Ruwaard *et al.* [12.39] elaborated on the possibilities of incorporating green issues when analysing industrial supply chains and more generally on the value of using operations research (OR) models and techniques in GSCM research. Van Hoek [12.40] presented a categorisation of green approaches and suggested the value-seeking approach as the most relevant in greening the supply chain. Zhu and Sarkis [12.41] examined the relationships between GSCM practice as well as environmental and economic performance. They evaluated the general relationships between specific GSCM practices and performance using moderated hierarchical regression analysis. Georgiadis and Vlachos [12.42] examined the impact of environmental issues on the long-term behaviour of a single product supply chain with product recovery. The behaviour of the system was analysed through a dynamic simulation model based on the principles of the system dynamics methodology. Sheu *et al.* [12.43] formulated a linear multi-objective programming model that systematically optimised the operations of both integrated logistics and corresponding used-product reverse logistics in a given green supply chain. Factors such as the used-product return ratio and corresponding subsidies from a governmental organisation for reverse logistics were considered in the model formulation. Vlachos *et al.* [12.44] tackled the development of efficient capacity planning policies for remanufacturing facilities in reverse supply chains, taking into account not only economic but also environmental issues, such as the take-back obligation imposed by legislation and the 'green image' effect on customer demand.

Carlson [12.45] used weighted non-linear goal programming to discuss the economic impacts of material recycling on energy recovery facilities. Pohlen and Farris [12.14] identified a number of fundamental functions, including collection, separation, transitional processing, delivery and integration, within a typical reverse logistics channel in a plastic recycling case. Spengler *et al.* [12.18] discussed two cases, one for recycling building debris and one for the recycling of by-products in German steel industry. Johnson [12.46] described the reverse logistics systems for ferrous scrap in twelve North American manufacturing plants; examined the role of purchasing and other functions in the reverse logistics system; and assessed the contribution made by various departments. Barros *et al.* [12.22] proposed a two-level location model for a sand recycling problem and considered its optimisation using heuristic procedures. They formulated a mixed integer linear programming model to minimise the total cost of the network.

Krikke *et al.* [12.47] discussed a PC-monitor recycling case as a part of a broader pilot project at Roteb (the municipal waste company of Rotterdam in The Netherlands). They applied a two-step procedure for optimising a recovery strategy for durable consumer products in a multi-product situation. Tsoulfas *et al.* [12.33] performed an environmental analysis of the used SLI batteries sector, based on the logistics involved in the recovery process, and measured the environmental impact of such a process using life cycle analysis (LCA). Boon *et al.* [12.48] investigated the critical factors influencing the profitability of EOL processing of PCs. They also suggested suitable policies for both PC manufacturers and legislators to ensure that there was a viable PC recycling infrastructure. Khoei *et al.* [12.19] used the Taguchi method to optimise recycling processes and Hoyle [12.49] used technical–economical constraints analysis for the case of aluminium recycling. Degher [12.50] reported the take-back and recycling programmes at Hewlett-Packard Ltd and concluded that electronic manufacturers and government agencies should work together to better provide customers with environmentally responsible take-back and recycling programmes.

More recently, Spicer and Johnson [12.51] proposed the concept of *third-party demanufacturing*, which was defined as an extended producer responsibility approach in which private companies take up EOL responsibility for products on behalf of the original equipment manufacturers. Bufardi *et al.* [12.52] proposed a multi-criteria decision-aid approach to aid the decision maker in selecting the best compromise EOL alternative on the basis of his/her preferences and the performances of EOL alternatives with respect to the relevant environmental, social and economic criteria. Ravi *et al.* [12.53] presented an analytic network process (ANP)-based decision model to structure the problem of the conduct of reverse logistics for EOL computers in a hierarchical form and linked the determinants, dimensions and enablers of the reverse logistics, and the alternatives available to the decision maker for a computer industry.

Nagurney and Toyasaki [12.23] developed an integrated framework for the modelling of electronic waste RSC management. They formulated a multi-tiered, e-cycling network model (with the objective of profit maximisation), consisting of sources of electronic waste, recyclers, processors and consumers associated with the demand markets for the distinct products.

Listes and Dekker [12.54] presented a stochastic programming based approach to a case study on recycling sand from demolition waste in The Netherlands, by which a deterministic location model for product recovery network design was extended to explicitly account for the uncertainties. Wright *et al.* [12.21] illustrated how to improve the recyclability of fibre optic cable in a practical and relatively easy way, leading to both environmental and economic benefits. Pati *et al.* [12.55] presented a linear optimisation model for the paper industry to compare the total system cost of wood, as a raw material, with the recycling of waste paper. Bian and Yu [12.56] analysed various countries in the Asia Pacific region to determine their suitability in carrying out reverse logistics operations for an international electrical manufacturer using an analytic hierarchy process (AHP). Staikos and Rahimifard [12.57] also used AHP as a decision-making model to identify the most appropriate reuse, recovery and recycling option for post-consumer shoes.

Pagell *et al.* [12.58] presented a framework that highlighted the supply chain implications for firms forced into EOL product management where recycling was the only viable option. Queiruga *et al.* [12.59] applied PROMETHEE as a multi-criteria decision-making method for the selection of good alternatives for potential locations of WEEE recycling plants in Spain. Pati *et al.* [12.15] formulated a mixed integer goal programming model to assist in the proper management of the paper recycling logistics system in India and studied the inter-relationship between the multiple objectives of a recycled paper distribution network. In Ravi *et al.* [12.60], a combination of ANP and zero one goal programming was used as solution methodologies to deal with the problem related to the selection of feasible reverse logistics for EOL alternatives.

Kannan *et al.* [12.61] used AHP and fuzzy AHP as a multi-criteria decision-making (MCDM) model for selecting the collecting centre location in the reverse logistics. Gomes *et al.* [12.62] used THOR as a multi-criteria decision support system for ranking the alternatives and presented two cases where the decision makers had different preferences concerning the environmental investments at stake. In the first case, different methods of disposing of plastic waste were evaluated, while in the second, construction and demolition waste recycling facilities were submitted to a performance evaluation. Wadhwa *et al.* [12.63] made an attempt to bring fuzzy-based flexible MCDM and reverse logistics together as a well-suited group decision support tool for alternative selections. Kannan *et al.* [12.35] developed a multi-echelon, multi-period, multi-product closed-loop supply chain network model for the case of battery recycling, and decisions were made regarding material procurement, production, distribution, recycling and disposal using a genetic algorithm.

In summary, the knowledge gap revealed in the previous work on the reverse supply chain is the lack of analysis of the interactions among the battery recycling barriers. Therefore, the aim of this work is to identify the major barriers of battery recycling and present an ISM model for analysing the interactions among these barriers.

12.3 Description of Recycling Barriers

Disposal of used batteries is a serious environmental issue faced both by the government and by battery manufacturing industries. Increased attention has been given by the government in recent years to handle this problem in a safe and hygienic manner, and recycling has been identified as a viable option to tackle this problem. Also, a high level of battery recycling will be extremely useful in reducing the amount of lead dumped in the environment. Thus, it is very important to identify the battery recycling barriers and understand the mutual relationships among these barriers. The battery recycling barriers described in Table 12.1 are identified from the literature and the opinions of experts in the industry.

Table 12.1. Descriptions of the battery recycling barriers

Barriers	Descriptions
1. Lack of customer interest/motivation	Not getting an exchange offer (*i.e.* cash discount price) for used batteries returned; not getting a personal motivational reward from recycling
2. Lack of storage space	Space constraints often make recycling prohibitive for some establishment
3. Lack of knowledge	Not understanding the real benefits of recycling (*i.e.* not accepting that there is an environmental and economical benefit)
4. Inadequate collection points	Participation in recycling was impacted by the presence of number of collection points. In order to enhance the level of customer service, it is necessary to provide an adequate number of collection points
5. Inconvenient collecting centres	Poor participation in recycling schemes was usually a result of inconvenient collecting centres. There is a need to provide convenient collection facilities for easy access to customer locations
6. Lack of effective communication	The manufacturing company should pay particular attention to effective communications with distributors, retailers and customers. Use media (television, radio and newspaper) advertising; regular leaflets for effective communication about the potential benefits of battery recycling programs. The more that people see recycling as effective, the more likely they are to participate
7. Lack of political will	Fiscal policies can make a significant contribution to the successful implementation of a battery recycling program, as was recognised by the implementation of landfill tax. Treatments other than reduce, re-use and recycling should attract a tax
8. Lack of suitable recycling plant sites	There are often conflicts between citizens and local body officials with regard to the site of recycling facility
9. Vehicle access problems	This problem is mainly because of the distance between the collection points and the recycling plant; time; and labour
10. Financial constraints	Unwilling/unable to provide recycling services at a reasonable cost; stating recycling markets are unstable/unprofitable. The breakdown of expenditure is not always clear, even by practitioners. Proper finances and system-to-system coordination is an important factor

12.4 Interpretive Structural Modelling

Interpretive structural modelling (ISM) is proposed as a solution methodology to analyse the interactions among the battery recycling barriers. This section explains the details and various steps involved in the ISM methodology. ISM is an interactive learning process in which a set of different but directly related elements are structured into a comprehensive systemic model. The model thus formed portrays the structure of a complex issue or problem, system or field of study in a carefully designed pattern implying graphics as well as words. The basic idea of ISM is to use experts' practical experience and knowledge to decompose a complicated system into several sub-systems (elements) and construct a multi-level structural model. The ISM methodology helps to impose order and direction on the complexity of relationships among the elements of a system [12.64, 12.65].

Saxena *et al.* [12.66, 12.67] have presented the results of an ISM application to 'energy conservation in the Indian cement industry', and have developed direct relationship matrices for the key factors, objectives, and activities for energy conservation. Mandal and Deshmukh [12.68] used the ISM methodology to analyse some of the important criteria on vendor selection and have shown the inter-relationships for the criteria and their levels. These criteria have also been categorised depending on their driver power and dependence. Sharma *et al.* [12.69] used the ISM methodology to develop a hierarchy of actions that are required to achieve the future objectives of waste management in India. Singh *et al.* [12.70] used the ISM methodology to categorise variables for implementing knowledge management in manufacturing industries. Ravi *et al.* [12.53] employed an ISM-based approach to model the reverse logistics variables typically found in computer hardware supply chains. These variables had been categorised under *enablers* and *results*. The main objectives were: to identify and rank the variables of reverse logistics activities in the computer hardware industry; to find out the interaction among the variables identified; and to understand the managerial implications of this research.

Huang *et al.* [12.71] proposed a method of integrating ISM and ANP to analyse subsystems' interdependence and feedback relationships. Ravi and Shankar [12.72] identified eleven barriers to reverse logistics in the automobile industry and used the ISM methodology to analyse the interaction among these barriers. Kannan and Haq [12.73] used the ISM methodology to analyse the interactions among the criteria and sub-criteria that influence supplier selection in the built-to-order supply chain environment. Kannan *et al.* [12.74] analysed the interaction of criteria that were used to select the green suppliers who addressed the environmental performance using ISM, and the effectiveness of the model was illustrated using an automobile company.

Singh and Kant [12.75] also used the ISM methodology to evolve mutual relationships among the identified knowledge management barriers. Vivek *et al.* [12.76] used the same ISM methodology to represent the interrelationships among core, transaction and relationship specific investments for the case of offshoring. Raj *et al.* [12.77] used the ISM approach to understand the mutual interaction of the enablers that help in the implementation of flexible manufacturing systems (FMS) and identify the *driving enablers* (*i.e.* that influence the other enablers) and the *dependent enablers* (*i.e.* that are influenced by others). Kannan *et al.* [12.78] used ISM and fuzzy TOPSIS as a hybrid approach for the analysis and selection of a third-party reverse logistics provider.

The ISM methodology is interpretive from the fact that the judgment of the group decides whether and how the variables are related. It is structural, too, on the basis of relationship; an overall structure is extracted from the complex set of variables. It is a modelling technique in which the specific relationships of the variables and the overall structure of the system under consideration are portrayed in a digraph model. ISM is primarily intended as a group learning process, but it can also be used individually.

The steps involved in ISM are shown in Figure 12.1 and summarised below:

Step 1: Variables considered for the system under consideration are listed.
Step 2: From the variables identified in step 1, a contextual relationship is established among the variables with respect to which pairs of variables would be examined.
Step 3: A structural self-interaction matrix (SSIM) is developed for variables, which indicates pair-wise relationships among variables of the system under consideration.
Step 4: A reachability matrix is developed from the SSIM and the matrix is checked for transitivity. The transitivity of the contextual relation is a basic assumption made in ISM. It states that if a variable *A* is related to *B* and *B* is related to *C*, then *A* is necessarily related to *C*.
Step 5: The reachability matrix obtained in step 4 is partitioned into different levels.
Step 6: Based on the relationships given above in the reachability matrix, a directed graph is drawn and the transitive links are removed.
Step 7: The resultant digraph is converted into an ISM, by replacing variable nodes with statements.
Step 8: The ISM model developed in step 7 is reviewed to check for conceptual inconsistency and necessary modifications are made.

12.5 Case Study

The proposed decision-making methodology is applied to the battery manufacturing industry in the southern part of India. The purpose of this study is to assess the current battery recycling practices, identifying major barriers to its ineffectiveness and inefficiency, and to gain some suggestions and recommendations to improve the recycling systems. Ten barriers, as given in Table 12.1, were identified for analysing the interactions among the battery recycling barriers. Once the variables (barriers) are listed for analysing the interactions, it is essential to establish the contextual relationship among the variables for developing the structural self-interaction matrix (SSIM).

12.5.1 Structural Self-interaction Matrix

The ISM methodology suggests the use of expert opinions based on various management techniques, such as brain storming, nominal technique, *etc.*, in developing the contextual relationship between the variables. Thus, in this research into identifying the contextual relationship between the battery recycling barriers, three experts, two from the industry and one from academia, were consulted. For analysing the barriers, a contextual relationship of a 'leads to' type is chosen. This means that one variable leads to another variable. Based on this, the contextual relationship between the variables is developed.

Keeping in mind the contextual relationship for each variable, the existence of a relationship between any two barriers (*i* and *j*) and the associated direction of the

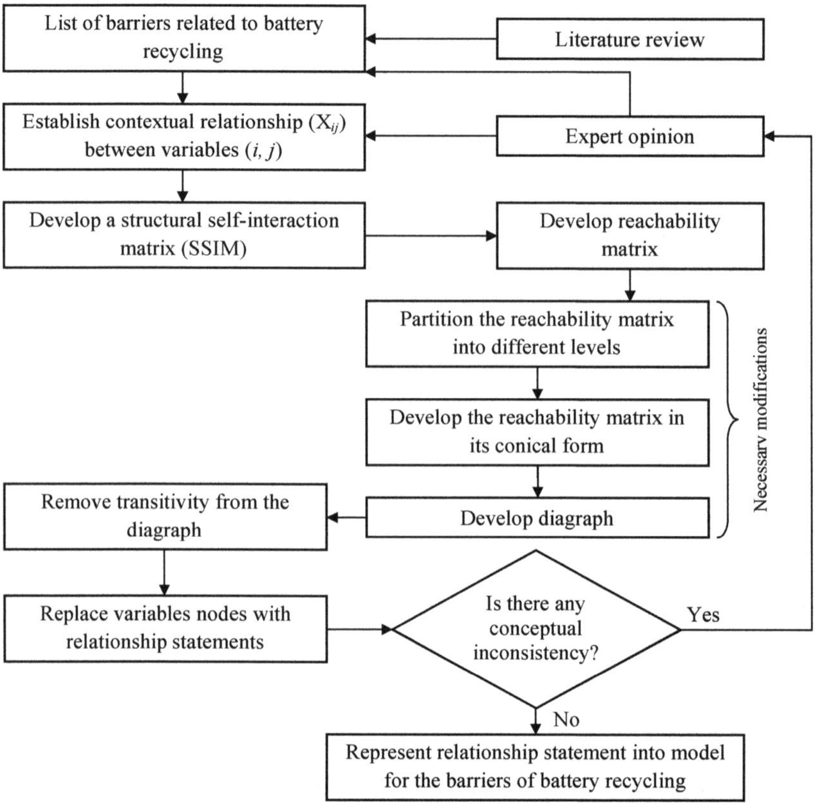

Figure 12.1. Flow diagram for preparing the ISM model

relation is questioned. Four symbols are used to denote the direction of relationship between the barriers (i and j):

V: criterion i will help alleviate criterion j;
A: criterion j will be alleviated by criterion i;
X: criteria i and j will help achieve each other; and
O: criteria i and j are unrelated.

The SSIM for the barriers to municipal solid waste recycling is given in Table 12.2. The following explains the use of the symbols V, A, X and O in the SSIM.

The *lack of storage space* barrier will help alleviate the *vehicle access problems* barrier, so the relationship of ‘V’ is denoted for barriers 2 and 9 in the SSIM. The *lack of customer interest/motivation* barrier can be alleviated by the *lack of effective communication* barrier. Thus, the relationship between these barriers is denoted by ‘A’ in the SSIM. The *inconvenient collecting centres* barrier and the *inadequate collection points* barrier help achieve each other. Thus, the relationship between these barriers is denoted by ‘X’ in the SSIM. No relationship exists between the *lack of effective communication* barrier and the *lack of suitable recycling plant sites* barrier and hence their relationship is denoted by ‘O’ in the SSIM.

Table 12.2. Structural self-interaction matrix (SSIM)

Barriers	10	9	8	7	6	5	4	3	2	1
1. Lack of customer interest/motivation	A	O	O	A	A	V	V	A	V	
2. Lack of storage space	A	V	X	A	A	V	V	A		
3. Lack of knowledge	V	V	V	A	V	V	V			
4. Inadequate collection points	A	V	A	A	A	X				
5. Inconvenient collecting centres	A	V	A	A	A					
6. Lack of effective communication	A	O	O	A						
7. Lack of political will	V	V	V							
8. Lack of suitable recycling plant sites	A	V								
9. Vehicle access problems	A									
10. Financial constraints										

12.5.2 Reachability Matrix

The SSIM is transformed into a binary matrix, called the initial reachability matrix, by substituting 1 and 0 for V, A, X, O as required. The rules for the substitution of 1s and 0s are as follows:

- if the (*i*, *j*) entry in the SSIM is V, then the (*i*, *j*) entry in the reachability matrix becomes 1 and the (*j*, *i*) entry becomes 0;
- if the (*i*, *j*) entry in the SSIM is A, then the (*i*, *j*) entry in the reachability matrix becomes 0 and the (*j*, *i*) entry becomes 1;
- if the (*i*, *j*) entry in the SSIM is X, then the entries for both (*i*, *j*) and (*j*, *i*) in the reachability matrix becomes 1;
- if the (*i*, *j*) entry in the SSIM is O, then the entries for both (*i*, *j*) and (*j*, *i*) in the reachability matrix becomes 0.

Following these rules, the initial reachability matrix for the barriers is given in Table 12.3.

Table 12.3. Initial reachability matrix

Barriers	1	2	3	4	5	6	7	8	9	10
1. Lack of customer interest/motivation	1	1	0	1	1	0	0	0	0	0
2. Lack of storage space	0	1	0	1	1	0	0	1	1	0
3. Lack of knowledge	1	1	1	1	1	1	0	1	1	1
4. Inadequate collection points	0	0	0	1	1	0	0	0	1	0
5. Inconvenient collecting centres	0	0	0	1	1	0	0	0	1	0
6. Lack of effective communication	1	1	0	1	1	1	0	0	0	0
7. Lack of political will	1	1	1	1	1	1	1	1	1	1
8. Lack of suitable recycling plant sites	0	1	0	1	1	0	0	1	1	0
9. Vehicle access problems	0	0	0	0	0	0	0	0	1	0
10. Financial constraints	1	1	0	1	1	1	0	1	1	1

The final reachability matrix for the barriers, shown in Table 12.4, is obtained by incorporating the transitivities from step 4 of the ISM methodology. In this table, the driving power and dependence of each barrier are also shown. The driving power of a particular barrier is the total number of barriers (including itself) that it may help achieve. The dependence is the total number of barriers that may help achieve it. These driving power and dependencies will be used in the analysis of *impact matrix cross-reference multiplication applied to a classification* (MICMAC), where the barriers are classified into four groups, *i.e.* autonomous, dependent, linkage and independent (driver) barriers.

Table 12.4. Final reachability matrix

Barriers	1	2	3	4	5	6	7	8	9	10	*Driver power* ↓
1. Lack of customer interest/motivation	1	1	0	1	1	0	0	1	1	0	6
2. Lack of storage space	0	1	0	1	1	0	0	1	1	0	5
3. Lack of knowledge	1	1	1	1	1	1	0	1	1	1	9
4. Inadequate collection points	0	0	0	1	1	0	0	0	1	0	3
5. Inconvenient collecting centres	0	0	0	1	1	0	0	0	1	0	3
6. Lack of effective communication	1	1	0	1	1	1	0	1	1	0	7
7. Lack of political will	1	1	1	1	1	1	1	1	1	1	10
8. Lack of suitable recycling plant sites	0	1	0	1	1	0	0	1	1	0	5
9. Vehicle access problems	0	0	0	0	0	0	0	0	1	0	1
10. Financial constraints	1	1	0	1	1	1	0	1	1	1	8
Dependence power →	5	7	2	9	9	4	1	7	10	3	

Table 12.5. Level partitions for barriers – iteration 1

Barriers	Reachability set	Antecedent set	Intersection set	Level
1	1,2,4,5,8,9	1,3,6,7,10	1	
2	2,4,5,8,9	1,2,3,6,7,8,10	2,8	
3	1,2,3,4,5,6,8,9,10	3,7	3	
4	4,5,9	1,2,3,4,5,6,7,8,10	4,5	
5	4,5,9	1,2,3,4,5,6,7,8,10	4,5	
6	1,2,4,5,6,8,9	3,6,7,10	6	
7	1,2,3,4,5,6,7,8,9,10	7	7	
8	2,4,5,8,9	1,2,3,6,7,8,10	2,8	
9	9	1,2,3,4,5,6,7,8,9,10	9	**I**
10	1,2,4,5,6,8,9,10	3,7,10	10	

12.5.3 Level Partitions

The reachability and antecedent set [12.64] for each barrier are obtained from the final reachability matrix. The reachability set for a particular variable consists of the variable itself and the other variables, which it may help achieve. The antecedent set consists of the variable itself and the other variables, which may help in achieving them. Subsequently, the intersection of these sets is derived for all variables. The variable for which the reachability and the intersection sets are the same is given the top-level variable in the ISM hierarchy, as they would not help achieve any other variable above their own level. After the identification of the top-level element, it is discarded from the other remaining variables. From Table 12.5, it can be seen that the vehicle access problems barrier is found at level I. Thus, it would be positioned at the top of the ISM model (Figure 12.2). This iteration is continued until the levels corresponding to each variable are obtained. The identified levels aid in building the digraph and the final ISM model. The barriers, along with their reachability set, antecedent set, intersection set and the levels, are shown in Tables 12.5–12.11.

Table 12.6. Level partitions for barriers – iteration 2

Barriers	Reachability set	Antecedent set	Intersection set	Level
1	1,2,4,5,8	1,3,6,7,10	1	
2	2,4,5,8	1,2,3,6,7,8,10	2,8	
3	1,2,3,4,5,6,8,10	3,7	3	
4	4,5	1,2,3,4,5,6,7,8,10	4,5	**II**
5	4,5	1,2,3,4,5,6,7,8,10	4,5	**II**
6	1,2,4,5,6,8	3,6,7,10	6	
7	1,2,3,4,5,6,7,8,10	7	7	
8	2,4,5,8	1,2,3,6,7,8,10	2,8	
10	1,2,4,5,6,8,10	3,7,10	10	

Table 12.7. Level partitions for barriers – iteration 3

Barriers	Reachability set	Antecedent set	Intersection set	Level
1	1,2,8	1,3,6,7,10	1	
2	2,8	1,2,3,6,7,8,10	2,8	**III**
3	1,2,3,6,8,10	3,7	3	
6	1,2,6,8	3,6,7,10	6	
7	1,2,3,6,7,8,10	7	7	
8	2,8	1,2,3,6,7,8,10	2,8	**III**
10	1,2,6,8,10	3,7,10	10	

Table 12.8. Level partitions for barriers – iteration 4

Barriers	Reachability set	Antecedent set	Intersection set	Level
1	1	1,3,6,7,10	1	**IV**
3	1,3,6,10	3,7	3	
6	1,6	3,6,7,10	6	
7	1,3,6,7,10	7	7	
10	1,6,10	3,7,10	10	

Table 12.9. Level partitions for barriers – iteration 5

Barriers	Reachability set	Antecedent set	Intersection set	Level
3	3,6,10	3,7	3	
6	6	3,6,7,10	6	**V**
7	3,6,7,10	7	7	
10	6,10	3,7,10	10	

Table 12.10. Level partitions for barriers – iteration 6

Barriers	Reachability set	Antecedent set	Intersection set	Level
3	3,10	3,7	3	
7	3,7,10	7	7	
10	10	3,7,10	10	**VI**

Table 12.11. Level partitions for barriers – iteration 7

Barriers	Reachability set	Antecedent set	Intersection set	Level
3	3	3,7	3	**VII**
7	3,7	7	7	**VIII**

12.6 Formation of the ISM-based Model

From the final reachability matrix, the structural model is generated. The relationship between the barriers j and i is shown by an arrow pointing from i to j. This resulting graph is called a digraph. Removing the transitivities as described in the ISM methodology, the digraph is finally converted into the ISM model as shown in Figure 12.2. This figure reveals that lack of political will is a very significant barrier for battery recycling as it forms the base of the ISM hierarchy. The vehicle access problem is the barrier on which the effectiveness of the battery recycling depends. This barrier appears at the top of the hierarchy. More details of the full ISM model for the barriers are given in Figure 12.2.

12.7 MICMAC Analysis

The MICMAC principle is based on the multiplication properties of matrices. The objective of the MICMAC analysis is to analyse the driver power and the dependence power of the variables [12.68]. The variables are classified into four clusters (Figure 12.3). The first cluster consists of the 'autonomous barriers', which have weak driver power and weak dependence power. These barriers are relatively disconnected from the system, with which they have only few links that may be strong. The second cluster consists of the dependent barriers, which have weak driver power but strong dependence. The third cluster has the linkage barriers, which have strong driver power and also strong dependence. These barriers are unstable in the fact that any action on these barriers will have an effect on others and also a feedback on themselves. The fourth cluster includes the independent barriers, which have strong driver power but weak dependence.

It is observed that a variable with a very strong driver power, called the key variable, falls into the category of independent or linkage criteria. The driver power and dependence power of each of these barriers are shown in Table 12.4. Subsequently, the diagram of driver power vs. dependence power for the barriers is constructed as shown in Figure 12.3. As an illustration, it is observed from Table 12.4 that a lack of knowledge barrier has a driver power of 9 and a dependence power of 2. Therefore, in Figure 12.3, it is positioned corresponding to a driver power of 9 and a dependence power of 2.

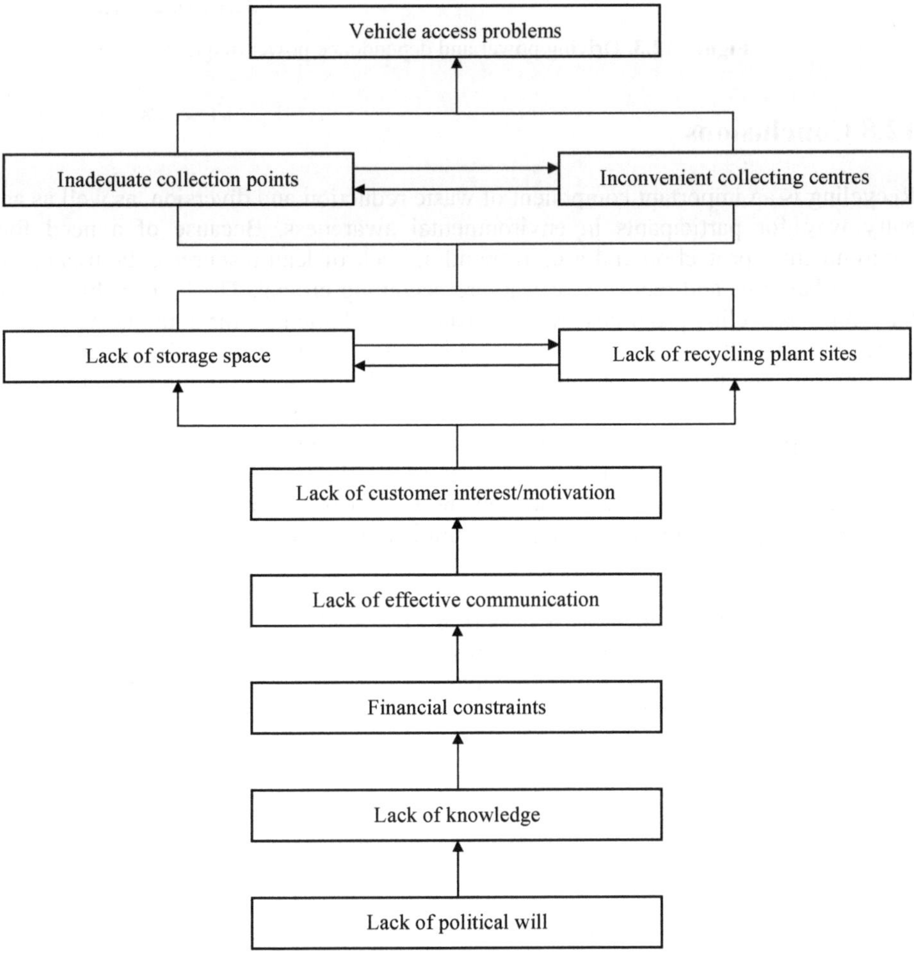

Figure 12.2. ISM-based model for the battery recycling barriers

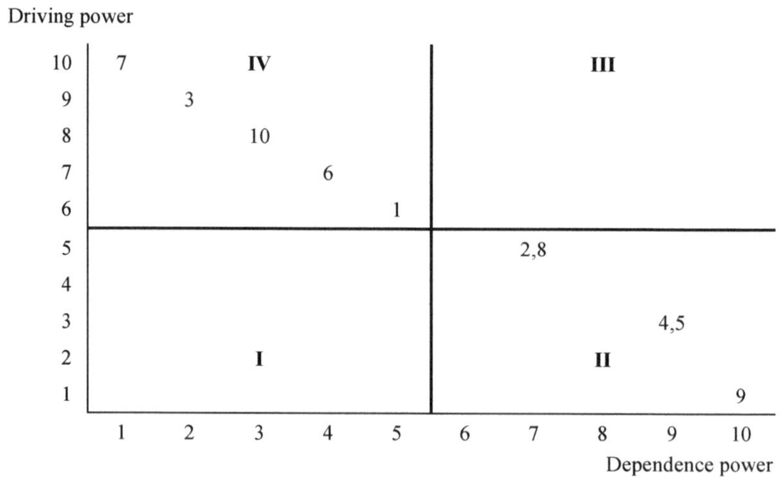

Figure 12.3. Driving power and dependence power diagram

12.8 Conclusions

Recycling is an important component of waste reduction and diversion, as well as an entry way for participants in environmental awareness. Because of a need for environmental protection and a corresponding lack of lead resources, the treatment of spent batteries and recovery of lead are becoming crucial. The barriers hindering the battery recycling programs create considerable challenges for both managers and policymakers in the battery manufacturing industry. Some of the major barriers have been highlighted here and put into an ISM model to analyse the interaction among the barriers. These barriers need to be overcome to ensure the success of battery recycling programmes. The driver–dependence diagram gives some valuable insights into the relative importance and interdependencies among the barriers. This can give better insights to the management so that they can proactively deal with these barriers. The important managerial implications emerging from this study are as follows:

- There is no autonomous barrier (see Figure 12.3). Autonomous barriers are weak drivers and weak dependents and do not have much influence on the system. The absence of any autonomous barriers in the present study indicates that all the barriers considered play a significant role.
- Dependent barriers are (i) lack of storage space (barrier 2), (ii) inadequate collection points (barrier 4), (iii) inconvenient collecting centres (barrier 5), (iv) lack of suitable recycling plant sites (barrier 8), and (v) vehicle access problems (barrier 9). These barriers are weak drivers but strongly depend on one another. Therefore, managers should take special care in handling these barriers.
- No barrier is found under the linkage element category possessing a strong driver power along with strong dependence. Therefore, among all ten selected battery recycling barriers, no barrier is unstable.

- It is further observed from Figure 12.3 that lack of customer interest/motivation (barrier 1), lack of knowledge (barrier 3), lack of effective communication (barrier 6), lack of political will (barrier 7), and financial constraints (barrier 10) are independent barriers, *i.e.* they have strong driver power and weak dependency on other barriers. They may be treated as the 'key barriers'. Management should place a high priority in tackling these barriers.

The levels of different barriers are important in better understanding their implications in the successful implementation of the battery recycling system. An insight into the ISM model indicates that barrier 9 (*i.e.* vehicle access problems) is the top-level barrier. This is the one that is most affected by the lower-level barriers. The second-level barriers (*i.e.* inadequate collection points, inconvenient collecting centres) and the third-level barriers (*i.e.* lack of storage space, lack of suitable recycling plant sites) are the operational level barriers that are essential for the successful operation of the recycling system. Lack of political will, lack of knowledge, financial constraints, lack of effective communication and lack of customer interest/motivation have the highest driver power and lowest dependence; hence, they appear as the bottom level of the hierarchy. This implies that a lack of political will, lack of knowledge and financial constraints play a significant role and work as the main drivers in the successful implementation of the battery recycling system. This fact is also very true from the practical point of view, because if management does not have a fiscal policy and knowledge for the implementation of battery recycling, none of the others will have any important significance. Therefore, the ISM methodology strengthens the practical views of the battery manufacturing industry and depicts a clear picture of the significance of the recycling barriers. In this way, different barriers can be identified and dealt with to ensure the successful implementation of the battery recycling system.

References

[12.1] Rogers, D.S. and Tibben-Lembke, R., 1999, *Going Backwards: Reverse Logistics Trends and Practices*, Reverse Logistics Executive Council, Reno, NV.

[12.2] Carter, C.R. and Ellram, L.M., 1998, "Reverse logistics: a review of the literature and framework for future investigation," *Journal of Business Logistics*, **19**(1), pp. 85–102.

[12.3] Beullens, P., 2004, "Reverse logistics in effective recovery of products from waste materials," *Reviews in Environmental Science and Bio/Technology*, **3**, pp. 283–306.

[12.4] Prahinski, C. and Kocabasoglu, C., 2006, "Empirical research opportunities in reverse supply chains," *Omega*, **34**(6), pp. 519–532.

[12.5] Srivastava, S.K., 2007, "Green supply-chain management: a state-of-the-art literature review," *International Journal of Management Reviews*, **9**(1), pp. 53–80.

[12.6] Rubio, S., Chamorro, A. and Miranda, F.J., 2008, "Characteristics of the research on reverse logistics (1995-2005)," *International Journal of Production Research*, **46**(4), pp. 1099–1120.

[12.7] Sasikumar, P. and Kannan, G., 2009, "Issues in reverse supply chain, part III: classification and simple analysis," *International Journal of Sustainable Engineering*, **2**(1), pp. 2–27.

[12.8] Thierry, M., Salomon, M., Van Nunen, J. and Van Wassenhove, L., 1995, "Strategic issues in product recovery management," *California Management Review*, **37**(2), pp. 114–135.

[12.9] Dekker, R., Fleischmann, M., Inderfurth, K. and Van Wassenhove, L.N., 2004, *Reverse Logistics – Quantitative Models for Closed Loop Supply Chains*, Springer, New York.

[12.10] Moyer, L. and Gupta, S.M., 1997, "Environmental concerns and recycling/ disassembly efforts in the electronics industry," *Journal of Electronics Manufacturing*, 7(1), pp. 1–22.

[12.11] Gungor, A. and Gupta, S.M., 1999, "Issues in environmentally conscious manufacturing and product recovery: a survey," *Computers and Industrial Engineering*, **36**(4), pp. 811–853.

[12.12] Johnson, M.R. and Wang, M.H., 1995, "Planning product disassembly for material recovery opportunities," *International Journal of Production Research*, **33**(11), pp. 3119–3142.

[12.13] Sasikumar, P. and Kannan, G., 2008, "Issues in reverse supply chain, part I: classification and simple analysis," *International Journal of Sustainable Engineering*, **1**(3), pp. 154–172.

[12.14] Pohlen, T.L. and Farris, M.T., 1992, "Reverse logistics in plastics recycling," *International Journal of Physical Distribution and Logistics Management*, **22**(7), pp. 35–47.

[12.15] Pati, R.K., Vrat, P. and Kumar, P., 2008, "A goal programming model for paper recycling system," *Omega*, **36**(3), pp. 405–417.

[12.16] González-Torre, P.L. and Adenso-Díaz, B., 2006, "Reverse logistics practices in the glass sector in Spain and Belgium," *International Business Review*, **15**(5), pp. 527–546.

[12.17] Ayres, R.U., 1997, "Metals recycling: economic and environmental implications," *Resources, Conservation and Recycling*, **21**, pp. 145–173.

[12.18] Spengler, T., Püchert, H., Penkuhn, T. and Rentz, O., 1997, "Environmental integrated production and recycling management," *European Journal of Operational Research*, **97** (2), pp. 308–326.

[12.19] Khoei, A.R., Masters, I. and Gethin, D.T., 2002, "Design optimization of aluminium recycling processes using Taguchi technique," *Journal of Materials Processing Technology*, **127**, pp. 96–106.

[12.20] Logozar, K., Radonjic, G. and Bastic, M., 2006, "Incorporation of reverse logistics model into in-plant recycling process: a case of aluminium industry," *Resources, Conservation and Recycling*, **49**(1), pp. 49–67.

[12.21] Wright, E., Azapagic, A., Stevens, G., Mellor, W. and Clift, R., 2005, "Improving recyclability by design: a case study of fibre optic cable," *Resources, Conservation and Recycling*, **44**(1), pp. 37–50.

[12.22] Barros, A.I., Dekker, R. and Scholten, V., 1998, "A two-level network for recycling sand: a case study," *European Journal of Operational Research*, **110**(2), pp. 199–214.

[12.23] Nagurney, A. and Toyasaki, F., 2005, "Reverse supply chain management and electronic waste recycling: a multitiered network equilibrium framework for e-cycling," *Transportation Research Part E: Logistics and Transportation Review*, **41**(1), pp. 1–28.

[12.24] Ammons, J.C., Realff, M.J. and Newton, D., 1999, "Carpet recycling: determining the reverse production system design," *Journal of Polymer-Plastics Technology and Engineering*, **38**(3), pp. 547–567.

[12.25] Louwers, D., Kip, B.J., Peters, E., Souren, F. and Flapper, S.D.P., 1999, "A facility location allocation model for reusing carpet material," *Computers and Industrial Engineering*, **36**(4), pp. 855–869.

[12.26] Realff, M.J., Ammons, J.C. and Newton, D.J., 2004, "Robust reverse production system design for carpet recycling," *IIE Transactions*, **36**(8), pp. 767–776.

[12.27] Biehl, M., Prater, E. and Realff, M.J., 2007, "Assessing performance and uncertainty in developing carpet reverse logistics systems," *Computers & Operations Research*, **34**(2), pp. 443–463.

[12.28] Jolly, R. and Rhin, C., 1994, "The recycling of lead batteries: production of lead and polypropylene," *Resources Conservation and Recycling*, **10**, pp. 137–143.

[12.29] Muzi, A., 1995, "Collection of spent batteries in Rome," *Journal of Power Sources*, **57**, pp. 19–21.

[12.30] Phillips, M.J. and Lim, S.S., 1998, "Secondary lead production in Malaysia," *Journal of Power Sources*, **73**, pp. 11–16.

[12.31] Frost, P.C., 1999, "Developments in lead–acid batteries: a lead producer's perspective," *Journal of Power Sources*, **78**, pp. 256–266.

[12.32] Zabaniotou, A., Kouskoumvekaki, E. and Sanopoulos, D., 1999, "Recycling of spent lead/acid batteries: the case of Greece," *Resources, Conservation and Recycling*, **25**(3–4), pp. 301–317.

[12.33] Tsoulfas, G.T., Pappis, C.T. and Minner, S., 2002, "An environmental analysis of the reverse supply chain of SLI batteries," *Resources, Conservation and Recycling*, **36**(2), pp. 135–154.

[12.34] Daniel, S.E., Pappis, C.P. and Voutsinas, T.G., 2003, "Applying life cycle inventory to reverse supply chains: a case study of lead recovery from batteries," *Resources, Conservation and Recycling*, **37**, pp. 251–281.

[12.35] Kannan, G., Sasikumar, P. and Devika, K., 2010, "A genetic algorithm approach for solving a closed loop supply chain model: a case of battery recycling," *Applied Mathematical Modelling*, **34**(3), pp. 655–670.

[12.36] Espinosa, D.C.R., Bernardes, A.M. and Tenório, A.S., 2004, "Brazilian policy on battery disposal and its practical effects on battery recycling," *Journal of Power Sources*, **137**, pp. 134–139.

[12.37] Andrews, D., Raychaudhuri, A. and Frias, C., 2000, "Environmentally sound technologies for recycling secondary lead," *Journal of Power Sources*, **88**, pp. 124–129.

[12.38] Bernardes, A.M., Espinosa, D.C.R. and Tenório, J.A.S., 2004, "Recycling of batteries: a review of current processes and technologies," *Journal of Power Sources*, **130**, pp. 291–298.

[12.39] Bloemhof-Ruwaard, J.M., Van Beek, P., Hordijk, L. and Van Wassenhove, L.N., 1995, "Interactions between operational research and environmental management," *European Journal of Operational Research*, **85**(2), pp. 229–243.

[12.40] Van Hoek, R.I., 1999, "From reversed logistics to green supply chains," *Supply Chain Management*, **4**(3), pp. 129–135.

[12.41] Zhu, Q. and Sarkis, J., 2004, "Relationships between operational practices and performance among early adopters of green supply chain management practices in Chinese manufacturing enterprises," *Journal of Operations Management*, **22**(3), pp. 265–289.

[12.42] Georgiadis, P. and Vlachos, D., 2004, "The effect of environmental parameters on product recovery," *European Journal of Operational Research*, **157**(2), pp. 449–464.

[12.43] Sheu, J.B., Chou, Y.H. and Hu, C.C., 2005, "An integrated logistics operational model for green-supply chain management," *Transportation Research Part E: Logistics and Transportation Review*, **41**(4), pp. 287–313.

[12.44] Vlachos, D., Georgiadis, P. and Iakovou, E., 2007, "A system dynamics model for dynamic capacity planning of remanufacturing in closed-loop supply chains," *Computers & Operations Research*, **34**, pp. 367–394.

[12.45] Carlson, R.L., 1985, "The impact of materials recycling programs on energy recovery economics," *Journal of Resource Management and Technology*, **15**(1), pp. 28–36.

[12.46] Johnson, P.F., 1998, "Managing value in reverse logistics systems," *Transportation Research Part E: Logistics and Transportation Review*, **34**(3), pp. 217–227.

[12.47] Krikke, H.R., Van Harten, A. and Schuur, P.C., 1999, "Business case Roteb: recovery strategies for monitors," *Computers & Industrial Engineering*, **36**(4), pp. 739–757.

[12.48] Boon, J.E., Isaacs, J.A. and Gupta, S.M., 2002, "Economic sensitivity for end of life planning and processing of personal computers," *Journal of Electronics Manufacturing*, **11**(1), pp. 81–93.

[12.49] Hoyle, G., 1995, "Recycling opportunities in the UK for aluminium-bodied motor cars," *Resources, Conservation and Recycling*, **15**, pp. 181–191.

[12.50] Degher, A., 2002, "HP's worldwide take back and recycling programs: lessons on improving program implementation," In *Proceedings of IEEE International Symposium on Electronics and the Environment*, pp. 224–227.

[12.51] Spicer, A.J. and Johnson, M.R., 2004, "Third-party demanufacturing as a solution for extended producer responsibility," *Journal of Cleaner Production*, **12**(1), pp. 37–45.

[12.52] Bufardi, A., Gheorghe, R., Kiritsis, D. and Xirouchakis, P., 2004, "Multi-criteria decision-aid approach for product end-of-life alternative selection," *International Journal of Production Research*, **42**, pp. 3139–3157.

[12.53] Ravi, V., Shankar, R. and Tiwari, M.K., 2005, "Analyzing alternatives in reverse logistics for end-of-life computers: ANP and balanced scorecard approach," *Computers & Industrial Engineering*, **48**(2), pp. 327–356.

[12.54] Listes, O. and Dekker, R., 2005, "A stochastic approach to a case study for product recovery network design," *European Journal of Operational Research*, **160**(1), pp. 268–287.

[12.55] Pati, R.K., Vrat, P. and Kumar, P., 2006, "Economic analysis of paper recycling vis-à-vis wood as raw material," *International Journal of Production Economics*, **103**, pp. 489–508.

[12.56] Bian, W. and Yu, M., 2006, "Location analysis of reverse logistics operations for an international electrical manufacturer in Asia Pacific region using the analytic hierarchy process," *International Journal of Services Operations and Informatics*, **1**(1–2), pp. 187–201.

[12.57] Staikos, T. and Rahimifard, S., 2007, "A decision-making model for waste management in the footwear industry," *International Journal of Production Research*, **45**(18), pp. 4403–4422.

[12.58] Pagell, M., Wu, Z. and Murthy, N.N., 2007, "The supply chain implications of recycling," *Business Horizons*, **50**, pp. 133–143.

[12.59] Queiruga, D., Walther, G., Gonzalez-Benito, J. and Spengler, T., 2008, "Evaluation of sites for the location of WEEE recycling plants in Spain," *Waste Management*, **28**, pp. 181–190.

[12.60] Ravi, V., Shankar, R. and Tiwari, M.K., 2008, "Selection of a reverse logistics project for end-of-life computers: ANP and goal programming approach," *International Journal of Production Research*, **46**(17), pp. 4849–4870.

[12.61] Kannan, G., Haq, A.N. and Sasikumar, P., 2008, "An application of the analytical hierarchy process and fuzzy analytical hierarchy process in the selection of collecting centre location for the reverse logistics multi-criteria decision-making supply chain model," *International Journal of Management and Decision Making*, **9**(4), pp. 350–365.

[12.62] Gomes, C.F.S., Nunes, K.R.A., Xavier, L.H., Cardoso, R. and Valle, R., 2008, "Multi-criteria decision making applied to waste recycling in Brazil," *Omega*, **36**, pp. 395–404.

[12.63] Wadhwa, S., Madaan, J. and Chan, F.T.S., 2009, "Flexible decision modeling of reverse logistics system: a value adding MCDM approach for alternative selection," *Robotics and Computer-Integrated Manufacturing*, **25**(2), pp. 460–469.

[12.64] Warfield, J.W., 1974, "Developing interconnected matrices in structural modelling," *IEEE Transactions on Systems, Men and Cybernetics*, **4**(1), pp. 51–81.

[12.65] Sage, A.P., 1977, *Interpretive Structural Modeling: Methodology for Large-scale Systems*, McGraw-Hill, New York.

[12.66] Saxena, J.P., Sushil and Vrat, P., 1990, "The impact of indirect relationships in classification of variables – a MICMAC analysis for energy conservation," *System Research*, **7**(4), pp. 245–253.

[12.67] Saxena, J.P., Sushil and Vrat, P., 1992, "Scenario building: a critical study of energy conservation in the Indian cement industry," *Technological Forecasting and Social Change*, **41**(2), pp. 121–146.

[12.68] Mandal, A. and Deshmukh, S.G., 1994, "Vendor selection using interpretive structural modelling (ISM)," *International Journal of Operations and Production Management*, **14**(6), pp. 52–59.

[12.69] Sharma, H.D., Gupta. A.D. and Sushil, 1995, "The objectives of waste management in India: a futures inquiry," *Technological Forecasting and Social Change*, **48**(3), pp. 285–309.

[12.70] Singh, M.D., Shankar, R., Narain, R. and Agarwal, A., 2003, "An interpretive structural modeling of knowledge management in engineering industries," *Journal of Advances in Management Research*, **1**(1), pp. 28–40.

[12.71] Huang, J.J., Tzeng, G.H. and Ong, C.S., 2005, "Multidimensional data in multidimensional scaling using the analytic network process," *Pattern Recognition Letters*, **26**(6), pp. 755–767.

[12.72] Ravi, V. and Shankar, R., 2005, "Analysis of interactions among the barriers of reverse logistics," *Technological Forecasting and Social Change*, **72**(8), pp. 1011–1029.

[12.73] Kannan, G. and Haq, A.N., 2007, "Analysis of interactions of criteria and sub-criteria for the selection of supplier in the built-in-order supply chain environment", *International Journal of Production Research*, **45**(17), pp. 3831–3852.

[12.74] Kannan, G., Haq, A.N., Sasikumar, P. and Arunachalam, S., 2008, "Analysis and selection of green suppliers using interpretative structural modeling and analytic hierarchy process," *International Journal of Management and Decision Making*, **9**(2), pp. 163–182.

[12.75] Singh, M.D. and Kant, R., 2008, "Knowledge management barriers: an interpretive structural modelling approach," *International Journal of Management Science and Engineering Management*, **3**(2), pp. 141–150.

[12.76] Vivek, S.D., Banwet, D.K. and Shankar, R., 2008, "Analysis of interactions among core, transaction and relationship-specific investments: the case of offshoring," *Journal of Operations Management*, **26**, pp. 180–197.

[12.77] Raj, T., Shankar, R. and Suhaib, M., 2008, "An ISM approach for modelling the enablers of flexible manufacturing system: the case for India," *International Journal of Production Research*, **46**(24), pp. 6883–6912.

[12.78] Kannan, G., Pokharel, S. and SasiKumar, P. 2009, "A hybrid approach using ISM and fuzzy TOPSIS for the selection of reverse logistics provider," *Resources, Conservation and Recycling*, **54**(1), pp. 28–36.

13

Design of Reverse Supply Chains in Support of Agile Closed-loop Logistics Networks

Anastasios Xanthopoulos and Eleftherios Iakovou

Industrial Management Division, Department of Mechanical Engineering
Aristotle University of Thessaloniki, 54124 Thessaloniki, P.O. Box 461, Greece
Emails: axanthop@auth.gr; eiakovou@auth.gr

Abstract
Reverse logistics is a key supply chain management discipline addressing the need for environmentally conscious manufacturing and processing of the end-of-life products. As the recovery processes are being recognised as a new value-added profit centre, the design of reverse logistics is receiving increased attention and scrutiny. In this chapter, we first present a comprehensive up-to-date literature review on the optimal design of reverse logistics and closed-loop supply chain networks. The chapter builds upon the general concepts that were developed by previous works, while extending them by presenting an integrated decision-support methodological approach for the optimal configuration of reverse supply chain networks in support of agile closed-loop supply chains. The proposed decision-making methodology provides a valuable strategic generalised model to decision makers that can be applied to various business environments. Finally, useful managerial insights regarding the implementation of the proposed solution methodology and sensitivity analysis are discussed, while specific directions for future research are provided.

13.1 Introduction: Motivation and Concepts

Sustainable development requires the environmental performance of supply chain management processes to be continually improving. Reverse logistics is a key supply chain management discipline, addressing the need for the environmentally conscious manufacturing and processing of end-of-life (EOL) products. Over-consumption and ever-shorter lifecycles of electrical and electronic products have as a consequence caused the accumulation of large volumes of waste products. On the other hand, EOL, end-of-lease, warranty, obsolete and overstocked products' returns usually have significant salvage value. As the recovery processes of the EOL products are now being recognised as a new value-added profit centre, and the environmental regulatory interventions regarding their uncontrollable disposal are getting ever stricter, reverse logistics are receiving increased attention. Although only a small part of today's business function is dedicated to recovery operations, the business world is starting to realise their potential [13.1].

Under this framework, the manufacturing industry has attempted to optimally exploit products reaching the end of their useful life. The objectives of the manufacturing companies are now threefold as they strive to (i) develop efficient forward supply chains, (ii) comply with the established regulatory interventions for products' recovery by designing the new products and handling the returns in the most environmentally friendly manner (*e.g.* the European Community Directive on waste electrical and electronic equipment – WEEE [13.2]), and (iii) increase profitability. Thus, the paradigm of the exclusive traditional forward character of supply chain management is not necessarily valid anymore, while the need for a comprehensive and optimal configuration of the closed-loop supply chain (CLSC) network per case has clearly emerged on the corporate agenda.

In this context, we first present a comprehensive up-to-date literature review of previous research efforts related to the optimal design of reverse logistics and CLSC networks. Second, an analytical multiple product and period mixed-integer linear programming (MILP) model is presented to address the important real-world problem of the optimal configuration of reverse supply chain networks. Developing a reverse logistics network entails its proper integration into the existing forward logistics channel, so that an agile, cost-effective and profitable closed-loop network emerges. Towards this, a large number of issues must be tackled optimally. For example, it should be investigated whether the network under development will lead to a significant transformation of the forward supply network, and whether this reverse logistics network will be exclusively privately structured or if the manufacturers will resort to outsourcing and to what degree. The decisions regarding the type, location, number and capacity of the collection, sorting, warehouse, disassembly and recovery facilities are at the heart of this problem.

The proposed optimisation model is rather comprehensive since it aims to address a large number of problem instances, in order to provide decision makers with a valuable methodology that can be applied to various business environments; thus, it addresses the 'superset' of all relevant decisions that need to be made. For instance, the reverse logistics network under configuration may be both remanufacturing- and recycling-driven, depending on each occasion. Furthermore, the proposed model holds for both the case of one manufacturer and the case of two or more manufacturers that may wish to collaborate by developing common recovery facilities.

The presented work was originally motivated by our two-year involvement with a research grant funded by the Greek Ministry of National Education and Religious Affairs under the title: 'Optimum management of industrial products at the end of their useful life'. The intention was the development of analytical methodological approaches for the optimisation of the recovery/environmental management of the EOL electrical and electronic products, using as prototypes real modem network terminals developed by one of the consortium's partners [13.3, 13.4].

The remainder of the chapter is organised as follows. In Section 13.2, we present a comprehensive up-to-date literature review on research works tackling the reverse logistics network design problem. Section 13.3 provides the problem definition as well as the main modelling assumptions. Section 13.4 deals with the formulation of the proposed MILP model, while its solution behaviour is extensively discussed. Moreover, alternative ways of conducting sensitivity analysis and significant managerial insights are provided. Next, in Section 13.5, specific extensions of the proposed model are presented, while directions for future research are given. Finally, in Section 13.6, we conclude by summarising the value of this work.

13.2 Design of Reverse Logistics Networks: a Literature Review

The optimal configuration of reverse logistics networks is rapidly evolving research field. Financial, environmental and regulatory motivations are forcing companies to redesign their forward supply channels so as to accommodate the recovery processes of EOL products. The vast majority of previous research efforts employ mathematical programming techniques. More particularly, the first research steps dealt with the appropriate adaptation of the corresponding quantitative models from the forward supply chain problem. Today, the relevant works found in the literature can generally be classified into three categories:

- research papers dealing with the configuration of an independent reverse logistics network;
- papers that aim at optimising the configuration of a reverse logistics network, while simultaneously taking into account to some extent the synergies with the existing forward supply chain; and
- manuscripts dealing with the joint configuration of forward and reverse supply chains (CLSC network design problem).

Below, using these classifications, we present an up-to-date comprehensive review of the research papers relevant to this topic.

13.2.1 Independent Reverse Logistics Networks

One of the first research papers covering the independent reverse logistics network design problem is that of Gottinger [13.5], who presented a single-product, single-period MILP location-allocation model with recycling and incineration as the examined recovery alternatives. Caruso *et al.* [13.6] developed a quantitative multi-criteria location model for an urban solid waste management system that determines the number, location and capacity of the waste disposal and recycling plants to be opened. Berger and Debaillie [13.7] and Realff *et al.* [13.8] proposed alternative single-period cost-minimisation MILP models in order to support the reverse logistics network design decision-making processes. The latter work is associated with the development of carpet recycling networks. Later, Realff *et al.* [13.9] extended their work by proposing a more elaborate multi-period model. In parallel, Louwers *et al.* [13.10] presented a deterministic nonlinear facility location-allocation model for the collection, pre-processing and redistribution of carpet waste, while taking into account the relevant depreciation costs. Chang and Wei [13.11] proposed a fuzzy multi-objective nonlinear integer programming model for determining the location of collection centres in a specific geographic area. Jayaraman *et al.* [13.12, 13.13] introduced MILP models for the reverse logistics network configuration problem, while in the latter case [13.13] a mixed heuristic concentration and expansion/greedy solution algorithm was proposed. Finally, Min *et al.* [13.14] addressed the problem of determining the number and location of centralised collection centres, by presenting a nonlinear mixed integer programming model and a genetic algorithm for its solution.

13.2.2 Configuration of Reverse Logistics Networks by Considering the Synergies with the Forward Channel

A number of research papers dealing with the reverse logistics network design problem systematically consider its synergies with the forward supply channel. In this context, Kroon and Vrijens [13.15] developed a single-period, cost-minimisation MILP model for designing a logistics system for returnable containers by taking into account the decisions concerning the number and the location of the warehouses/ redistribution centres. Spengler *et al.* [13.16] and Barros *et al.* [13.17] developed similar approaches for the optimisation of the location planning of recycling installations for the by-products of the steel industry in Germany and for the configuration of sand recycling networks, respectively. Listes and Dekker [13.18] extended the work of Barros *et al.* [13.17] by proposing a stochastic programming-based approach. Krikke *et al.* [13.19] developed a two-phased methodology in which first the recovery strategy of the returned products is determined, and then the reverse logistics network is designed optimally through a MILP model.

More recently, Beamon and Fernandes [13.20] developed a multi-period single-product MILP model that addresses the following questions: which warehouses and collection centres should be opened, which warehouses should have sorting capabilities and how much material should be transported between each pair of sites? The net present values of the employed costs are also taken into account. In parallel, Pochampally and Gupta [13.21] proposed a three-phased analytical approach for the design of reverse supply chains. In the first phase the set of products to be reprocessed are selected, in the second phase the potential locations of the recovery facilities are identified via the analytic hierarchy process, and finally the sourcing and deployment plans are addressed through a discrete location model. Min *et al.* [13.22] and Ko and Evans [13.23] presented mixed-integer non-linear programming models accompanied by genetic solution algorithms for the location and allocation problem of collection centres, and for the design of a dynamic integrated distribution network in a CLSC, respectively. Finally, Wang *et al.* [13.24] presented a two-phased location-inventory model for first determining the location of the collection/sorting centres, and then for finding an optimal replenishment policy.

13.2.3 CLSC Networks

One of the most comprehensive research works for the CLSC network design problem is that of Fleischmann [13.25], who developed a cost minimisation single-period and single-product MILP model for the optimal development of new plants, warehouses and reverse-activities centres, while considering the coordination issues between the forward and the reverse supply channels. Later, Salema *et al.* [13.26] extended the work of Fleischmann by considering capacity limitations, uncertainty on product demands and returns in a scenario-based manner, and the multi-product case. Lieckens and Vandaele [13.27] addressed the single-product and single-period problem by including queuing characteristics to allow for stochastic lead-times, while they provided a genetic solution algorithm. In addition, Marin and Pelegrin [13.28], Krikke *et al.* [13.29], Listes [13.30], Lu and Bostel [13.31] and Sahyouni *et al.* [13.32] proposed interesting modelling approaches for the CLSC network design problem.

13.2.4 Literature Review Insights

The conducted literature review reveals the existence of a considerable number of reverse logistics network design analytical models, the main characteristics of which are illustrated in Table 13.1. The first step in this research field was to properly adapt the respective forward logistics models to the case of a recovery network, but progressively the modelling efforts became more sophisticated as modelling aspects were also incorporated. However, very few of the existing models consider the multi-period problem, the relevant environmental and legislative aspects, the maximisation of profit, and the time value of money. On the other hand, most papers do consider the development of general recovery facilities, in which the total of the various recovery processes can occur (considering economies of scale). Furthermore, a few particular modelling aspects are not addressed at all, including: (i) the outsourcing issues, (ii) the existence of state reverse logistics systems, (iii) the development of disassembly centres with a recycling and/or remanufacturing orientation, (iv) the potential marketing gains from the adoption of an ecological profile, and (v) the backordering and lost sales issues.

Moreover, only a small portion of the literature body addresses the supply and demand uncertainty in recovery operations, and especially in a scenario-based manner. Concluding, in spite of the significant developments in reverse logistics network design modelling, there is a clear lack of analytical approaches that capture the complexity and the multitude of issues that need to be taken into account for any meaningful decision-making methodology. The present chapter builds upon the general concepts that were developed by previous works, while extending them by presenting an integrated decision-support methodological approach and not merely a new optimisation model.

13.3 System Description

13.3.1 Problem Definition

The efficient management of the returned EOL products, so that the profitability is increased and the interdependencies with the traditional forward channel are tackled properly, represents a major challenge and a very important issue for the international manufacturing industry. Today, companies that acknowledge the strategic importance of reverse logistics strive to properly configure their CLSCs. In order for

Table 13.1. Main characteristics of the reverse logistics network design models

	Reference number [13.*]													
Parameters	5	6	7	8	9	10	11	12	13	14	15	16	17	18
Deterministic demand	×	×	×	×	×	×		×	×	×	×	×	×	
Stochastic demand														×
Deterministic returns	×	×	×	×	×	×		×	×	×	×	×	×	
Stochastic returns														×
Single product	×	×	×				×		×	×	×			
Multiple products				×	×	×		×				×	×	×
Single period	×	×	×	×		×	×	×	×	×	×	×	×	×
Multiple periods					×									
State reverse logistics system														
Outsourcing														
General recovery centres	×		×			×			×	×				
Collection centres				×	×		×		×	×		×	×	×
Warehouses			×		×			×			×	×	×	×
Disassembly centres														
Remanufacturing centres								×						
Recycling centres		×		×	×							×	×	×
Environmental issues		×												
Regulatory issues														
Time value of money						×								
Capacity constraints	×	×	×	×	×	×	×	×	×	×		×	×	×
Lead-time														
Quality aspects													×	×
Service level														
Backorders														
Uncollected EOL products														
Marketing gains														
Cost minimisation	×	×	×			×		×	×	×	×	×	×	
Profit maximisation				×	×									×

Table 13.1. Main characteristics of the reverse logistics network design models (continued)

	Reference number [13.*]													
Parameters	19	20	21	22	23	24	25	26	27	28	29	30	31	32
Deterministic demand	×	×	×	×	×	×	×		×	×	×		×	×
Stochastic demand								×				×		
Deterministic returns	×	×	×	×	×	×	×		×	×	×		×	×
Stochastic returns								×				×		
Single product	×	×		×		×	×		×	×		×	×	×
Multiple products			×		×			×			×			
Single period	×		×			×	×	×	×	×	×	×	×	×
Multiple periods		×		×	×									
State reverse logistics system														
Outsourcing					×									
General recovery centres	×		×		×		×	×	×	×	×	×		
Collection centres		×	×	×		×							×	×
Warehouses		×			×		×	×						
Disassembly centres														
Remanufacturing centres													×	
Recycling centres														
Environmental issues							×				×			
Regulatory issues							×				×			
Time value of money		×												
Capacity constraints	×	×	×		×			×	×	×	×	×		
Lead-time			×						×					
Quality aspects		×	×								×	×	×	
Service level							×		×					
Backorders														
Uncollected EOL products							×		×			×		
Marketing gains														
Cost minimisation	×	×	×	×	×		×	×		×	×		×	×
Profit maximisation									×			×		

a reverse logistics network to be flexible and efficient, it is essential that all the related functional areas that affect or can be affected by the recovery operations should be taken into account. According to Fleischmann [13.33], there are five possible classes of reverse logistics networks: (i) dedicated remanufacturing networks, (ii) recycling-driven networks for material recovery, (iii) reuse networks, (iv) regulatory-driven networks, and (v) manufacturers' or third-party reverse logistics (3PRL) providers' networks. To this effect, a strategic analytical quantitative decision-making methodology is proposed for the electrical and electronic products industry, which considers explicitly all of the above network structures through a large number of realistic problem instances. More specifically, we treat this problem as a generalised multi-period, multi-product MILP model that addresses the optimisation of the related decision-making processes (*i.e.* type, number, location and capacity of the facilities under development, as well as product flows decisions). Additionally, regulatory and environmentally related constraints are also taken into account.

Figure 13.1 depicts the considered alternative options of configuring a reverse logistics network for electrical and electronic products. The development of such a recovery network presupposes the existence of the forward supply chain, which may be transformed properly so that an efficient and agile CLSC network will finally result. Indeed, this case is of interest to all manufacturing firms that wish to get involved in product recovery operations. Decision makers should thoroughly consider the financial ramifications of the alternatives to developing a reverse logistics network. Such alternatives include:

- the case where a firm joins the national/state reverse logistics system;
- the proper transformation of the existing forward supply channel facilities in order to accommodate the reverse logistics activities;
- the development of new facilities dedicated to reverse logistics processes;
- the collaboration with third-party reverse logistics (3PRL) providers, so as to outsource the whole or a part of the recovery processes;
- a combination of the previous alternatives.

The collection of the returned EOL products constitutes the starting point of every reverse supply chain. To this end, manufacturers could set up private collection centres. These centres may consist of new facilities dedicated to the above reverse logistics processes and/or can be established by adapting the facilities of the forward supply chain (for instance, the outlet stores of electrical and electronic products can also serve as initial collection points). Sorting processes can also take place in collection centres. More specifically, the recoverability of each EOL product can be identified through quality control techniques, so as to facilitate the subsequent recovery processes. Products of high quality can be refurbished for reuse, remanufactured or recycled. On the other hand, products of poor quality that are unsuitable for any recovery processes are disposed of to sanitary landfills. The manufacturers also have the choice of whether resort to outsourcing. 3PRL providers can undertake all the collection and sorting processes or they can operate complementarily to manufacturers. It is very important for the decision makers to determine the optimal contractual agreement with the 3PRL providers, in terms of cost and quantity commitments.

Considering the recovery processes, manufacturers can develop new refurbishing, remanufacturing and recycling facilities or appropriately transform the existing forward supply network facilities to this purpose. An EOL product needs to be first disassembled before being remanufactured or recycled: *i.e.* non-destructive disassembly processes for the products to be remanufactured, and destructive disassembly processes for the products that have to be recycled. Moreover, manufacturers have the option of outsourcing the respective reverse logistics processes to 3PRL providers. Outsourcing in reverse logistics is a newly established international practice with a constantly increasing potential [13.34] and a field of growing research interest [13.23, 13.35–13.37]. The reasons that entice companies to resort to outsourcing can vary: (i) many manufacturers focus only on delivering new products to the end-users and want to exploit a 3PRL provider's core competency, (ii) the investment cost of developing reverse logistics facilities is prohibitive for firms to undertake it, and (iii) limited EOL returns limit reverse logistics to only a small part of the company's business-related processes with poor profitability potential.

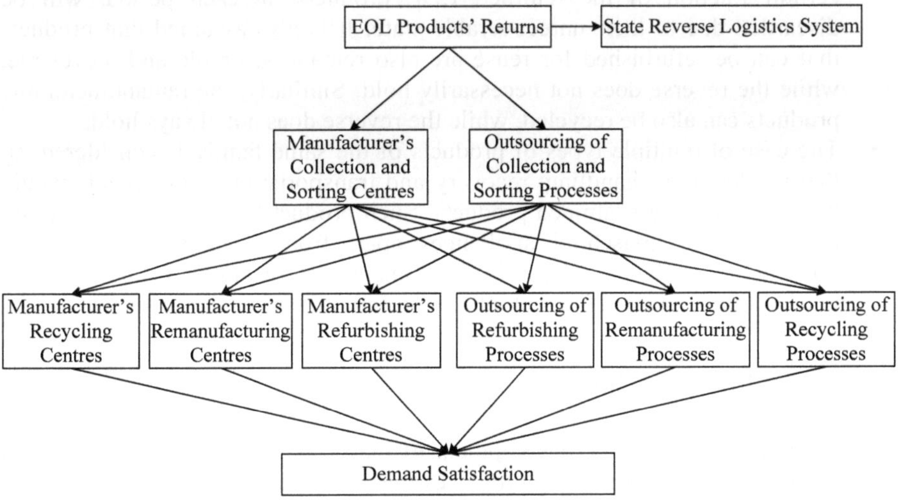

Figure 13.1. Structure of the integrated reverse logistics network design problem

Apart from developing reverse logistics centres and outsourcing the recovery processes, a manufacturer has the alternative of entering into the national/state recovery system for waste electrical and electronic equipment (provided that such a system exists in his/her home country). This is common practice especially among European Union countries, whereby the state recovery system undertakes the collection and recovery processes of certain types of products on behalf of the manufacturer for a specified fee. The state reverse logistics system may undertake the recovery operations of some or all of the different types of products that a manufacturer fabricates. For these types of products the manufacturer will neither be responsible for their recovery operations (WEEE directive) [13.2], nor would he/she have to satisfy a specific demand in the recovered products.

13.3.2 Major Modelling Assumptions

In this subsection, we elaborate on the main modelling assumptions for the investigating problem:

- The time span of the examined strategic problem depends on the amortisation time of the facilities under development and on the duration of the contractual agreements with the 3PRL providers. Usually the planning horizon can span an interval of four to six years, and the base planning period (unit time-step) can be three to six months.
- The present values of all the employed costs are taken into account. When computing the fixed investment costs, the manufacturing costs during the construction period prior to the start of the planning horizon should also be calculated. The employed interest rate is the company's specific cost of capital.
- The variable transportation costs from the source to a destination site are incorporated into the unit variable costs of recovering EOL products.
- Not all the returned EOL products are in good condition to be recovered. A certain fraction of the returned EOL products, in each period, will be discarded due to their unrecoverable state. It is also assumed that products that can be refurbished for reuse are also remanufacturable and recyclable, while the reverse does not necessarily hold. Similarly, the remanufacturable products can also be recycled, while the reverse does not always hold.
- The case of multiple types of products of the same family is considered, so that the facilities, handling, recovery and transportation requirements would be similar for the examined products. Such product families include cooling devices, large household appliances apart from cooling devices, small household appliances, lighting equipment, information technology and telecommunication equipment, other consumer or medical equipment, *etc*.

13.4 Model Formulation

In this section, we present the development of the proposed strategic MILP model for addressing the decision-making processes related to the design of reverse supply chains in support of agile CLSC networks. The development of the proposed model, which addresses a relatively large number of problem instances, was motivated by a large number of manufacturing companies (of electrical and electronic products) that face the problem under investigation.

13.4.1 Nomenclature

The notation of the employed sets of indices is provided in Table 13.2. The indices relating to the development of collection and recovery facilities are tied to investment scenarios relating to the number, type, location, and capacity of the examining facilities. These facilities can be either new and/or result from the appropriate transformation of the forward channel facilities. On the other hand, the contractual outsourcing agreements are related to alternative scenarios of quantity commitments and costs.

In Tables 13.3–13.5 we provide the definitions of the decision variables, and of the general and cost parameters, respectively.

Table 13.2. Definition of the employed indices

Notation	Description
$i = 1,\ldots,I$	Types of products under consideration
$t = 1,\ldots,T$	Set of periods that constitute the strategic planning horizon
m	Manufacturer
$ma = 1,\ldots,MA$	Alternative scenarios for developing collection and sorting installations
$mb = 1,\ldots, MB$	Alternative scenarios for developing refurbishing installations
$mc = 1,\ldots,MC$	Alternative scenarios for developing remanufacturing installations
$md = 1,\ldots,MD$	Alternative scenarios for developing recycling installations
p	3PRL providers
$pa = 1,\ldots,PA$	Alternative scenarios of outsourcing the collection and sorting processes
$pb = 1,\ldots,PB$	Alternative scenarios of outsourcing the refurbishing processes
$pc = 1,\ldots,PC$	Alternative scenarios of outsourcing the remanufacturing processes
$pd = 1,\ldots,PD$	Alternative scenarios of outsourcing the recycling processes
s	State reverse logistics system

Table 13.3. Definition of the decision variables

Notation	Description
$X^{s}_{i,t}$	Type *i* EOL products that the state reverse logistics system collect/handle in period *t*
$XA^{m}_{i,t,ma}$ / $XA^{p}_{i,t,pa}$	Type *i* EOL products that are collected and sorted in type *ma* manufacturer's centre or by a 3PRL provider under contract type *pa*, respectively, in period *t*
$XB^{mm}_{i,t,ma,mb}$ / $XB^{pm}_{i,t,pa,mb}$	Type *i* EOL products that are refurbished in type *mb* centre in period *t*; the products stem from type *ma* manufacturer's site and/or from a 3PRL provider under contract type *pa*
$XB^{mp}_{i,t,ma,pb}$ / $XB^{pp}_{i,t,pa,pb}$	Type *i* EOL products that are refurbished in period *t* by a 3PRL provider under contract type *pb*; the products stem from type *ma* manufacturer's site and/or from a 3PRL provider under contract type *pa*
$XBB_{i,t}$ / $XBC_{i,t}$ / $XBD_{i,t}$	Backorders of type *i* refurbished, remanufactured, and recycled products, respectively, in period *t*
$XC^{mm}_{i,t,ma,mc}$ / $XC^{pm}_{i,t,pa,mc}$	Type *i* EOL products that are remanufactured in type *mc* centre in period *t*; the products stem from type *ma* manufacturer's site and/or from a 3PRL provider under contract type *pa*
$XC^{mp}_{i,t,ma,pc}$ / $XC^{pp}_{i,t,pa,pc}$	Type *i* EOL products that are remanufactured in period *t* by a 3PRL provider under contract type *pc*; the products stem from type *ma* manufacturer's site and/or from a 3PRL provider under contract type *pa*
$XD^{mm}_{i,t,ma,md}$ / $XD^{pm}_{i,t,pa,md}$	Type *i* EOL products that are recycled in type *md* centre in period *t*; the products stem from type *ma* manufacturer's site and/or from a 3PRL provider under contract type *pa*

Table 13.3. Definition of the decision variables (continued)

Notation	Description
$XD^{mp}_{i,t,ma,pd}$ / $XD^{pp}_{i,t,pa,pd}$	Type *i* EOL products that are recycled in period *t* by a 3PRL provider under contract type *pd*; the products stem from type *ma* manufacturer's site and/or from a 3PRL provider under contract type *pa*
$XI^{m}_{i,t,ma}$ / $XI^{p}_{i,t,pa}$	On-hand inventory in type *i* EOL products stored in type *ma* manufacturer's facility or by a 3PRL provider under contract type *pa*, respectively, in period *t*
$XLB_{i,t}$ / $XLC_{i,t}$ / $XLD_{i,t}$	Lost sales in type *i* refurbished, remanufactured, and recycled products, respectively, in period *t*
$XVB^{m}_{i,t}$ / $XVC^{m}_{i,t}$ / $XVD^{m}_{i,t}$	Type *i* EOL products that are refurbished, remanufactured, and recycled, respectively, by the manufacturer in period *t*
$XVB^{p}_{i,t}$ / $XVC^{p}_{i,t}$ / $XVD^{p}_{i,t}$	Type *i* EOL products that are refurbished, remanufactured, and recycled, respectively, by the 3PRL providers in period *t*
$XW^{m}_{i,t,ma}$ / $XW^{p}_{i,t,pa}$	Type *i* EOL products that are discarded in period *t* due to their poor quality, from type *ma* manufacturer's facility or from 3PRL provider's facility under contract type *pa*, respectively
Y^{s}_{i}	Boolean indicating whether the manufacturer will enter the state reverse logistics system for type *i* EOL products or not
YA^{m}_{ma} / YB^{m}_{mb} / YC^{m}_{mc} / YD^{m}_{md}	Boolean indicating the development of type *ma*/*mb*/*mc*/*md* collection/refurbishing/remanufacturing/recycling centres
YA^{p}_{pa} / YB^{p}_{pb} / YC^{p}_{pc} / YD^{p}_{pd}	Boolean indicating that the collection/refurbishing/ remanufacturing/recycling processes will be outsourced to a 3PRL provider under contract type *pa*/*pb*/*pc*/*pd*

Table 13.4. Definition of the general parameters of the model

Notation	Description
CA^{m}_{ma} / CI^{m}_{ma}	Collection and sorting, and storage capacity, respectively, of type *ma* facility
CA^{p}_{pa} / CI^{p}_{pa}	Maximum quantity of EOL products that a 3PRL provider under contract type *pa* is committed to collect and sort in a period, and store, respectively
$CB^{m}_{i,mb}$ / $CC^{m}_{i,mc}$ / $CD^{m}_{i,md}$	Refurbishing/remanufacturing/recycling capacity of type *mb*/*mc*/*md* facility for type *i* EOL products
$CB^{p}_{i,pb}$ / $CC^{p}_{i,pc}$ / $CD^{p}_{i,pd}$	Maximum quantity of type *i* EOL products that a 3PRL provider under contract type *pb*/*pc*/*pd* is committed to refurbish/ remanufacture/recycle in a period
$DB_{i,t}$ / $DC_{i,t}$ / $DD_{i,t}$	Demand for type *i* refurbished/remanufactured/recycled EOL products in period *t*
FB_{i} / FC_{i} / FD_{i}	Maximum acceptable percentage of unsatisfied demand in type *i* refurbished/remanufactured/recycled EOL products
ir	Interest rate
L_{i}	Minimum percentage of type *i* returned EOL products that is possible (by the legislation) to not be recovered
M	A very large positive number
QB_{i} / QC_{i} / QD_{i}	Maximum percentage of type *i* collected EOL products that are reusable, remanufacturable, and recyclable, respectively
$R_{i,t}$	Type *i* EOL returned products in period *t*
V_{i}	Volume of type *i* product

Table 13.5. Definition of the cost parameters of the model

Notation	Description
c_i^s	Fee that the manufacturer pays to the state reverse logistics system per type *i* product
$ca_{i,ma}^m$ / $ca_{i,pa}^p$	Unit collection and sorting costs of type *i* EOL products in type *ma* manufacturer's centre and for a 3PRL provider under contract type *pa*, respectively
$cb_{i,ma,mb}^{mm}$ / $cb_{i,pa,mb}^{pm}$	Unit refurbishing cost of type *i* products in type *mb* facilities; the products stem from type *ma* manufacturer's site and/or from a 3PRL provider under contract type *pa*, respectively
$cb_{i,ma,pb}^{mp}$ / $cb_{i,pa,pb}^{pp}$	Unit refurbishing cost of type *i* products for a 3PRL provider under contract type *pb*; the products stem from type *ma* manufacturer's site and/or from a 3PRL provider under contract type *pa*, respectively
cbb_i / cbc_i / cbd_i	Unit backorder cost per type *i* refurbished/remanufactured/recycled products per period
$cc_{i,ma,mc}^{mm}$ / $cc_{i,pa,mc}^{pm}$	Unit disassembly and remanufacturing cost of type *i* products in type *mc* facilities; the products stem from type *ma* manufacturer's site and/or from a 3PRL provider under contract type *pa*, respectively
$cc_{i,t,ma,pc}^{mp}$ / $cc_{i,t,pa,pc}^{pp}$	Unit disassembly and remanufacturing cost of type *i* products for a 3PRL provider under contract type *pc*; the products stem from type *ma* manufacturer's site and/or from a 3PRL provider under contract type *pa*, respectively
$cd_{i,ma,md}^{mm}$ / $cd_{i,pa,md}^{pm}$	Unit disassembly and recycling cost of type *i* products in type *md* facilities; the products stem from type *ma* manufacturer's site and/or from a 3PRL provider under contract type *pa*, respectively
$cd_{i,ma,pd}^{mp}$ / $cd_{i,pa,pd}^{pp}$	Unit disassembly and recycling cost of type *i* products for a 3PRL provider under contract type *pd*; the products stem from type *ma* manufacturer's site and/or from a 3PRL provider under contract type *pa*, respectively
$ci_{i,ma}^m$ / $ci_{i,pa}^p$	Unit holding cost per type *i* EOL product in type *ma* manufacturer's centre and for a 3PRL provider under contract type *pa*, respectively
clb_i / clc_i / cld_i	Unit penalty cost for the unmet demand in refurbished/remanufactured/recycled type *i* EOL products
vb_i / vc_i / vd_i	Mean revenues per type *i* EOL product being refurbished/remanufactured/recycled
$cw_{i,ma}^m$ / $cw_{i,pa}^p$	Unit discarding cost of type *i* EOL products from type *ma* manufacturer's centre and from a 3PRL provider under contract type *pa*, respectively
ka_{ma}^m / kb_{mb}^m / kc_{mc}^m / kd_{md}^m	Fixed cost of developing collection/refurbishing/remanufacturing/recycling centres under scenario *ma*/*mb*/*mc*/*md*
ka_{pa}^p / kb_{pb}^p / kc_{pc}^p / kd_{pd}^p	Fixed cost of agreement with a 3PRL provider under contract type *pa*/*pb*/*pc*/*pd* for the collection/refurbishing/remanufacturing/recycling processes

13.4.2 Optimisation Model

The developed objective function and multiple groups of constraints of the proposed MILP model are explicitly presented in this subsection.

Objective Function

Maximise:

$$\sum_i \sum_t \begin{bmatrix} vb_i \cdot \left(XVB^m_{i,t} + XVB^p_{i,t}\right) + vc_i \cdot \left(XVC^m_{i,t} + XVC^p_{i,t}\right) + \\ vd_i \cdot \left(XVD^m_{i,t} + XVD^p_{i,t}\right) - c^s_i \cdot X^s_{i,t} - cbb_i \cdot XBB_{i,t} - \\ cbc_i \cdot XBC_{i,t} - cbd_i \cdot XBD_{i,t} - clb_i \cdot XLB_{i,t} - \\ clc_i \cdot XLC_{i,t} - cld_i \cdot XLD_{i,t} \end{bmatrix} \cdot (1+ir)^{-t}$$

$$-\sum_i \sum_t \sum_{ma} \begin{bmatrix} ca^m_{i,ma} \cdot XA^m_{i,t,ma} + ci^m_{i,ma} \cdot XI^m_{i,t,ma} + cw^m_{i,ma} \cdot XW^m_{i,t,ma} + \\ \sum_{mb}\left(cb^{mm}_{i,ma,mb} \cdot XB^{mm}_{i,t,ma,mb}\right) + \sum_{pb}\left(cb^{mp}_{i,ma,pb} \cdot XB^{mp}_{i,t,ma,pb}\right) + \\ \sum_{mc}\left(cc^{mm}_{i,ma,mc} \cdot XC^{mm}_{i,t,ma,mc}\right) + \sum_{pc}\left(cc^{mp}_{i,ma,pc} \cdot XC^{mp}_{i,t,ma,pc}\right) + \\ \sum_{md}\left(cd^{mm}_{i,ma,md} \cdot XD^{mm}_{i,t,ma,md}\right) + \sum_{pd}\left(cd^{mp}_{i,ma,pd} \cdot XD^{mp}_{i,t,ma,pd}\right) \end{bmatrix} \cdot (1+ir)^{-t}$$

$$-\sum_i \sum_t \sum_{pa} \begin{bmatrix} ca^p_{i,pa} \cdot XA^p_{i,t,pa} + ci^p_{i,pa} \cdot XI^p_{i,t,pa} + cw^p_{i,pa} \cdot XW^p_{i,t,pa} + \\ \sum_{mb}\left(cb^{pm}_{i,pa,mb} \cdot XB^{pm}_{i,t,pa,mb}\right) + \sum_{pb}\left(cb^{pp}_{i,pa,pb} \cdot XB^{pp}_{i,t,pa,pb}\right) + \\ \sum_{mc}\left(cc^{pm}_{i,pa,mc} \cdot XC^{pm}_{i,t,pa,mc}\right) + \sum_{pc}\left(cc^{pp}_{i,pa,pc} \cdot XC^{pp}_{i,t,pa,pc}\right) + \\ \sum_{md}\left(cd^{pm}_{i,pa,md} \cdot XD^{pm}_{i,t,pa,md}\right) + \sum_{pd}\left(cd^{pp}_{i,pa,pd} \cdot XD^{pp}_{i,t,pa,pd}\right) \end{bmatrix} \cdot (1+ir)^{-t}$$

$$-\sum_{ma}\left(ka^m_{ma} \cdot YA^m_{ma}\right) - \sum_{pa}\left(ka^p_{pa} \cdot YA^p_{pa}\right) - \sum_{mb}\left(kb^m_{mb} \cdot YB^m_{mb}\right) - \sum_{pb}\left(kb^p_{pb} \cdot YB^p_{pb}\right)$$

$$-\sum_{mc}\left(kc^m_{mc} \cdot YC^m_{mc}\right) - \sum_{pc}\left(kc^p_{pc} \cdot YC^p_{pc}\right) - \sum_{md}\left(kd^m_{md} \cdot YD^m_{md}\right) - \sum_{pd}\left(kd^p_{pd} \cdot YD^p_{pd}\right)$$

The objective function enables the optimisation of the reverse logistics network design problem through the maximisation of the recovered value from the EOL products minus the investment, outsourcing, collection, sorting, recovery, transportation and operational costs.

Constraints

Capacity constraints:

$$X^s_{i,t} \le M \cdot Y^s_i, \forall i,t \tag{13.1}$$

$$\sum_{i} XA^{m}_{i,t,ma} \leq CA^{m}_{ma} \cdot YA^{m}_{ma}, \forall t, ma \tag{13.2}$$

$$\sum_{i} \left(XI^{m}_{i,t,ma} \cdot V_{i} \right) \leq CI^{m}_{ma} \cdot YA^{m}_{ma}, \forall t, ma \tag{13.3}$$

$$\sum_{ma} XB^{mm}_{i,t,ma,mb} + \sum_{pa} XB^{pm}_{i,t,pa,mb} \leq CB^{m}_{i,mb} \cdot YB^{m}_{mb}, \forall i, t, mb \tag{13.4}$$

$$\sum_{ma} XC^{mm}_{i,t,ma,mc} + \sum_{pa} XC^{pm}_{i,t,pa,mc} \leq CC^{m}_{i,mc} \cdot YC^{m}_{mc}, \forall i, t, mc \tag{13.5}$$

$$\sum_{ma} XD^{mm}_{i,t,ma,md} + \sum_{pa} XD^{pm}_{i,t,pa,md} \leq CD^{m}_{i,md} \cdot YD^{m}_{md}, \forall i, t, md \tag{13.6}$$

$$\sum_{i} XA^{p}_{i,t,pa} \leq CA^{p}_{pa} \cdot YA^{p}_{pa}, \forall t, pa \tag{13.7}$$

$$\sum_{i} \left(XI^{p}_{i,t,pa} \cdot V_{i} \right) \leq CI^{p}_{pa} \cdot YA^{p}_{pa}, \forall t, pa \tag{13.8}$$

$$\sum_{ma} XB^{mp}_{i,t,ma,pb} + \sum_{pa} XB^{pp}_{i,t,pa,pb} \leq CB^{p}_{i,pb} \cdot YB^{p}_{pb}, \forall i, t, pb \tag{13.9}$$

$$\sum_{ma} XC^{mp}_{i,t,ma,pc} + \sum_{pa} XC^{pp}_{i,t,pa,pc} \leq CC^{p}_{i,pc} \cdot YC^{p}_{pc}, \forall i, t, pc \tag{13.10}$$

$$\sum_{ma} XD^{mp}_{i,t,ma,pd} + \sum_{pa} XD^{pp}_{i,t,pa,pd} \leq CD^{p}_{i,pd} \cdot YD^{p}_{pd}, \forall i, t, pd \tag{13.11}$$

Inequalities (13.1)–(13.11) provide the various capacity constraints for the reverse logistics facilities under development, as well as the upper bounds for the EOL products that a 3PRL provider will be committed to manage. Specifically, constraint (13.1) is applicable for the case where the manufacturer enters the national reverse logistics system for a subset of or all of his/her products. Constraints (13.2)–(13.6) represent the capacity restrictions for the manufacturer's owned facilities: the collection and sorting, warehouse, refurbishing, remanufacturing and recycling facilities. Similarly, constraints (13.7)–(13.11) capture the quantities of the EOL products that the 3PRL providers will be responsible for handling (for the collection and sorting, warehousing, refurbishing, remanufacturing and recycling processes). These quantity commitments are case-specific and depend on the contractual agreement between the manufacturer and the third-party provider. Finally, the collection and sorting capacities, (13.2) and (13.7), depend on the total number of the returned products, irrespectively of the types of products considered, while the warehouse capacities, (13.3) and (13.8), are volume rather than product specific.

Balance constraints:

$$X^{s}_{i,t} + \sum_{ma} XA^{m}_{i,t,ma} + \sum_{pa} XA^{p}_{i,t,pa} = R_{i,t}, \forall i, t \tag{13.12}$$

$$XA_{i,t,ma}^{m} + XI_{i,t-1,ma}^{m} = XI_{i,t,ma}^{m} + \sum_{mb} XB_{i,t,ma,mb}^{mm} + \sum_{pb} XB_{i,t,ma,pb}^{mp} + \sum_{mc} XC_{i,t,ma,mc}^{mm} + \sum_{pc} XC_{i,t,ma,pc}^{mp} + \sum_{md} XD_{i,t,ma,md}^{mm} + \sum_{pd} XD_{i,t,ma,pd}^{mp} + XW_{i,t,ma}^{m}, \forall i,t,ma \tag{13.13}$$

$$XA_{i,t,pa}^{p} + XI_{i,t-1,pa}^{p} = XI_{i,t,pa}^{p} + \sum_{mb} XB_{i,t,pa,mb}^{pm} + \sum_{pb} XB_{i,t,pa,pb}^{pp} + \sum_{mc} XC_{i,t,pa,mc}^{pm} + \sum_{pc} XC_{i,t,pa,pc}^{pp} + \sum_{md} XD_{i,t,pa,md}^{pm} + \sum_{pd} XD_{i,t,pa,pd}^{pp} + XW_{i,t,pa}^{p}, \forall i,t,pa \tag{13.14}$$

$$XI_{i,0,ma}^{m}, XI_{i,T,ma}^{m}, XI_{i,0,pa}^{p}, XI_{i,T,pa}^{p} = 0, \forall i, ma, pa \tag{13.15}$$

$$\sum_{ma}\sum_{mb} XB_{i,t,ma,mb}^{mm} + \sum_{ma}\sum_{pb} XB_{i,t,ma,pb}^{mp} = XVB_{i,t}^{m}, \forall i,t \tag{13.16}$$

$$\sum_{ma}\sum_{mc} XC_{i,t,ma,mc}^{mm} + \sum_{ma}\sum_{pc} XC_{i,t,ma,pc}^{mp} = XVC_{i,t}^{m}, \forall i,t \tag{13.17}$$

$$\sum_{ma}\sum_{md} XD_{i,t,ma,md}^{mm} + \sum_{ma}\sum_{pd} XD_{i,t,ma,pd}^{mp} = XVD_{i,t}^{m}, \forall i,t \tag{13.18}$$

$$\sum_{pa}\sum_{mb} XB_{i,t,pa,mb}^{pm} + \sum_{pa}\sum_{pb} XB_{i,t,pa,pb}^{pp} = XVB_{i,t}^{p}, \forall i,t \tag{13.19}$$

$$\sum_{pa}\sum_{mc} XC_{i,t,pa,mc}^{pm} + \sum_{pa}\sum_{pc} XC_{i,t,pa,pc}^{pp} = XVC_{i,t}^{p}, \forall i,t \tag{13.20}$$

$$\sum_{pa}\sum_{md} XD_{i,t,pa,md}^{pm} + \sum_{pa}\sum_{pd} XD_{i,t,pa,pd}^{pp} = XVD_{i,t}^{p}, \forall i,t \tag{13.21}$$

The set of constraints, (13.12)–(13.21), constitutes the classical balance equations from the point of collecting the EOL products up to the point of recovering value from them. More specifically, Equation 13.12 describes the allocation of the returned products to the manufacturer, to the 3PRL providers, and to the national reverse logistics system. The collected products are then allocated to the alternative recovery centres, according to equalities (13.13) and (13.14). Equation 13.15 defines the starting and ending conditions for the on-hand inventory in collected EOL products throughout the planning horizon. Normally, these conditions can take either zero or positive values, depending on the specific instance. Equations 13.16–13.21 are auxiliary constraints and provide the totals of refurbished, remanufactured and recycled products in manufacturer's and 3PRL providers' facilities, respectively.

Quality constraints:

$$XVB_{i,t}^{m} \leq QB_i \cdot \left(\sum_{ma} XA_{i,t,ma}^{m} + XI_{i,t-1,ma}^{m} - XI_{i,t,ma}^{m} \right), \forall i,t \tag{13.22}$$

$$XVB_{i,t}^{m} + XVC_{i,t}^{m} \leq QC_i \cdot \left(\sum_{ma} XA_{i,t,ma}^{m} + XI_{i,t-1,ma}^{m} - XI_{i,t,ma}^{m} \right), \forall i,t \tag{13.23}$$

$$XVB_{i,t}^{m} + XVC_{i,t}^{m} + XVD_{i,t}^{m} \leq QD_i \cdot \left(\sum_{ma} XA_{i,t,ma}^{m} + XI_{i,t-1,ma}^{m} - XI_{i,t,ma}^{m} \right), \forall i,t \tag{13.24}$$

$$XVB_{i,t}^{p} \leq QB_i \cdot \left(\sum_{pa} XA_{i,t,pa}^{p} + XI_{i,t-1,pa}^{p} - XI_{i,t,pa}^{p} \right), \forall i,t \tag{13.25}$$

$$XVB_{i,t}^{p} + XVC_{i,t}^{p} \leq QC_i \cdot \left(\sum_{pa} XA_{i,t,pa}^{p} + XI_{i,t-1,pa}^{p} - XI_{i,t,pa}^{p} \right), \forall i,t \tag{13.26}$$

$$XVB_{i,t}^{p} + XVC_{i,t}^{p} + XVD_{i,t}^{p} \leq QD_i \cdot \left(\sum_{pa} XA_{i,t,pa}^{p} + XI_{i,t-1,pa}^{p} - XI_{i,t,pa}^{p} \right), \forall i,t \tag{13.27}$$

Inequalities (13.22)–(13.27) indicate that not all the EOL products will be in a satisfactory condition/quality for recovery purposes. Only a certain fraction of the collected EOL products in each period, along with the on-hand inventory of the previous period minus the products being inventoried in the current period, will be reusable, remanufacturable and recyclable. In general, it is assumed that the major percentage of the returned products will be recyclable, a lesser percentage remanufacturable and an even lesser part reusable. Products that can be reused can also be remanufactured and recycled, while products that are remanufacturable are also recyclable. In both cases, the reverse conditions do not always hold.

Demand satisfaction:

$$XVB_{i,t}^{m} + XVB_{i,t}^{p} - XBB_{i,t-1} + XBB_{i,t} \geq DB_{i,t} - DB_{i,t} \cdot Y_i^{s} - XLB_{i,t}, \forall i,t \tag{13.28}$$

$$XVC_{i,t}^{m} + XVC_{i,t}^{p} - XBC_{i,t-1} + XBC_{i,t} \geq DC_{i,t} - DC_{i,t} \cdot Y_i^{s} - XLC_{i,t}, \forall i,t \tag{13.29}$$

$$XVD_{i,t}^{m} + XVD_{i,t}^{p} - XBD_{i,t-1} + XBD_{i,t} \geq DD_{i,t} - DD_{i,t} \cdot Y_i^{s} - XLD_{i,t}, \forall i,t \tag{13.30}$$

$$XBB_{i,0}, XBC_{i,0}, XBD_{i,0}, XBB_{i,T}, XBC_{i,T}, XBD_{i,T} = 0, \forall i \tag{13.31}$$

Inequalities (13.28)–(13.30) capture the satisfaction of the demand by refurbished, remanufactured and recycled products, respectively. It is assumed that when the national reverse logistics system undertakes the reverse logistics processes of specific types of products on behalf of the manufacturer, then the manufacturer will no longer be responsible for the recovery operations of these products, nor has he/she to satisfy a specific demand for them (WEEE directive) [13.2]. The right-hand side of the inequalities (see the use of Y_i^s binary variables) refers to the latter case. Finally, the starting and ending values of the backorders in recovered EOL products are determined from Equation 13.31 and they could have either zero or positive values.

Regulatory-type constraints:

$$\begin{aligned} & XVB_{i,t}^{m} + XVB_{i,t}^{p} + XVC_{i,t}^{m} + XVC_{i,t}^{p} + \\ & XVD_{i,t}^{m} + XVD_{i,t}^{p} \geq L_i \cdot R_{i,t} \cdot \left(1 - Y_i^s\right), \forall i,t \end{aligned} \tag{13.32}$$

$$\begin{aligned} & \sum_t \left(\sum_{ma} XA_{i,t,ma}^{m} + \sum_{pa} XA_{i,t,pa}^{p} \right) + \\ & \sum_t \begin{pmatrix} XVB_{i,t}^{m} + XVB_{i,t}^{p} + XVC_{i,t}^{m} + \\ XVC_{i,t}^{p} + XVD_{i,t}^{m} + XVD_{i,t}^{p} \end{pmatrix} \leq M \cdot \left(1 - Y_i^s\right), \forall i \end{aligned} \tag{13.33}$$

Inequalities (13.32) concern specific regulatory-type constraints that are effective in many European Union countries (WEEE directive) [13.2]. More specifically, in case where the manufacturer does not enter into the national reverse logistics system, he/she will be responsible for recovering at least a minimum weight percentage of the returned products. This group of constraints can be easily adjusted, so as to accommodate alternative regulatory restrictions that are effective in other countries. Furthermore, constraints (13.33) ensure that the manufacturer will not be responsible for the collection and recovery processes of those products that the national reverse logistics system handles.

Environmental-based constraints:

$$XLB_{i,t} \leq FB_i \cdot DB_{i,t}, \forall i,t \tag{13.34}$$

$$XLC_{i,t} \leq FC_i \cdot DC_{i,t}, \forall i,t \tag{13.35}$$

$$XLD_{i,t} \leq FD_i \cdot DD_{i,t}, \forall i,t \tag{13.36}$$

Constraints (13.34)–(13.36) provide the upper acceptable bounds for the unsatisfied demand in refurbished, remanufactured and recycled products, respectively. These constraints have an environmental rationale, while aiming indirectly to minimise the uncontrollable disposal of the EOL products.

Finally, the variables related to the investments in reverse logistics facilities and to the agreements with the 3PRL providers are binary, while the rest are positive continuous variables.

13.4.3 Solution Performance

The proposed MILP model can be solved through any of the commercially available mathematical programming software. In order to test the solution behaviour of the model, a number of different instances were developed. All the problems were solved using standard branch and bound (B&B) techniques through the commercially available CPLEX 9.1 solver. The solution performance of the proposed MILP model for small- and medium-scale problems is reasonably acceptable: for ten types of products, five time periods, and four alternative scenarios for each network configuration option, the solution time is approximately 2.5 min. On the other hand, for large-scale problems (*e.g.* twenty types of products, ten time periods, and eight alternative scenarios for each network configuration option), the computational time is about an hour. Generally, as the size of the model increases the computational time increases significantly. However, considering the fact that the proposed model is a strategic-design one (and thus the problem will need to be solved infrequently), the resulting computational times are quite satisfactory.

The major determinants/drivers of the size of the proposed model are the number of considered products types and considered time periods. The need to obtain high-quality solutions for modest computational effort remains a priority in mathematical programming. To this effect, below we propose a few specific approaches for reducing the size of the problem and the solution time:

- By properly aggregating the examined different types of products, we can drastically limit their number and therefore the size of the problem. For instance, the grouping procedure can be based on particular design of the products (size, volume, weight, *etc.*), the potential similarities at the implementation of their recovery processes, and on the volume of their returns.
- An initial evaluation of the different network structures under investigation can be achieved through the development of an aggregated analytical model. This model could be a preliminary one and in a short period of time would provide significant insights on the particular network structures that are more 'appealing'. In this case, the initial alternative network structure scenarios for the comprehensive model will be drastically limited, and the scale of the model will be significantly reduced.
- It is not always necessary to explore all the B&B nodes of a problem in order to obtain an optimal or near-optimal solution. When restricting the solution time to a specific time limit (significantly smaller than the total computational time), the resulting solution may still happen to be the optimal one or has an insignificantly small optimality gap, something common in MILP modelling. In this way, the use of heuristic solution algorithms can be 'substituted' when necessary.
- Although it is not within the scope of this chapter to present heuristic solution algorithms, and given the fact that there exists a plethora of efficient solution algorithms relevant to our problem in the literature body, the reader is referred to the detailed overviews of such algorithms and solution techniques

of Mirchandani and Francis [13.38], Daskin [13.39] and Labbe and Louveaux [13.40].

13.4.4 Sensitivity Analysis and Managerial Insights

The conduct of sensitivity analysis is pivotal for the real-world applicability of any integrated decision-making methodology. Therefore in this subsection we discuss the different ways of conducting sensitivity analysis, while providing some useful managerial insights that can result from the solution of the proposed model.

Generally, if the proposed model is a linear-programming model then classical sensitivity analysis can be implemented. However, for an MILP problem the most popular way of performing such an analysis is through fixing the binary variables to their optimal values and running the resulting linear model [13.41]. By doing so and by taking account of the changes in the continuous variables, one can obtain significant insights and information.

The precise estimation of the input data has a critical role in the whole analysis. The cost and general parameters should be estimated as accurately as possible. However, there may exist some parameters with innate stochasticity. For instance, the volume and the quality of the returned EOL products in each period may be variable and thus they would follow specific probability distributions. In order to tackle this challenge, we can generate a large number of random instances through Monte Carlo simulation, and the resulting solutions can be statistically processed in order to obtain useful managerial insights and be guided to more robust decisions. This sensitivity analysis approach facilitates the design of proper statistical experiments, and as well as the evaluation of various 'what-if' scenarios.

A significant insight that results from the solution of the model involves the identification of the system parameters that mainly define the structure of the network under development. For example, when the regulatory constraints are demanding, and/or when the investment costs for developing recovery facilities are quite high, and/or when the EOL returns are limited, then the options of outsourcing and of joining the national reverse logistics system may be more appealing than the development of reverse logistics facilities. Moreover, for large and increased volumes of EOL returns of good quality, the development of recovery facilities may be more profitable than the other two alternatives.

Additional interesting insights can result from answering the following questions and from the proper conduct of sensitivity analysis:

- What is the effect of a considerable increase or decrease in the returns of EOL products on the optimal solution?
- What is the impact of specific changes in the unit variable costs of recovering an EOL product on the optimal configuration of the reverse logistics network?
- What is the impact of stricter regulatory-type constraints on the final decisions?
- What are the consequences of a considerable increase in low-quality EOL products?
- Are the decision-making processes sensitive to the value of the interest rate?

- Does the case of time-increasing capacities instead of constant ones affect the optimal network structure?
- Are specific shifts in the unit variable costs related to how demanding a regulatory constraint is?

13.5 Extensions and Future Research Directions

13.5.1 Model Extensions

The proposed methodological approach can be easily adjusted in order to accommodate a plethora of additional or alternative problem instances. Towards this effect, we present below additional issues that can be included in the developed MILP model:

- *Specific number of facilities* – in many business environments, it is preferable to develop a specific number of reverse logistics facilities, instead of allowing the model to decide the optimal number of the facilities to be opened. For instance, let *n* be the maximum number of collection and sorting facilities that the manufacturer wishes to develop. In such case the following constraint should be inserted to the model:

$$\sum_{ma} YA^{m}_{ma} \leq n \tag{13.37}$$

  Analogous constraints can be added for the refurbishing, remanufacturing and recycling facilities.

- *Implementation of the same recovery policy across all the types of products examined* – in order to limit the complexity of the network under development and potentially to further improve the control and monitoring of it, a uniform recovery policy can be implemented across all product types. More specifically, all the products can be processed through the manufacturer's owned facilities, or all the reverse logistics processes can be outsourced, or the manufacturer may enter into the national reverse logistics system. To this effect, Y^{s}_{i} variables are substituted by a single binary variable Y^{s} representing all the products. Moreover, the following constraints should be added:

$$Y^{s} + YA^{m}_{ma} \leq 1, \forall ma \tag{13.38}$$

$$Y^{s} + YA^{p}_{pa} \leq 1, \forall pa \tag{13.39}$$

$$YA^{m}_{ma} + YA^{p}_{pa} \leq 1, \forall ma, pa \tag{13.40}$$

$$YB^{m}_{mb} + YB^{p}_{pb} \leq 1, \forall mb, pb \tag{13.41}$$

$$YB^m_{mb} + YC^p_{pc} \leq 1, \forall mb, pc \tag{13.42}$$

$$YB^m_{mb} + YD^p_{pd} \leq 1, \forall mb, pd \tag{13.43}$$

$$YC^m_{mc} + YB^p_{pb} \leq 1, \forall mc, pb \tag{13.44}$$

$$YC^m_{mc} + YC^p_{pc} \leq 1, \forall mc, pc \tag{13.45}$$

$$YC^m_{mc} + YD^p_{pd} \leq 1, \forall mc, pd \tag{13.46}$$

$$YD^m_{md} + YB^p_{pb} \leq 1, \forall md, pb \tag{13.47}$$

$$YD^m_{md} + YC^p_{pc} \leq 1, \forall md, pc \tag{13.48}$$

$$YD^m_{md} + YD^p_{pd} \leq 1, \forall md, pd \tag{13.49}$$

- *Establishing new facilities and capacity utilisation* – constraints that limit the potential build-up of new facilities to only the ones with capacity utilisation ratio larger than a pre-specified threshold can be reasonably added per case to the model. For instance, if *TD* is the capacity utilisation threshold for the recycling facilities, then the following constraints should be added to the model:

$$\sum_i \sum_t \sum_{ma} XD^{mm}_{i,t,ma,md} + \sum_i \sum_t \sum_{pa} XD^{pm}_{i,t,pa,md} \geq TD \cdot \sum_i \sum_t CD^m_{i,md} \cdot YD^m_{md}, \forall md \tag{13.50}$$

  Analogous constraints can be added for the collection, refurbishing and remanufacturing facilities.

- *Common refurbishing, remanufacturing and recycling facilities* – the proposed model can also accommodate in two different ways (implicitly and explicitly) the development of common collection, refurbishing, remanufacturing and recycling facilities, so as to exploit the employing economies of scale. Firstly without any changes, when the indices *ma*, *mb*, *mc* and *md* take the same value then the respective facilities under development are to be co-developed within the same installation. For example, when $ma=mb=mc=md=1$ and the corresponding binary variables take non-zero values, then the collection, refurbishing, remanufacturing and recycling facilities will be accommodated under the same installation/location. On the other hand, the integrated recovery facilities can also be inserted into the model in a more explicit way. More particularly, proper continuous and binary variables that indicate the development of integrated reverse logistics facilities could be added. The model can be easily transformed in order to accommodate this additional case.

- *Selection of recovery technology* – without any change, for each recovery process we can make a selection among alternative facilities of varying technology, and as a result of different capacity, investment and variable costs.
- *Time-varying capacities* – the capacities considered so far by the proposed model are constant. However, we can also consider the case where extra capacity is installed over time. For instance, instead of examining alternative recovery facilities of constant capacity, we can also examine in parallel scenarios for facilities with time-increasing capacities; this can be an interesting extension of the proposed model for when the volume of the returned EOL products is time-increasing.
- *Alternative regulatory requirements* – apart from considering the regulatory restrictions of the WEEE directive, alternative regulatory constraints can be easily accommodated into the proposed model.
- *Different recovery options* – the case of different recovery options can be easily incorporated into the proposed MILP model. Additional variables can be added that will indicate, for instance, alternative remanufacturing and recycling options (stemming from different disassembly depths) with different revenues and variable costs.
- *Collaboration issues* – the presented model can also hold for the case that two or more manufacturers wish to collaborate by developing common recovery facilities. It is not unusual that the investment costs for developing recovery facilities can be very high [13.35]. As a consequence, some manufacturers may wish to configure a common reverse logistics network. The presented model can also take into account 'as is' this alternative problem, while the allocation of the investment and variable recovery costs among the manufacturers can be pre-specified.

13.5.2 Future Research

The development of analytical methodological approaches for the strategic design/ configuration of reverse and agile closed-loop logistics networks is a research field with several promising research avenues. Below, specific directions for future research are provided:

- An interesting direction for future research lies in the development of a holistic methodological approach that will jointly treat the forward and reverse supply chain network configuration problems, by relaxing the assumption of the pre-existence of the forward supply channel. This case involves the joint design of the forward and the reverse supply chains (CLSC network) right from the start, by considering their synergies.
- The assignment of the returned EOL products to specific collection and recovery centres and the determination of the geographic area that a specific facility covers are two significant realistic allocation aspects that can be taken into account.
- It seems worthwhile to systematically explore the potential of geographical information systems (GIS) in the examined problem.

- The inclusion of vehicle routing aspects in the forward supply chain network configuration problem is a relatively new and very promising stream of research. As a result, the inclusion of vehicle routing issues in the design of reverse logistics networks seems to have a significant potential.
- Finally, an important direction for future research stems from employing game theory in order to systematically take into account the synergies with pre-existing competitive facilities in the examined geographic area, and also to strengthen the stochastic nature of the new models.

13.6 Conclusions

The design of reverse logistics supply chains has emerged as a significant issue in the executive agenda of many manufacturing firms. Because of the high-value content of the returned EOL products, especially if they are recovered quickly, reverse logistics can be turned into a significant corporate competitive advantage. In this framework the present chapter builds upon the general concepts that were developed by previous research works, extending them first by the presentation of a comprehensive up-to-date literature review of related papers dealing with the reverse logistics network design problem, and second by proposing a generic analytical decision-making methodology for the optimal design/configuration of a reverse logistics network with forward supply channel synergies. The proposed methodology constitutes an integrated decision-support framework and not merely a new optimisation model. On the other hand, the presented model can accommodate a large number of different problem instances. Certain modelling aspects such as outsourcing, state reverse logistics systems, regulatory constraints, the time value of money, and the case of both remanufacturing- and recycling-driven networks, which are (almost) ignored by previous research works, are explicitly taken into account. Moreover, the solution performance of the model seems to be satisfactory both for medium- and large-scale problems. In addition, a novel sensitivity analysis type methodology is proposed based on Monte Carlo simulation. Finally, significant potential managerial insights are discussed regarding the optimal structure of the network under development, while many useful extensions of the proposed model and directions for future research are provided.

References

[13.1] Thierry, M., Salomon, M., van Nunen, J.A.E.E. and van Wassenhove, L.N., 1995, "Strategic issues in product recovery management," *California Management Review*, **37**(2), pp. 114–135.

[13.2] European Council, 2003, *Directive 2002/96 on Waste Electrical and Electronic Equipment*.

[13.3] Koroneos, C., Moussiopoulos, N., Achillas, H., Dompros, A., Bouzakis, K.-D., Michailidis, N., Iakovou, E. and Xanthopoulos, A., 2005, "Implementation of recycling aspects' integrated product policy in electrical and electronic equipment," In *Proceedings of 2nd International Conference on Manufacturing Engineering*, pp. 855–865.

[13.4] Xanthopoulos, A. and Iakovou, E., 2006, "A methodological framework for the efficient configuration of a reverse logistics network," In *Proceedings of Protection and Restoration of the Environment VIII*, pp. 358:1–358:8.

[13.5] Gottinger, H.W., 1988, "A computational model for solid waste management with application," *European Journal of Operational Research*, **35**, pp. 350–364.

[13.6] Caruso, C., Colorni, A. and Paruccini, M., 1993, "The regional urban solid waste management system: a modelling approach," *European Journal of Operational Research*, **70**, pp. 16–30.

[13.7] Berger, T. and Debaillie, B., 1997, *Location of Disassembly Centres for Reuse to Extend an Existing Distribution Network*, Master's Thesis, University of Leuven.

[13.8] Realff, M., Ammons, J. and Newton, D., 1999, "Carpet recycling: determining the reverse production system design," *The Journal of Polymer-Plastics Technology and Engineering*, **38**, pp. 547–567.

[13.9] Realff, M., Ammons, J. and Newton, D., 2004, "Robust reverse production system design for carpet recycling," *IIE Transactions*, **36**, pp. 767–776.

[13.10] Louwers, D., Kip, B.J., Peters, E., Souren, F. and Flapper, S.D.P., 1999, "A facility location-allocation model for reusing carpet materials," *Computers and Industrial Engineering*, **36**, pp. 855–869.

[13.11] Chang, N.-B. and Wei, Y.L., 2000, "Sitting recycling drop-off stations in urban area by genetic algorithm-based fuzzy multiobjective nonlinear integer programming modelling," *FUZZY Sets and Systems*, **114**, pp. 133–149.

[13.12] Jayaraman, V., Guide Jr, V.D.R. and Srivastava, R., 1999, "A closed-loop logistics model for remanufacturing," *Journal of the Operational Research Society*, **50**, pp. 497–508.

[13.13] Jayaraman, V., Patterson, R.A. and Rolland, E., 2003, "The design of reverse distribution networks: models and solution procedures," *European Journal of Operational Research*, **150**, pp. 128–149.

[13.14] Min, H., Ko, H.J. and Ko, C.S., 2006, "A genetic algorithm approach to developing the multi-echelon reverse logistics network for product returns," *Omega*, **34**, pp. 56–69.

[13.15] Kroon, L. and Vrijens, G., 1995, "Returnable containers: an example of reverse logistics," *International Journal of Physical Distribution and Logistics Management*, **25**(2), pp. 56–68.

[13.16] Spengler, T., Püchert, H., Penkuhn, T. and Rentz, O., 1997, "Environmental integrated production and recycling management," *European Journal of Operational Research*, **97**, pp. 308–326.

[13.17] Barros, A.I., Dekker, R. and Scholten, V., 1998, "A two-level network for recycling sand: a case study," *European Journal of Operational Research*, **110**, pp. 199–214.

[13.18] Listes, O. and Dekker, R., 2005, "A stochastic approach to a case study for product recovery network design," *European Journal of Operational Research*, **160**, pp. 268–287.

[13.19] Krikke, H.R., Van Harten, A. and Schuur, P.C., 1999, "Business case Océ: reverse logistics network redesign for copiers," *OR-Spektrum*, **21**(3), pp. 381–409.

[13.20] Beamon, B.M. and Fernandes, C., 2004, "Supply chain network configuration for product recovery," *Production Planning and Control*, **15**(3), pp. 270–281.

[13.21] Pochampally, K.K. and Gupta, S.M., 2005, "Strategic planning of a reverse supply chain network," *International Journal of Integrated Supply Chain Management*, **1**(4), pp. 421–441.

[13.22] Min, H., Ko, C.S. and Ko, H.J., 2006, "The spatial and temporal consolidation of returned products in a closed-loop supply chain network," *Computers & Industrial Engineering*, **51**, pp. 309–320.

[13.23] Ko, H.J. and Evans, G.W., 2007, "A genetic algorithm-based heuristic for the dynamic integrated forward/reverse logistics network for 3PLs," *Computers and Operations Research*, **34**, pp. 346–366.
[13.24] Wang, Z., Yao, D.-Q. and Huang, P., 2007, "A new location-inventory policy with reverse logistics applied to B2C e-markets of China," *International Journal of Production Economics*, **107**, pp. 350–363.
[13.25] Fleischmann, M., 2000, *Quantitative Models for Reverse Logistics*, Doctoral Thesis, Erasmus University Rotterdam.
[13.26] Salema, M.I.G., Barbosa-Povoa, A.P. and Novais, A.Q., 2007, "An optimization model for the design of a capacitated multi-product reverse logistics network with uncertainty," *European Journal of Operational Research*, **179**, pp. 1063–1077.
[13.27] Lieckens, K. and Vandaele, N., 2007, "Reverse logistics network design with stochastic lead times," *Computers and Operations Research*, **34**, pp. 395–416.
[13.28] Marin, A. and Pelegrin, B., 1998, "The return plant location problem: modelling and resolution," *European Journal of Operational Research*, **104**, pp. 375–392.
[13.29] Krikke, H.R., Bloemhof-Ruwaard, J.M. and van Wassenhove, L.N., 2003, "Concurrent product and closed-loop supply chain design with an application to refrigerators," *International Journal of Production Research*, **41**(16), pp. 3689–3719.
[13.30] Listes, O., 2007, "A generic stochastic model for supply and return network design," *Computers and Operations Research*, **34**, pp. 417–442.
[13.31] Lu, Z. and Bostel, N., 2007, "A facility location model for logistics systems including reverse flows: the case of remanufacturing activities," *Computers and Operations Research*, **34**, pp. 299–323.
[13.32] Sahyouni, K., Savaskan, R.C. and Daskin, M.S., 2007, "A facility location model for bidirectional flows," *Transportation Science*, **41**(4), pp. 484–499.
[13.33] Fleischmann, M., 2003, "Reverse logistics network structures and design," In *Business Aspects of Closed-Loop Supply Chains*, van Wassenhove, L.N. and Guide Jr., V.D.R. (eds.), Carnegie Mellon University Press, Pittsburgh, pp. 117–148.
[13.34] EyeforTransport, 2005, *European 3PL Market Survey Report*, 3rd EyeforTransport European 3PL Summit.
[13.35] Malone, R., 2005, *Reverse Side of Logistics: The Business of Returns*, Forbes.com, http://www.forbes.com/.
[13.36] Rupnow, P., 2003, *Reverse Logistics Services for 3PL's*, Reverse Logistics Professional, http://www.reverselogisticsprofessional.com/.
[13.37] Savaskan, R.C., Bhattacharya, S. and van Wassenhove, L.N., 2004, "Closed-loop supply chain models with product remanufacturing," *Management Science*, **50**(2), pp. 239–252.
[13.38] Mirchandani, P.B. and Francis, R.L., 1989, *Discrete Location Theory*, Wiley, New York.
[13.39] Daskin, M.S., 1995, *Network and Discrete Location, Models, Algorithms, and Application*, Wiley-Interscience, New York.
[13.40] Labbe, M. and Louveaux, F., 1997, "Location problems," In *Annotated Bibliographies in Combinatorial Optimization*, Wiley, New York, pp. 261–282.
[13.41] Bloemhof-Ruwaard, J.M., Krikke, H. and van Wassenhove, L.N., 2004, "OR model for eco-eco closed-loop supply chain optimization," In *Reverse Logistics Quantitative Models for Closed-Loop Supply Chains*, Springer, Heidelberg, pp. 357–372.

14

The Evolution of Logistics Service Providers and the Role of Internet-based Applications in Facilitating Global Operations

Aristides Matopoulos[1] and Eleni-Maria Papadopoulou[2]

[1] City College, Business Administration and Economics Department
International Faculty of the University of Sheffield
3 Leontos Sofou Street, 54626 Thessaloniki, Greece
Email: amatopoulos@city.academic.gr

[2] Department of Applied Informatics, University of Macedonia
156 Egnatia Street, 540 06, Thessaloniki, Greece
Email: elpap@uom.gr

Abstract
The need for global logistics services has increased dramatically and become extremely complex and dynamic as a result of a number of changes in manufacturing and in industrial production. In response, the logistics industry is changing in a variety of ways, including mergers to form integrated transportation service providers, outsourcing and increased use of information technology. The aim of this chapter is to provide an overview of the evolution and the most important trends in the logistics services provider (LSP) industry. Specific emphasis will be given to the role of Internet-based applications. Within this context, the chapter will also present the role of logistics e-marketplaces. In particular, based on the secondary research of currently existing logistics on-line marketplaces, an analysis and classification of them is provided with the aim of identifying service gaps. The analysis reveals that logistics electronic marketplaces, despite the increased range of services currently offered, still face limitations with reference to integrated customs links or translation services, which both reduce the efficiency of global operations.

14.1 Introduction

The changes we have witnessed in industrial production in the new economy have been shaped to a great extend by long-term economic and logistics trends. Markets have become global in scale, and companies have outsourced part, if not all, of their production. Not surprisingly, these new models of production and distribution are changing the demand for transportation, but also the nature of the services offered, with shippers putting increasing emphasis on other attributes – particularly, speedy, reliable deliveries.

In parallel to these changes, the need for global logistics services has dramatically increased and has become extremely complex and dynamic. Indeed, several entities are currently involved in a product's distribution, often characterised by separate, and sometimes conflicting, goals. Transportation, for example, is an extremely complex activity, given the number of participating agents [14.1], as well as the magnitude of interactions needed in order to achieve distribution on a global scale. The complexity further increases due to the pluralism of physical, information and communication structures, the difference in dynamics regarding the flows of cargo, cash and information, and the differentiation in property adaptation, regarding the decision-making processes followed by each company [14.2].

In response, the structure of the logistics industry is changing in a variety of ways, including mergers to form integrated transportation service providers, outsourcing and increased use of information technology. The aim of this chapter is to provide an overview of the logistics services provider (LSP) sector with specific emphasis being paid to the role of Internet-based applications. Electronic business has, in theory, great potential to reshape markets, and with them, the demand for logistics services. Within this context, the chapter will present the role and the current implementation of e-business applications in the logistics industry.

In the next section, a discussion of the evolution and the most important trends in the LSP industry is provided. In Section 14.3, the emergence of electronic marketplaces is presented together with an analysis of currently existing electronic logistics marketplaces, while in the final section conclusions are drawn and key issues for further research are identified.

14.2 Logistics Service Providers: Evolution and Major Trends

14.2.1 LSPs: Context and Types

Logistics service providers (LSPs) are companies, often acting as intermediaries, that undertake the execution of logistics-related activities that have been traditionally kept in-house [14.3, 14.4], also tendering for the dissemination of accurate and timely information among supply chain partners. The most common types of LSPs are: carriers, third-party logistics providers (3PLs), international freight forwarders (IFF), non-vessel-operating common carriers (NVOCC) and fourth-party logistics providers (4PLs).

Carriers refer to companies that possess the means to accomplish goods transportations, and can be specialised in ocean carriage, air-/rail-shipments and inland haulage. The services that they normally provide include inbound and outbound transportation, door-to-door transportation, contract delivery, transport administration, documentation processes, shipment scheduling, tracking and tracing [14.5]. The intermediation of freight forwarders or 3PLs is not always necessary, thus allowing the carriers to involve directly in the buyer–supplier relationship with long-term contracts, trying to deploy strategic planning, aiming at the establishment of a fruitful partnership [14.6].

In contrast to carriers, 3PLs do not perform only transportation-related activities, but instead are more involved in a wide variety of services categorised into the areas of transportation, warehousing, inventory management, order processing, information systems and packaging [14.7, 14.8]. IFFs and NVOCCs possess logistical expertise, holding a prominent position in international trade. Acting as non-asset-based intermediaries, they provide consultative information regarding the selection of the appropriate transport mode and routing, they monitor the shipments through track and trace mechanisms, they handle commercial documentation, cargo insurance and customs clearance [14.9], and in some occasions they enter the field of the 3PLs by offering packing and warehousing services [14.10].

NVOCCs differ from freight forwarders in that they arrange the consolidation of partial shipments from multiple origins to a common destination into a single container. Their functions include purchasing transportation services from vessel-operating common carriers for resale, payment of port-to-port or multi-modal transportation charges, issuing bills of lading, arranging and paying for inland haulage on door-to-door transportation, *etc.* [14.11]. Finally, 4PLs act as 'pure brokers' [14.12], aiming at coordination of resources and synchronisation [14.13] of the supply chain members, in order to respond to the customer specifications and leverage the supply chain into a value chain [14.14]. According to Stefansson [14.5], most 4PLs are non-asset owners that undertake administrative processes, leaving the 3PLs to handle the physical ones.

14.2.2 Evolution and Characteristics of the LSP Market

The LSP market is a very dynamic one and has evolved over the decades, in an attempt to follow customers' pace and fulfil their constantly changing expectations [14.15]. Shippers'/consignees' competitive advantage is based on the provision of differentiated, innovative and quality services offered by LSPs [14.16], thus imposing significant pressure on them to continually evolve. An additional factor that drives the evolution of LSPs is the ever-expanding scale and scope of operations [14.17], as a consequence of the increased international competition [14.18]. For example, Berglund *et al.* [14.19] distinguished the three waves of entrants in the 3PL sector, namely the 'traditional third-party logistics providers', the 'network players' and the 'systems-based players'. The first category emerged during the 1980s providing the basic services of transportation and warehousing (*e.g.* Exel in UK, Frans Maas in the Netherlands). The second category appeared in the early 1990s, with the evolution of parcel and express services, based on global air express networks (*e.g.* DHL, TNT, UPS). The third wave came from the areas of information technology (IT), management consultancy and financial services in close cooperation with representatives of the first two categories. The involvement of shippers is also evident in several occasions (*e.g.* Andersen Consulting, Geologistics). Similarly, for ocean container carriers, the first voyage with a containership in 1956 [14.20] was followed by a rapid diffusion of the containerised cargo concept, leading to the reorganisation of general (dry) cargo traffic [14.21] (containerisation, inter-port competition and port selection).

In parallel to the evolution of the LSP industry and the changes mentioned above, the sector has also witnessed changes with respect to the criteria used in the selection process. Low-cost services can no longer be considered a competitive advantage if the performance specifications are not satisfied, such as on-time shipments and deliveries, financial stability, creative management, problem solving ability [14.22], reputation, prompt response to requests [14.23], reliability, traceability, smooth information sharing [14.24] and successful reverse logistics strategies [14.25].

Finally, the LSP market was also affected by the emergence of the Internet after the mid 1990s. There is no doubt that the LSP industry relies heavily on technology, being transformed in the last decade into an extremely information-intensive sector, given the hundreds of thousands of orders, deliveries, shipping notices or receipt documentations that are exchanged annually. Within this context, it is no surprise that the sector is one of the leaders, in Europe for example, in the use of ICT (information and communication technology)-based applications [14.26].

14.2.3 Major Trends

Outsourcing

The issues of survival and future growth constitute the initial drivers of outsourcing non-core functions to providers that possess the relevant expertise [14.23], so that operating costs are decreased and capital investments are avoided. Today, the pool of drivers has expanded significantly towards a more strategic direction [14.27], aiming to respond to the global competitive pressure, develop supply chain partnerships, reengineer business processes, achieve operational flexibility, penetrate new markets, access updated technology, share risks, optimise inventory levels and lead times, improve customer service and overall service quality, and broaden the services portfolio [14.15, 14.16, 14.19, 14.28, 14.29]. The risks mainly refer to service or quality issues, loss of control over the outsourced functions, lack of trust and proper communication mechanisms, inability of management to communicate the notion of outsourcing to employees, financial affairs, and no actual value added. Furthermore, information asymmetry, inadequate knowledge and lack of innovations in IT can also lead to distorted relationships [14.29–14.32]. The cooperation with efficient logistics service providers is a prerequisite; in order to properly exploit the advantages of outsourcing and deviate from the associated risks, LSPs must, therefore, be carefully determined based on a partner 'selection and negotiation' process [14.33], taking into consideration issues such as capacity constraints, streamlining of production with carriers' schedules, *etc.*, in order to establish lean logistics processes [14.34].

Global Sourcing

Trent and Monczka [14.35] defined global sourcing as the process of 'worldwide integration of engineering, operations, and procurement centres within the upstream portion of a firm's supply chain'. According to Zeng [14.36], this trend represents an integral part of the outsourcing process.

Kotabe *et al.* [14.37] recorded three waves of global sourcing, during the last 15 to 20 years. The first wave that started in the mid-1980s mainly focused on the establishment of manufacturing operations globally, aiming to achieve reductions in labour costs. The second wave that began in the early 1990s refers to the outsourcing of IT development to specialist providers, *e.g.* EDS, Accenture. The third wave started in recent years, focusing on business process outsourcing that includes accounting services, human resource management, call centres, *etc.*

The trend of global sourcing can be considered beneficial to companies that implement efficient sourcing strategies, in terms of cost, quality, supplier responsiveness, technology [14.35] and availability [14.38]. Cho and Kang [14.38] further identified the risks of global sourcing to be categorised in the clusters of logistics support, cultural differences and regulations. This trend has altered the way of conducting business, affecting both manufacturing and distribution strategies. Production and distribution are deployed through a network of coordinated partners, supported by shared information and communication technologies, thus empowering the dynamic nature of supply chains [14.39]. Although, according to Petersen *et al.* [14.40], the primary reason for global sourcing is unit price reduction, attention must be drawn to the estimation of logistics costs and their actual contribution to the total cost, in order to properly evaluate the decision for global sourcing [14.41].

Containerisation

The transfer of production facilities to low-cost countries, being the major driver of globalisation, has exemplified containerised ship transportation as the most familiar and economic solution [14.42], not representing a mode of transport, but rather a type of packaging that prevents pilferage, contamination and moisture that could be caused due to prolonged transit times. The numbers are just astonishing. According to the data published for 2009 by AXS-Alphaliner (http://www.axs-alphaliner.com/), there are 5,951 ships active on liner trades, accounting for 13,555,585 TEU (20-foot equivalent unit) and 180,876,929 TDW (tons dead weight). In addition, according to the AXS-Alphaliner data, the number of TEUs in the last decade has tripled. Figure 14.1 presents the evolution of the TEU numbers and the estimated growth in the last decade.

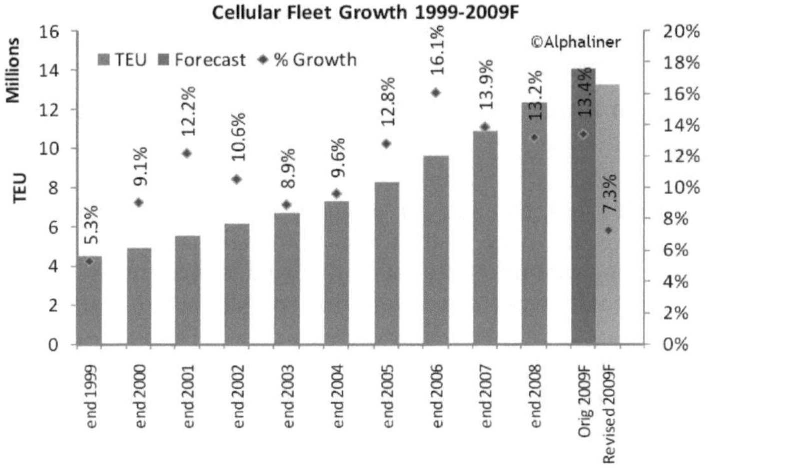

Figure 14.1. Evolution of TEU growth in the last decade (http://www.axs-alphaliner.com/)

The impact of containerisation is obvious in the modification of the trading patterns and commercial practices, the routing options, the size of vessels and terminals. One of its benefits is the economies of scale achieved, due to the design of vessels with increased capacity, which result in reduced transportation cost [14.43]. Pedersen [14.44] identified security of cargo and simplification of the transhipment process as additional benefits of containerisation that enhance the notion of intermodal door-to-door services. The concept of consolidation has emerged through containerisation as a value added service. Shippers/consignees that sell/purchase small batches of products to/from an overseas origin/destination are able to contact shipment consolidators or co-loaders that plan, arrange and optimise the movement of partial cargo [14.45], by co-loading individual less-than container load (LCL) shipments into a single container [14.46].

Transport Provider Integration

Globalisation, containerisation and outsourcing have led to the evolution of networked organisations in the field of transport and logistics services, facilitating the prompt and streamlined flow of cargo. New organisational network patterns emerge nowadays through mergers, strategic alliances, joint ventures, acquisitions, and partnerships that include mega carriers (operating global chains), niche firms (focusing on special markets and/or special commodities), and sub-suppliers (operating as sub-contractors to mega carriers and niche firms, based on their logistics competencies) [14.47]. An example of alliance in ocean carriage is the Sino-Japanese Alliance, consisting of Cosco, K-Line, Hanjin and Yang Ming Line. In the field of 3PL industry, the logistics provider Tibbett & Britten was acquired by Exel in 2004. The main objectives for LSPs to consolidate include the achievement of economies of scale and scope, penetration of new product/service markets, penetration of new geographical markets, leverage of transport chains through more intense control of global traffic flows, increase of company size, in order to invest in physical and technological infrastructure, enhancement of customer service by providing value-added services as well as competition with global 3PLs [14.24, 14.48, 14.49].

14.3 Evolution and Current State of Electronic Marketplaces in Logistics

ICT-based applications are able to promote new organisational forms on the markets, changing the way both transactions are executed and cooperation/ relationships among enterprises are established and implemented. Marketplaces and much more electronic marketplaces can play an extremely important role towards this direction.

14.3.1 Electronic Marketplaces and Logistics: Concept, Context and Evolution

A marketplace is a place where buyers and suppliers are met. Coordinating the supply and the demand and facilitating the transactions are both central tasks in a marketplace. Analogously, an electronic marketplace (EMP) is a place where the above-mentioned tasks are supported by electronic means [14.50]. In contrast to what most people believe, EMPs are not post-Internet 'inventions'. Their existence goes long before the Internet's boom. As Wheatley [14.51] argues, precursor to the currently existing Internet-based marketplaces was the development of electronic marketplaces for agricultural products (*e.g.* eggs, cotton, hog, cattle) in the 1970s or the early 1980s. In the following decades, in the mid-nineties, in the early stages of the Internet boom, marketplaces still had very simple business models centring on simply providing open marketplaces for listing or holding auctions for products or services. As a result, only a few marketplaces managed to survive, mainly due to low profits [14.51].

Christiaanse and Markus [14.52] proposed that marketplaces should be classified in two broad categories: transaction exchanges (where services or products are cleared) and collaboration exchanges (where integration and data-flow platforms exist). There is no doubt that, in the last decade, a new wave of marketplaces, falling in the second category, has emerged where market relationships can be built, meaning sharing information and fostering mutual cooperation rather than just doing transactions. Within this context, an EMP may thus represent a new environment in which intermediaries provide services that did not exist before [14.53] and where supply chain operations management [14.54] as well as relationships management are heavily affected by the available e-business applications.

14.3.2 Electronic Logistics Marketplaces: an Overview

As supply chains have become more global, more expanded and thus more complex, the need for the use of electronic marketplaces for logistics has become evident. Following the concept of electronic marketplace, an electronic logistics marketplace (ELM) is a place where the buyers and suppliers of logistics-related services meet. The development of such markets was expected to contribute much in the integration and improvement of the supply chain [14.55]. In the era of e-business, the existence of on-line freight marketplaces leverages the interaction among shippers/consignees and logistics service providers to a level where the disintermediation of physical contact with 3PLs simplifies the process. On the other hand, the virtual participation of multiple third parties verifies the pluralism of offers and finally leads to the selection of the most suitable solution.

According to Skjøtt-Larsen *et al.* [14.56], two types of ELMs have emerged since the late 1990s: open and closed systems. Open ELMs allow shippers and carriers to use their services with no barriers to entry. On the other hand, in closed ELMs, more closed and vertically integrated relationships are created [14.57]. Gudmundsson [14.58] distinguished two major types of ELM: horizontal, where a relatively homogeneous group of sellers participate, and vertical, where different groups of sellers and peripheral services integrate over the marketplace, such as air cargo, trucking, express services, shipping lines, freight forwarders, third-party warehousing, insurance, customs, *etc.* Finally, Regan and Song [14.59] suggested the following categories of online logistics providers: (a) *on-line freight marketplaces*, which are further distinguished in spot markets (active and passive), RFQ (request for quotation) and auction sites and exchanges. (Spot markets are used by shippers and carriers to post available equipment, loads or capacity, so that they are further matched with the needs of other members. In the second case, auctions or RFQs are enabled through an appropriate application. The exchange may also provide spot market and auction capabilities, but they far exceed the tasks of the previous categories, in that they must provide value-added, creative services and actively participate in the participants' processes.) (b) *Application service providers* (ASPs), which are considered technology enablers for the field of logistics, also developing spot/auction markets or even exchanges. (c) *Purchasing consolidation sites*, which provide the opportunity to small- or medium-sized companies to purchase supplies or equipment on site at premium rates. (d) *Web-based infomediaries*, which provide valuable information to the members of the logistics industry.

This subsection aims to provide a classification of currently existing on-line logistics service providers, based on the dimensions proposed by Gudmundsson and Walczuk [14.55]. Table 14.1 presents an analysis of currently existing logistics e-marketplaces, with a brief presentation of each marketplace as follows.

- *JCtrans.Net* is an open-access website, aiming to become a neutral platform to benefit all types of logistics participants through broadening their networks. The bulletin board includes postings of agents regarding their interests in overseas cooperation, container-/airfreight-/express-rates, as well as freight rates from China. The results can be filtered according to the customised preferences. Useful links are also developed, providing users with information regarding airlines, ports, shipping lines, embassies, and banks, or providing them with tools such as time-/currency-/ temperature-convertors, *etc.* Additional professional tools for cargo tracking are of major importance. Moreover, information regarding logistics exhibitions and conferences, as well as case studies and industry news, shipping statistics and indices help to expand users' knowledge.
- *uShip* is one of the largest shipping marketplaces, providing its members with a wide variety of supportive services that accompany the core ones of posting and auctions. In their effort to be creative and to offer value-added services, the developers of uShip provide users with a suite of shipping tools, such as providers' directories and a shipping price estimator that can be used to estimate the cost of shipping before posting the load, based on previous comparable shipments. Furthermore, the 'Book It Now' application allows the members to automatically book a shipment once the predefined price is reached. An additional value added service is the 'Bid Price Evaluator', an application that compares the bids received across the marketplace and advises on the chances of receiving any lower bids.
- *Freight Saver Online* is a free online freight quote service for regional, domestic and international shipments, providing discounted rates, due to contracts with multiple freight carriers and international agents. An RFQ can be completed online and submitted to the network of providers. Furthermore, a 'Transport Link' exists that includes numerous logistics providers for direct contact.
- *Bid Freight Logistics, Inc.* is a B2B web-based exchange that provides a technology-enabled platform for logistics participants to transact at optimal

Table 14.1. An overview of logistics e-marketplaces

Company name	JCtrans.Net [14.60]	uShip [14.61]	Freight Saver Online [14.62]	Bid Freight Logistics, Inc. [14.63]	TruckersB2B, Inc. [14.64]	eModal [14.65]
Type of online logistics provider	Online freight marketplace (spot market)	Online freight marketplace (exchange)	Online freight marketplace (RFQ)	Internet-based logistics manager (ASP)	Co-op	Infomediary
Launch year	2000	2003	–	–	2000	–
Partnerships	CIFA	e-Bay	Freight Broker Logistics LTD	Accuship.com, IBM, MARSH, Procert.com	Arrow Truck Sales, Advance Business Capital, EFS, *etc.*	US Terminals
Schedules	Based on availability	Not fixed	No	Case dependent	No	Upon request
Online bookings	No	Yes	No	Yes	Yes	–
Geographical coverage	Global	Global	Global	North America	USA	USA
Track & trace mechanism	Yes	Yes	No	Yes	No	Yes
Reports & statistics	No	Yes	No	Yes (press)	No	Yes
Customs links	Yes	No	No	No	No	No
Translation capabilities	Limited	No	No	No	No	No
Search for lowest rate	Yes	Yes	Yes	Yes	Yes	No
Suppliers/buyers/logistics agents' directives	Yes	Yes	Yes	Yes (upon registration)	Yes	Yes (upon registration)
Central billing	No	No	No	Yes	No	Yes
Security	Closed system for rates	Free access	Free access	Closed system	Closed system	Closed system
Orientation	Online platform for global freight forwarders	Reverse auctions marketplace/shipping platform	Quote service (Imp./exp. forwarding, customs brokerage, D2D transportation, shipping and freight management, warehousing & distribution)	Applications for total market (spot market, RFP, reverse auctions, load tendering, customer/carrier portal for order interface)	B2B membership program that provides savings to small and medium sized companies in terms of equipment, insurance, technology maintenance services, *etc.*	Port information service provider (e-handling of delivery orders, post and search for empty containers, *etc.*)

cost. Links with the industry news and general information are also provided. The company's product 'Bid Freight Network' represents a private lane for conducting business, also representing a valuable knowledge base for its members. The status of shipment postings can be viewed, along with a review of carrier bids, thus facilitating the selection process. The carriers' profile is also registered into the system for shippers' reference. The carriers can bid on specific loads through the system's analytical template, providing shippers with exact cost structure. Bid Freight Logistics, Inc. gives member carriers the opportunity to access value-added services, such as links to value-added web sites, electronic advertising and access to Bid Freight Customer directives, *etc.*

- *TruckersB2B, Inc.* comprises a B2B membership program beneficial to small- and medium-sized companies. Members enjoy benefits, such as industry-wide savings based on dedicated consultants, driver retention programs, access to 'Members Only Website', news and industry-based information. The company supplies members with discounts on fuel, tires and equipment, as well as on services, such as insurance, maintenance, financial or legal issues.
- *eModal* provides useful information regarding specific ports to the logistics community. Users can view the status of imported containers and arrange meetings for export ones. The partner-terminals have agreed to allow users pay the fees through this portal, so that prompt pick-up is facilitated. Through the 'Scheduler' application, appointments with terminals are arranged for pick-up or drop-off of containers. Through the 'eModal Trucker Check', the ability to create a driver list for prompt terminal access is verified. Port and industry news are also provided on site, keeping the users continuously informed.

14.4 Conclusions and Future Trends

Nowadays, both logistics service providers and the nature of logistics services offered are continuously changing. Companies are becoming more demanding and globalisation is putting enormous pressure on shippers and therefore on carriers. For example, the ability to trace shipments a few years ago was considered to be an added-value service, whereas nowadays it has become almost a commodity-type service. Within this very dynamic environment, ELMs can offer significant solutions to shippers. ELMs are far from their early stages of development and are increasingly becoming well-tested solutions for companies. Irrespectively, of the type of ELM, several services are offered, such as online bookings across global geographical coverage, track and trace mechanism, reports and statistics, central billing and security. However, there are still several issues that remain unresolved. The review of the analysis of currently existing ELMs revealed, for example, that *customs links* are absent. This is, however, of critical importance, as companies that operate in several countries spend a large amount of time untangling local procedures. Similarly, another important issue with reference to ELMs is the lack of *translation capabilities.* This is of major importance as it will provide the opportunity to companies from all over the word to plan and arrange logistics services on a worldwide basis.

Nowadays, more companies in contrast to the past are exposing themselves more often to more intense global sourcing practices and to global markets, which has created an increased need for fully integrated global solutions. At the same time, however, the need to bring simplicity, local expertise and transparency to logistics arrangements as well as customs compliance increases further. Consequently, the role and importance of ELMs is expected to increase in the near future as the solutions offered by ELMs can deal successfully with a number of problems in current global supply chains and operations.

References

[14.1] Wycisk, C., McKelvey, B. and Huelsmann, M., 2008, "Smart parts supply networks as complex adaptive systems: analysis and implications," *International Journal of Physical Distribution & Logistics Management*, **38**(2), pp. 108–125.

[14.2] Nilsson, F. and Waidringer, J., 2006, "Logistics management from a complexity perspective," *The International Journal of Logistics Management*, **17**(1), pp. 38–54.

[14.3] Lieb, R.C., Millen, R.A. and Van Wassenhove, L.N., 1993, "Third-party logistics services: a comparison of experienced American and European manufacturers," *International Journal of Physical Distribution & Logistics Management*, **23**(6), pp. 35–44.

[14.4] Coyle, J.J., Bardi, E.J. and Langley, C.J., 1996, *The Management of Business Logistics*, 6th edn., Western, St. Paul, MN.

[14.5] Stefansson, G., 2006, "Collaborative logistics management and the role of third-party service providers," *International Journal of Physical Distribution & Logistics Management*, **36**(2), pp. 76–92.

[14.6] Gentry, J.J., 1996, "Carrier involvement in buyer-supplier strategic partnerships," *International Journal of Physical Distribution & Logistics Management*, **26**(3), pp. 14–25.

[14.7] Lieb, R. and Randall, H.L., 1992, "The use of third-party logistics services by large American manufacturers," *Journal of Business Logistics*, **13**(2), pp. 29–42.

[14.8] Sink, H.L., Langley, C.J. and Gibson, B.J., 1996, "Buyer observations of the US third party logistics market," *International Journal of Physical Distribution & Logistics Management*, **26**(3), pp. 38–46.

[14.9] Babbar, S. and Prasad, S., 1998, "International purchasing, inventory management and logistics research – an assessment and agenda," *International Journal of Operations & Production Management*, **18**(1), pp. 6–36.

[14.10] Wong, P.C., Yan, H. and Bamford, C., 2008, "Evaluation of factors for carrier selection in the China Pearl River Delta," *Maritime Policy & Management*, **35**(1), pp. 27–52.

[14.11] Passas, N. and Jones, K., 2007, "The regulation of non-vessel-operating common carriers (NVOCC) and customs brokers – loopholes big enough to fit container ships," *Journal of Financial Crime*, **14**(1), pp. 84–93.

[14.12] Sarkis, J., Talluri, S. and Gunasekaran, A., 2007, "A strategic model for agile virtual enterprise partner selection," *International Journal of Operations & Production Management*, **27** (11), pp. 1213–1234.

[14.13] Sadler, I. and Gough, R., 2005, "Applying a strategic planning process to several supply chain partners," *Journal of Manufacturing Technology Management*, **16**(8), pp. 890–908.

[14.14] Win, A., 2008, "The value a 4PL provider can contribute to an organisation," *International Journal of Physical Distribution & Logistics Management*, **38**(9), pp. 674–684.

[14.15] Persson, G. and Virum, H., 2001, "Growth strategies for logistics service providers: a case study," *The International Journal of Logistics Management*, **12**(1), pp. 53–64.

[14.16] Wang, Q., Zantow, K., Lai, F. and Wang, X., 2006, "Strategic postures of third-party logistics providers in mainland China," *International Journal of Physical Distribution & Logistics Management*, **36**(10), pp. 793–819.

[14.17] Sohail, S., Austin, N.K. and Rushdi, M., 2004, "The use of third-party logistics services: evidence from a sub-Sahara African nation," *International Journal of Logistics: Research and Applications*, **7**(1), pp. 45–57.

[14.18] Hertz, S. and Alfredsson, M., 2003, "Strategic development of third party logistics providers," *Industrial Marketing Management*, **32**(2), pp. 139–149.

[14.19] Berglund, M., Van Laarhoven, P., Sharman, G. and Wandel, S., 1999, "Third party logistics: is there a future?," *International Journal of Logistics Management*, **10**(1), pp. 59–70.

[14.20] Talley, W.K., 2000, "Ocean container shipping: impacts of a technological improvement," *Journal of Economic Issues*, **34**(4), pp. 933–948.

[14.21] Slack, B., 1985, "Containerization, inter-port competition, and port selection," *Maritime Policy & Management*, **12**(4), pp. 293–303.

[14.22] McGinnis, M.A., Kochunny, C.M. and Ackerman, K.B., 1995, "Third party logistics choice," *The International Journal of Logistics Management*, **6**(2), pp. 93–102.

[14.23] Sink, H.L. and Langley, C.J., 1997, "A managerial framework for the acquisition of third-party logistics services," *Journal of Business Logistics*, **18**(2), pp. 163–89.

[14.24] Vasiliauskas, A.V. and Jakubauskas, G., 2007, "Principle and benefits of third party logistics approach when managing logistics supply chain," *Transport*, **22**(2), pp. 68–72.

[14.25] Meade, L. and Sarkis, J., 2002, "A conceptual model for selecting and evaluating third party reverse logistics services," *Supply Chain Management: an International Journal*, **7**(5), pp. 283–95.

[14.26] IDC, 2008, "RFID adoption and implications," *Study Report*, No. 07/2008, e-Business Watch.

[14.27] Skjoett-Larsen, T., 2000, "Third party logistics – from an interorganisational point of view," *International Journal of Physical Distribution & Logistics Management*, **30**(2), pp. 112–27.

[14.28] Jharkharia, S. and Shankar, R., 2007, "Selection of logistics service provider: an analytic network process (ANP) approach," *Omega*, **35**, pp. 274–289.

[14.29] Wilding, R. and Juriado, R., 2004, "Customer perceptions on logistics outsourcing in the European consumer goods industry," *International Journal of Physical Distribution & Logistics Management*, **34**(8), pp. 628–644.

[14.30] Wang, C. and Regan, A.C., 2003, "Risks and reduction measures in logistics outsourcing," In *Proceedings of TRB 2003 Annual Meeting* (CD-ROM).

[14.31] Min, H. and Zhou, G., 2001, "Supply chain modeling: past, present and future," *Computers & Industrial Engineering*, **43**(1–2), pp. 231–249.

[14.32] Van Laarhoven, P., Berglund, M. and Peters, M., 2000, "Third-party logistics in Europe – five years later," *International Journal of Physical Distribution & Logistics Management*, **30**(5), pp. 425–442.

[14.33] Banks, P.A., 2006, "Issues in supplier partner selection," *Journal of Enterprise Information Management*, **19**(3), pp. 262–276.

[14.34] Viswanadham, N. and Gaonkar, R.S., 2003, "Partner selection and synchronized planning in dynamic manufacturing networks," *IEEE Transactions on Robotics and Automation*, **19**(1), pp. 117–130.

[14.35] Trent, R.J. and Monczka, R.M., 2003, "Understanding integrated global sourcing," *International Journal of Physical Distribution & Logistics Management*, **33**(7), pp. 607–629.

[14.36] Zeng, A.Z., 2003, "Global sourcing: process and design for efficient management," *Supply Chain Management: An International Journal*, **8**(4), pp. 367–379.

[14.37] Kotabe, M., Mol, M.J. and Murray, J.Y., 2009, "Global sourcing strategy," In *Global Competitive Market Strategy*, SAGE Publications, Thousand Oaks, CA, pp. 288–302.

[14.38] Cho, J. and Kang, J., 2001, "Benefits and challenges of global sourcing: perceptions of US apparel retail firms," *International Marketing Review*, **18**(5), pp. 542–561.

[14.39] Hesse, M. and Rodrigue, J.P., 2004, "The transport geography of logistics and freight distribution," *Journal of Transport Geography*, **12**(3), pp. 171–184.

[14.40] Petersen, K.J., Frayer, D.J. and Scannel, T.V., 2000, "An empirical investigation of global sourcing strategy effectiveness," *Journal of Supply Chain Management*, **36**(2), pp. 29–38.

[14.41] Levy, D.L., 1995, "International sourcing and supply chain stability," *Journal of International Business Studies*, **26**(2), pp. 343–360.

[14.42] Pooler, H.V., Pooler, H.D. and Farney, S.D., 2004, "Transportation strategies to reduce logistics costs," In *Global Purchasing and Supply Management: Fulfill the Vision*, 2nd edition, Kluwer, Norwell, MA, pp. 349–372.

[14.43] Martin, J. and Thomas, B.J., 2001, "The container terminal community," *Maritime Policy & Management*, **28**(3), pp. 279–292.

[14.44] Pedersen, P.O., 2001, "Freight transport under globalisation and its impact on Africa," *Journal of Transport Geography*, **9**(2), pp. 85–99.

[14.45] Chow, K.H., Choy, K.L. and Lee, W.B., 2007, "A strategic knowledge-based planning system for freight forwarding industry," *Expert Systems with Applications*, **33**(4), pp. 936–954.

[14.46] Krajewska, M.A. and Kopfer, H., 2009, "Transportation planning in freight forwarding companies – Tabu search algorithm for the integrated operational transportation planning problem," *European Journal of Operational Research*, **197**(2), pp. 741–751.

[14.47] Lemoine, W. and Dagnæs, L., 2003, "Globalisation strategies and business organisation of a network of logistics service providers," *International Journal of Physical Distribution & Logistics Management*, **33**(3), pp. 209–228.

[14.48] Sohail, M.S. and Sohal, A.S., 2003, "The use of third party logistics services: a Malaysian perspective," *Technovation*, **23**(5), pp. 401–408.

[14.49] Carbone, V. and Stone, M.A., 2005, "Growth and relational strategies used by the European logistics service providers: rationale and outcomes," *Transportation Research Part E*, **41**(6), pp. 495–510.

[14.50] Bakos, Y., 1998, "The emerging role of electronic marketplaces on the Internet," *Communications of the ACM*, **41**(8), pp. 35–42.

[14.51] Wheatley, W.P., 2003, "Survival and ownership of Internet marketplaces for agriculture," In *Proceedings of the American Agricultural Economics Association Annual Meeting*, Montreal, Canada, 27–30 July.

[14.52] Christiaanse, E. and Markus, M.L., 2002, "Participation in collaboration electronic marketplaces," In *Proceedings of the 36th International Conference on System Sciences*, pp. 178–186.

[14.53] Rossignoli, C., 2009, "The contribution of transaction cost theory and other network-oriented techniques to digital markets," *Information Systems and E-Business Management*, **7**, pp. 57–79.

[14.54] Zhang, Y. and Bhattacharyya, S., 2009, "Analysis of B2B e-marketplaces: an operations perspective," *Information Systems and E-Business Management*, in press, doi:10.1007/s10257-008-0096-y.

[14.55] Gudmundsson, S.V. and Walczuk, R., 1999, "The development of electronic markets in logistics," *International Journal of Logistics Management*, **10**(2), pp. 99–113.

[14.56] Skjøtt-Larsen, T., Kotzab, H. and Grieger, M., 2003, "Electronic marketplaces and supply chain relationships," *Industrial Marketing Management*, **32**(3), pp. 199–210.

[14.57] Wang, Y., Potter, A. and Naim, M., 2007, "Electronic marketplaces for tailored logistics," *Industrial Management and Data Systems*, **107**(8), pp. 1170–1187.

[14.58] Gudmundsson, S.V., 2006, "A global electronic market (GEM) for logistics services and supply-chain management: the expert view," *World Review of Intermodal Transportation Research*, **1**(1), pp. 3–23.

[14.59] Regan, A. and Song, J., 2001, "An industry in transition: third party logistics in the information age," In *Proceedings of the 80th Annual Meeting of Transportation Research Board*, Washington DC, January.

[14.60] http://www.jctrans.net/

[14.61] http://www.uship.com/

[14.62] http://www.freightsaveronline.com/

[14.63] http://www.bidfreight.com/

[14.64] http://www.truckersb2b.com/

[14.65] http://www.emodal.com/

15

A Heuristic for Heterogeneous Capacitated Pick-up and Delivery Logistics Problems with Time Windows in Agile Manufacturing and the Distribution Supply Chain

P. Sivakumar[1], K. Ganesh[2], S. P. Nachiappan[3] and S. Arunachalam[4]

[1] Vickram College of Engineering, Madurai-Anna University, Tiruchirappalli, India
Email: rajasiva@yahoo.com

[2] Global Business Services – Global Delivery, IBM India Private Ltd., Mumbai, India
Emails: kog@iitm.ac.in; koganesh@yahoo.com

[3] Department of Mechanical Engineering, Thiagarajar College of Engineering, Madurai, India
Email: sp_nachi@yahoo.com

[4] School of Computing and Technology, University of East London, Essex, UK
Email: s.arunachalam@uel.ac.uk

Abstract

One type of decision of major importance that directly affects the performance of an agile manufacturing and distribution supply chain is the routing and scheduling of delivery trucks. Routing of vehicles leads to optimisation of logistics operations in enterprise networks. The present study addresses multiple-vehicle pick-up and delivery problems with time windows and heterogeneous capacitated vehicles (m-PDPTWH) for the application of blood bank logistics. The focus is to develop a heuristic to solve m-PDPTWH with the objective to maximise the number of requests assigned to vehicles routes and to minimise the total travel cost. Such a description scheme seems to be useful in the context of dynamic routing problems. The objective of this research is to provide a simple and fast meta-heuristic approach designed for the static case, before entrenching it in a dynamic context in future work. We choose simulated annealing (SA) as a search procedure to solve m-PDPTWH due to the reason that there is very limited literature on solving m-PDPTW using SA. Trials on a large number of test-problems have yielded encouraging results. The key contribution of the work is the development of a unified meta-heuristic to solve m-PDPTW and m-PDPTWH for large sized networks.

15.1 Introduction

In today's business world, transportation costs constitute more than half of the total logistics costs. This share has experienced a steady increase, since smaller, faster, more frequent, more on-time shipments are required as a result of trends such as:

- increased variability in consumers' demands;
- higher fuel costs – while fuel prices have eased slightly they are currently at or near all-time highs;
- constrained trucking capacity – the truck driver shortage has and will cause further price pressures;
- quest for total quality management;
- near-zero inventory production and distribution systems;
- expanded global sourcing that create new logistics challenges;
- sharp global-sized competition.

The benefit that may be achieved by reducing the transportation costs is of interest to the business at the micro level, and to the country at the macro level. Decreasing transportation costs can be achieved through better utilisation of resources such as people and vehicles. Utilising vehicles efficiently is through performing an efficient routing of a fleet of vehicles when they are to pick up/deliver goods from/to certain points.

The cost of health care and the efficient and appropriate utilisation of health care resources are of great public concern. Over the past decades, there have been increasing demands placed on both hospitals and transfusion services to reduce costs while maintaining and even improving the quality of patient care and services provided [15.1]. Hospitals use a variety of products in the treatment of their patients. Many of these products, most notably blood products, have a short life span and, therefore, their supply and inventory have to be managed carefully. Blood products are crucial for hospitals as they are required for surgeries and for the treatment of patients with chronic illnesses, *e.g.* cancer patients. As a consequence, blood products are delivered to hospitals on a regular basis in order to ensure that an adequate supply of the required blood products is available. Thus, a blood bank is faced with a situation in which customers (hospitals, clinics, medical institutes) requires regular deliveries of certain products (blood conserves) that they consume in different volumes.

Blood banks operate the routing of vehicles by the centralised logistics system. Blood banks include several regional blood banks and various types of vehicles of heterogeneous capacity. The required blood product should be picked up from one of the regional blood banks to be delivered to the destination. Any delivery policy should be such that no shortfalls of products occur to the customer, but at the same time spoilage of products has to be kept at a minimum. Vehicles pick up the blood products from the regional blood banks within the given time window and deliver it to hospitals within the stipulated time period. The situation is complicated by the fact that product usage varies over time. Of course, a blood bank also wants to minimise its delivery costs.

In India, most of the public and private blood bank delivery routes are planned manually; very few use routing software or geographic information system. Blood banks group the hospitals into regions, and fixed routes for visiting the hospitals in a region have emerged over time. Hospitals that have requested a delivery of blood products in the previous day are visited in the order of these fixed routes. As the need for blood products rises due to increased surgical activity and new treatments, new approaches are needed to maximise the utility of blood products collected and distributed.

15.2 Research Problem

A key issue in transportation is the cost-efficient management of a heterogeneous vehicle fleet providing pick-up/delivery services to a given set of customers with known demands. The collection/distribution system manager not only should decide on the number and types of vehicles to be used but also must specify which customers are serviced by which vehicle and what sequence to follow so as to minimise the transportation cost. Products to be delivered are loaded at the depot and picked-up products are transported back to the depot. Then, every vehicle route must start and finish at the assigned terminal and both vehicle capacity and working time constraints are to be satisfied. Moreover, each customer must be serviced by exactly one vehicle since split demand is not allowed. This class of logistic problems is usually known as the *vehicle routing problem* (VRP), and its objective is usually the minimisation of the overall distance travelled by the vehicles while servicing all the customers. The interest in VRP problems comes from its practical relevance as well as from the considerable difficulty to solve them exactly. In the field of combinatorial optimisation, the VRP is regarded as one of the most challenging problems. It is indeed non-deterministic polynomial (NP)-hard, so that the task of finding the best set of vehicle tours by solving the optimisation models is computationally prohibitive for real-world applications. As a result, different types of heuristic methodologies are usually applied. An example for a typical VRP is shown in Figure 15.1.

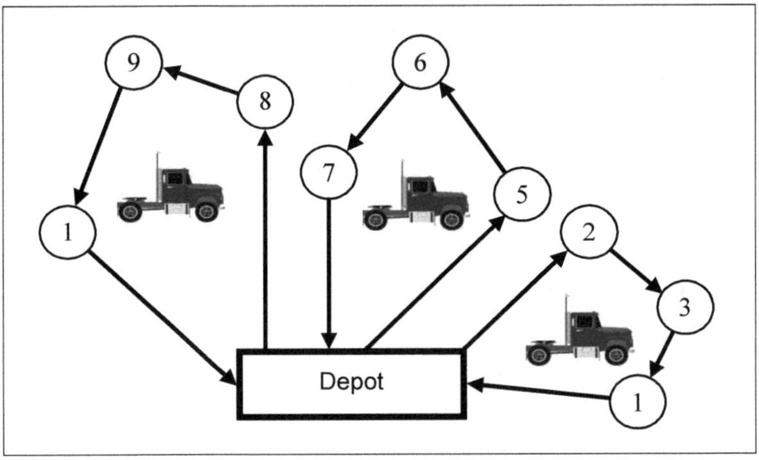

Figure 15.1. Pictorial representation of VRP

Several classes of vehicle routing problems have been studied in the literature. Although addressing different practical situations, they all focus on the common issue of efficiently managing a vehicle fleet for the purpose of serving a given set of customers. The most basic VRP is the *capacitated vehicle routing problem* (CVRP) that assumes a fixed fleet of vehicles of uniform capacity housed in a central depot. It is intrinsically a spatial problem with some capacity constraints. In addition to the geographic component, more realistic routing problems include a scheduling part by incorporating travel times between every pair of nodes, customer service times and the maximum tour duration as additional problem data. The *vehicle routing problem with time windows* (VRPTW) is a generalisation of the CVRP with the further complexity of time windows and other time data. In the VRPTW problem, each customer has an associated time window defined by the earliest and latest times to start the customer service. The depot may also have a time window defining the scheduling horizon. Time windows can be hard or soft. In the hard time window case, a vehicle arriving too early at the customer site is permitted to wait until the customer window is open. However, a vehicle is not permitted to arrive at the node after the latest service start time. In contrast, the soft time window case permits time window violations at the expense of a penalty cost.

Many core problems arising in logistics and transit involve pick-up and delivery in the routing along with the time windows. In the *pick-up and delivery problem with time windows* (PDPTW), each transportation request has its pick-up and delivery points, and the completion of servicing these points must be performed within a given time window. The difficulty in pick-up and delivery problems lies in the side-constraints. However, many practical applications naturally exhibit pick-up and delivery constraints in their modelling. This includes dial-a-ride problems, airline scheduling, bus routing, tractor-trailer problems, helicopter support of offshore oil field platforms, logistics and maintenance support. More generally, industrial vehicle routing problems are rarely pure and often feature side-constraints. Because of its practical relevance and its side-constraints, the PDPTW is a natural model to evaluate the robustness and scalability of various approaches with respect to side-constraints. This research focuses on multiple-vehicle pick-up and delivery problems with time windows and heterogeneous capacitated vehicles (m-PDPTWH).

The centralised blood bank logistics system has a fixed size fleet of m vehicles that receives a set of u requests. Each request consists of picking up a load, of a certain size, from some of the regional blood banks and to deliver it to a set of hospitals, with respect to the time windows associated with the regional blood banks and the hospitals. Since the heterogeneous fleet is finite, with finite capacity vehicles, some requests may not be assigned to a vehicle route without generating some delay in the time windows or exceeding the vehicle capacity. Then, such requests cannot be accepted and are rejected.

The classical VRP involves routing a fleet of vehicles, each visiting a set of nodes such that every node is visited exactly once and by exactly one vehicle, with the objective of minimising the total distance travelled by all the vehicles. It has been investigated exhaustively and has been proven to be *NP*-hard. From the background of the vehicle routing literature, it is understood that this problem can be viewed as m-PDPTWH, which is a generalisation of the well-studied VRPTW. Since m-PDPTWH is a generalisation of the VRPTW, it is at least as complex as the latter, which has been proven to be *NP*-hard [15.2]. m-PDPTWH can be described as follows: a set of transportation requests that is known in advance has to be satisfied by a given fleet of heterogeneous vehicles; each request is characterised by its pick-up location (origin), its delivery location (destination) and the size of the load that has to be transported from the origin to the destination; for each pick-up and delivery location, a time window of loading and unloading times are specified.

The load capacity, the maximum length of its operating interval, a start location and an end location are given for each vehicle. In order to fulfil the requests, a set of routes has to be planned such that each request is transported from its origin to its destination by exactly one vehicle. A reasonable objective function may use optimisation criteria such as, number of vehicles employed, total distance travelled, total schedule duration, or combinations of these. Basically, m-PDPTWH differs from the VRPTW by the additional precedence constraints, *i.e.* the restriction that the origin of each request has to be visited before the corresponding destination and with the set of heterogeneous capacitated vehicles.

The focus of the current research is to develop a heuristic to solve m-PDPTWH with the objective to maximise the number of requests assigned to vehicles routes among the u requests, and next to minimise the total travel cost. Such a description scheme seems to be useful in the context of dynamic routing problems. The purpose of this research is to provide a simple and fast heuristic designed for the static case, before entrenching it in a dynamic context in future work.

This chapter is organised as follows. Section 15.3 gives a review of previous work on m-PDPTWH and similar problems. In Section 15.4, the notations, problem representation, constraints and objectives are described. In Section 15.5, the proposed heuristic is explained in detail. Section 15.6 reports computational results obtained by the heuristic for available benchmark data sets. Section 15.7 presents the conclusion.

15.3 Literature Review

A careful analysis of literature reveals that there is no research on the variant m-PDPTWH. Since m-PDPTWH is an extension of the variant m-PDPTW, we present here a summary of literature of m-PDPTW and its other extensions in Table 15.1. A more recent book on vehicle routing [15.3] provides an excellent overview of techniques for solving vehicle routing problems.

Dynamic programming for PDPTW was attempted in [15.4–15.6] in the 1980s. A branch and price algorithm was then proposed for multiple-vehicle PDPTW [15.7, 15.8]. The branch and price algorithm is appropriate for instances in which each request occupies a relatively large percentage of vehicle capacity. By means of this algorithm, they were able to optimally solve two practical problems with 19 and 30 requests, respectively. The branch and price algorithm was also proposed by [15.9]. In this algorithm, authors have employed embedded heuristics and a special column management scheme to improve the search process. This algorithm has handled data-sets with 30 requests. In [15.10], an iterative procedure for m-PDPTW was proposed, whereas a clustering method clubbed with set partitioning procedure was introduced in [15.11] using the concept of a mini-cluster. An exact algorithm using the concept of a mini-cluster was proposed in [15.12]. Moreover, an insertion procedure was used in [15.13, 15.14]; arc exchange operators were developed for m-PDPTW in [15.15, 15.16]; and a user-defined request-oriented improvement method was developed in [15.17]. The first attempt using meta-heuristics was through the application of tabu search heuristic [15.18]. Later, a genetic algorithm for m-PDPTW was developed [15.19, 15.20]. Based on the inputs of [15.18], a reactive tabu search was proposed in [15.21]. Tabu search embedded with a partitioned insertion heuristic was developed in [15.22]. A hybrid approach combining tabu and simulated annealing was developed in [15.23]. Simulated annealing was further enhanced using large-scale neighbourhood search [15.24]. As competitive heuristics in [15.25], insertion-based heuristics was developed.

Table 15.1. Summary of literature in m-PDPTW

Methodology	Reference
Dynamic programming algorithm	[15.4], [15.5] and [15.6]
Integer programming formulation	[15.7] and [15.9]
Branch and price algorithm	[15.8]
Iterative procedure	[15.10]
Set partitioning algorithm via mini-cluster	[15.11]
Exact algorithm via mini-cluster	[15.12]
Insertion procedure	[15.13] and [15.14]
Arc exchange operators	[15.15] and [15.16]
Request-oriented improvement method	[15.17]
Tabu search heuristic	[15.18]
Genetic algorithm	[15.19] and [15.20]
Reactive tabu search	[15.21]
Tabu search and partitioned insertion heuristic	[15.22]
Tabu-embedded simulated annealing	[15.23]
Simulated annealing – large neighbourhood search	[15.24]
Insertion-based heuristic	[15.25]

15.4 Problem Description

The notation, problem representation, constraints and objectives are explained in this section.

15.4.1 Notations

Regional blood banks and hospitals are represented as customers, and the centralised logistics system is represented as a depot.

Customers:

u = total number of customers
(regional blood banks and hospitals considered in the problem)
R_i = request of customer i, where $i \in u$
R_i^p = pick-up node of customer i
R_i^d = delivery node of customer i
Q_i = size of the load to tranship from R_i^p to R_i^d
T_i^p = duration time of pick-up service
T_i^d = duration time of delivery service
$[E_i^p\ L_i^p]$ = time-window for pick-up start and finish
$[E_i^d\ L_i^d]$ = time-window for delivery start and finish

Vehicle and route:

v = heterogeneous number of fixed vehicles

For each vehicle V^f (where $f = 1, \ldots v$),

O^f = actual route associated with V^f
V_s^f = start depot of route O^f
V_e^f = end depot of route O^f
Q^f = vehicle capacity
$[E_s^f L_s^f]$ = time-window in which V^f must leave depot V_s^f
$[E_e^f L_e^f]$ = time-window in which V^f must enter depot V_e^f
O^I = imaginary route, where all the unallocated routes needs to be placed

Cost and time:

C_{lk} = travel cost between nodes l and k
D_{lk} = travel time between nodes l and k, where $l \neq k$

15.4.2 Problem Representation

The pictorial representation of m-PDPTWH is shown in Figure 15.2, where the request for each customer node is also depicted. The request includes delivery and pick-up, and each cluster indicates the coverage of one vehicle.

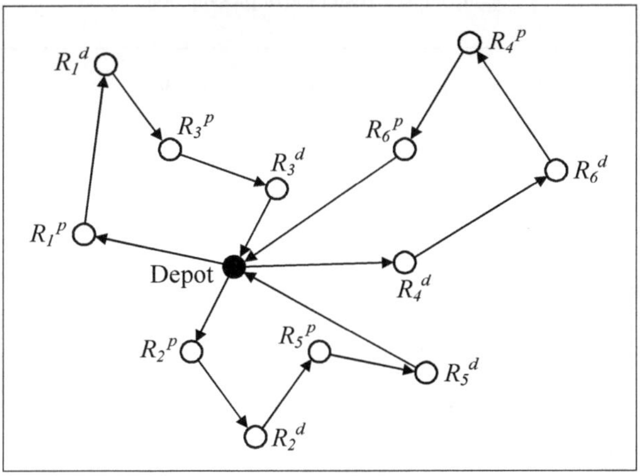

Figure 15.2. Pictorial representation of m-PDPTWH

A pictorial representation of time windows in m-PDPTWH is shown in Figure 15.3. The set of delivery and pick-up nodes now have time windows. The earliest and the latest time windows for each delivery and pick-up node are also depicted in the figure.

An example problem for m-PDPTWH is shown in Figure 15.4 and the routing results for the example are detailed in Figure 15.5. In the example, there are two vehicles, and the drivers of the vehicles are Bobby and Jo. This example includes travel time and travel distance. The drivers have to start around 8:00 AM and serve all the nodes that include both delivery and pick-up, and should return back before 7:00 PM to enjoy the party with their friends. Bobby's vehicle has a greater capacity

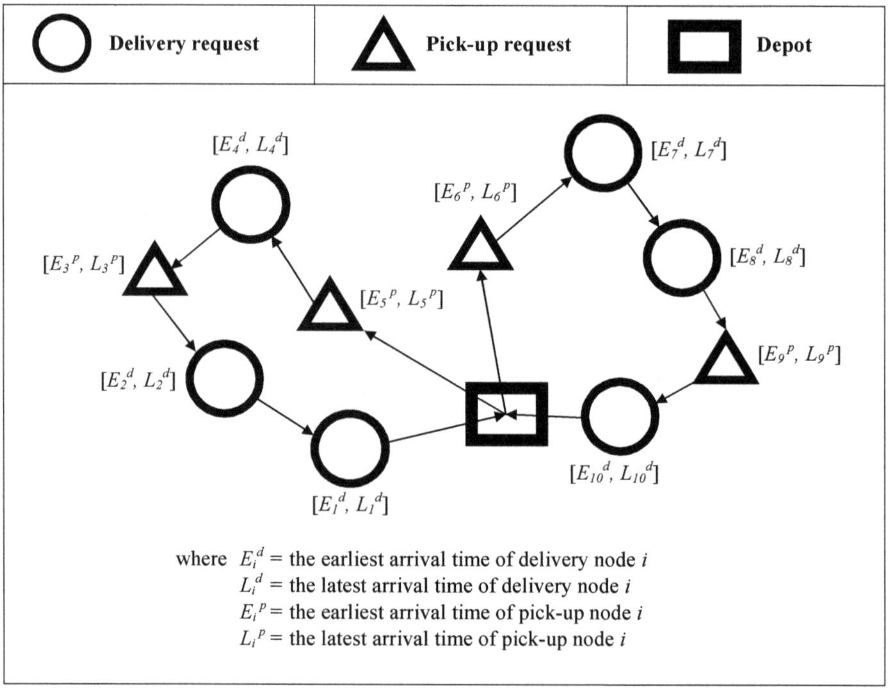

Figure 15.3. Pictorial representation of time windows in m-PDPTWH

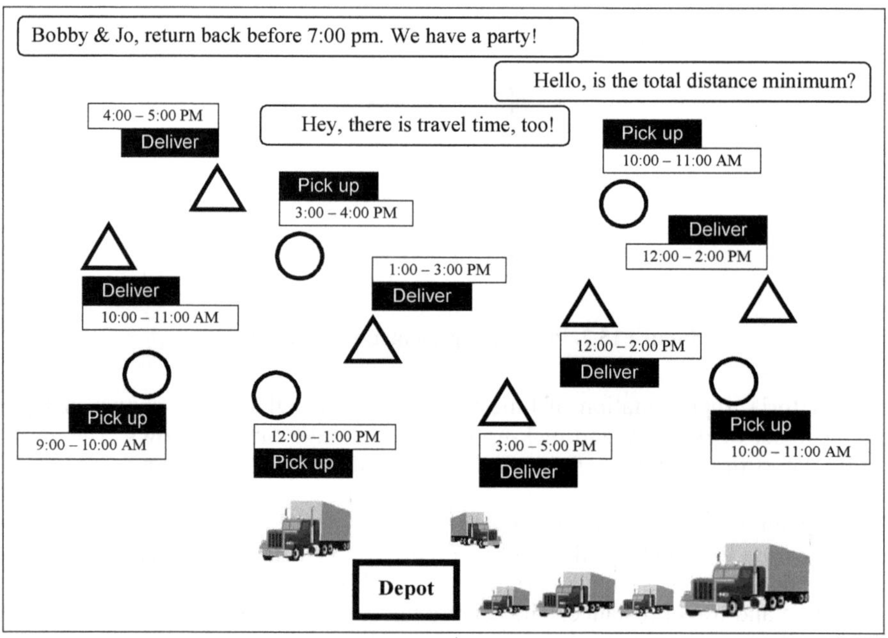

Figure 15.4. Example problem for m-PDPTWH

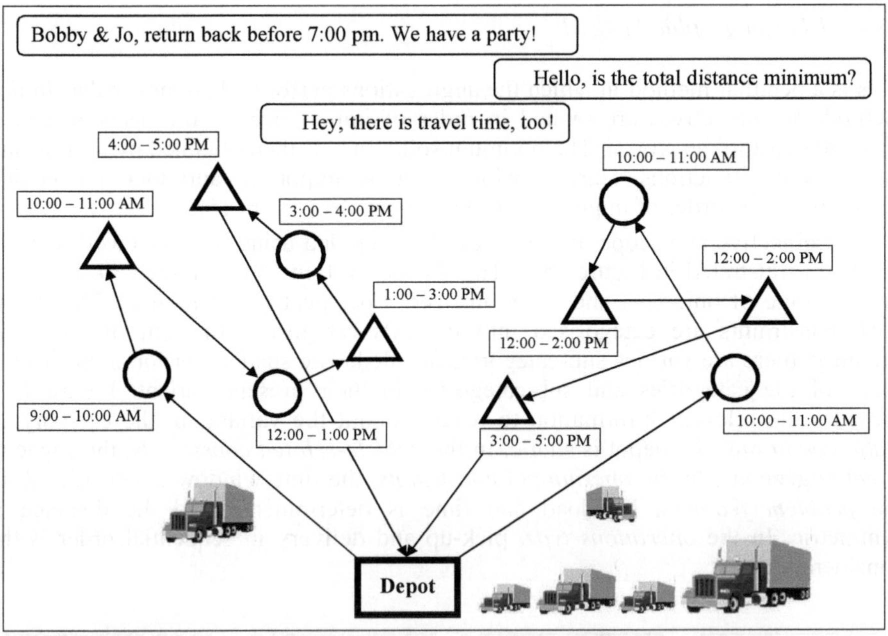

Figure 15.5. Routing results for the example problem of m-PDPTWH

than Jo's vehicle, and it is termed a heterogeneous vehicle. The objective is to find the shortest route for both drivers with respect to time and distance.

15.4.3 Problem Constraints

The focus is to design the route O^f for each vehicle V^f for m-PDPTWH by satisfying the following 5Ps constraints:

- parking constraint – vehicle starts and end at the depot;
- pairing constraint – pick-up and delivery requests from a customer must be served by the same vehicle;
- precedence constraint – a customer's pick-up request must be served before its delivery request;
- packing constraint – the total load of a vehicle must never exceed its capacity; and
- priority constraint – each customer is visited during the allotted time windows.

15.4.4 Problem Objective

The objective is to maximise the number of requests allotted to the actual route whilst minimising the total cost of the route. The objective will be treated in lexicographic order with penalty functions.

15.4.4.1 Lexicographic Method

This is a peculiar method in which the aggregations performed are not scalar. In this method, the objectives are ranked in order of importance by the decision maker (from the best to the worst). The optimum solution s^* is then obtained by minimising the objective functions, starting with the most important and then proceeding according to the order of importance of the objectives [15.26, 15.27].

The objective and scope of this research is detailed using the framework [15.28, 15.29] as illustrated in Figure 15.6. The shaded portion of the framework indicates the coverage of objective and scope in overall perspective. In Figure 15.6, all the VRP constraints are categorised into the 5Ps categories. In each of the main category, there are various sub-categories detailed. The shaded portion indicates the focus of the categories and sub-categories in their present variant. Figure 15.6 provides the relevant information to readers about the variant. In the category of *static conditions*, the depot is single. In the *vehicle-related constraints*, the capacity is heterogeneous. In the *operational constraints*, the time window is considered. In the *problem features*, the load and time is deterministic and the distance is symmetric. In the *operations type*, pick-up and delivery in sequential order is the consideration.

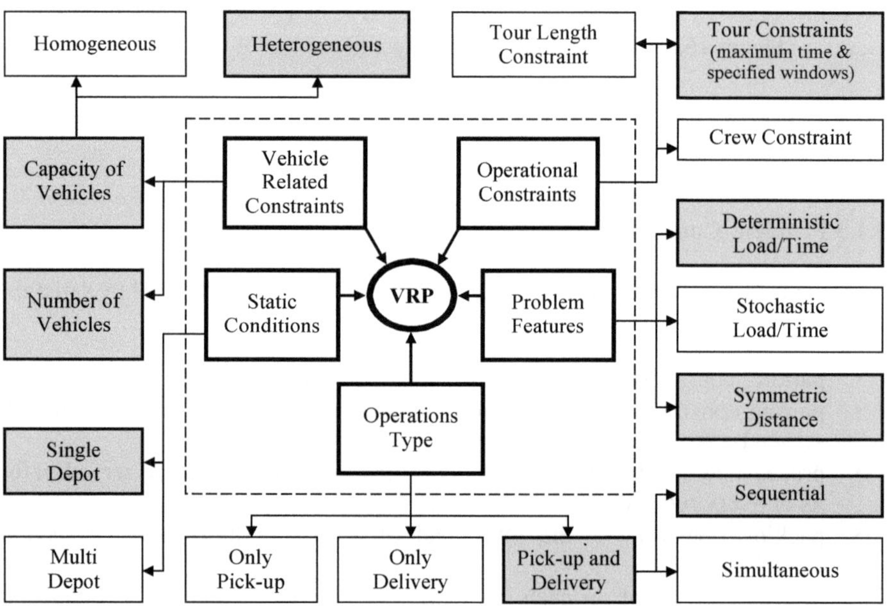

Figure 15.6. Indication of m-PDPTWH objective and scope

The chosen solution methodology to solve m-PDPTWH from the available methodologies in the literature is also detailed using the framework [15.28, 15.29] as shown in the Figure 15.7. The shaded portion of the framework indicates the chosen solution methodologies to solve m-PDPTWH. Among the various approaches, we leveraged SA along with the improvement heuristics for m-PDPTWH.

The other real-life applications of this variant m-PDPTWH include dial-a-ride problems, airline scheduling, bus routing, tractor–trailer problems, helicopter support of offshore oil field platforms, and logistics and maintenance support.

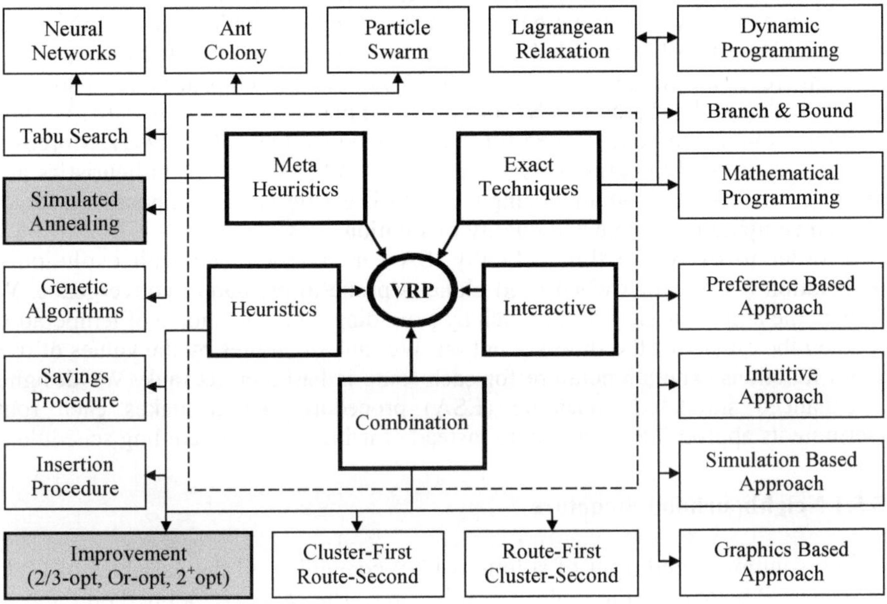

Figure 15.7. Indication of solution methodology for m-PDPTWH

15.5 Proposed Simulated Annealing for Solving m-PDPTWH

Problems of combinatorial optimisation are characterised by their well-structured problem definition as well as the huge number of action alternatives in practical application areas of reasonable size. Especially in areas such as routing, task allocation or scheduling, these types of problems often occur. Their advantage lies in the easy understanding of their action alternatives and their objective function. Therefore, an objective evaluation of the quality of action alternatives is possible in the context of combinatorial optimisation problems.

Utilising classical methods in operations research often fails due to the exponentially growing computational effort. Therefore, in practice, heuristics and meta-heuristics are commonly used even if they are unable to guarantee an optimal solution. Artificial intelligence heuristics, otherwise called meta-heuristic techniques that mimic natural processes developed over the last 30 years, have produced 'good' results in reasonable short runs for this class of optimisation problem. Even though bionic heuristics are much more flexible regarding modifications in the problem description when compared to classical problem-specific heuristics, they are often superior in their results. Those bionic heuristics have been developed following the principles of natural processes: in that sense, genetic algorithms (GAs) try to imitate the biological evolution of a species in order to achieve an almost optimal state whereas simulated annealing (SA) was initially inspired by the laws of thermodynamics in order to cool down a certain matter to its lowest energetic state.

Many optimisation problems of practical and theoretical importance consist of the search for the 'best' configuration of a set of variables to achieve some goals. They seem to divide naturally into two categories: solutions encoded with real-valued variables, and solutions encoded with discrete variables. Among the latter, we find a class of problems called *combinatorial optimisation* (CO) problems.

In this research, we choose SA [15.30, 15.31] as a search procedure to solve m-PDPTWH due to the reason that SA has provided better reason for m-PDPTW and it is believed that it can be leveraged for the extended variant m-PDPTWH. Moreover, there is very limited literature of solving m-PDPTW using any meta-heuristics and, therefore, we made first attempt using the meta-heuristic SA. SA imposes a trade-off between computational time and quality of solution.

In order to overcome this difficulty, SA can be combined with evolutionary computation (EC), which when used alone is prone to premature convergence. We enhance the ability of the SA approach by providing dynamic choice of temperature, based on the quality of the fitness function. We rank the routes by the values of their fitness functions. The temperature for each route is based on its rank. We designed an enhanced simulated annealing (ESA) procedure, which makes each route determine its appropriate temperature instead of using a uniform cooling schedule.

15.5.1 Neighbourhood Structure

For any solution z, and a set of solutions $G(z)$ containing all the variables of (R, V), the request R belongs to the route V in solution z. The shifts are defined on the basis of this variable set, removing and inserting a variable. Whenever a request is inserted into an actual route V^f, both the pick-up and delivery locations need to be inserted.

The ESA is embedded with Or-opt [15.32, 15.33] and used as a local search procedure to generate neighbourhoods. It is expected that the initial population of structured solutions from the Or-opt exchange evolves into high-quality solutions within a relatively small number of generations.

Or-opt, a well-known node exchange heuristic, removes a maximum of three consecutive nodes from a route and inserts them, in the same sequence, at another stretch of the same route. Or-opt can be considered a special case of 3-opt in which a chain of two or three consecutive nodes is shifted to a different part of the route. The preceding and succeeding nodes of the earlier route are now directly linked by an arc. The chain is inserted between some other pair of nodes, replacing the arc that linked them earlier (Figure 15.8). In [15.34], it is shown that Or-opt produces good solutions despite considering fewer exchanges than the 3-opt procedure; it also requires less computational time. After each set of exchanges, we check for its feasibility, compute the total distance travelled, compare it with the current best route and update the solution.

The Or-opt algorithm is given below:

Step 1: Consider an initial route and set $t = 1$ and $s = 3$.

Step 2: From the route, remove a chain of s consecutive vertices from position t to t+2, and tentatively insert the chain between all remaining pairs of consecutive vertices on the route.

Step 2.1: If one or more insertions bring about a decrease in the cost of the route, choose the new route based on the maximum reduction in cost.

Step 2.2: If no insertion decreases the cost of the route, set $t = t + 1$.

If $t < n + 1$, repeat step 2.

Step 3: Set $t = 1$ and $s = s - 1$. If $s > 0$, go to step 2; else, stop.

The Or-opt procedure for m-PDPTWH is also explained in Figure 15.9.

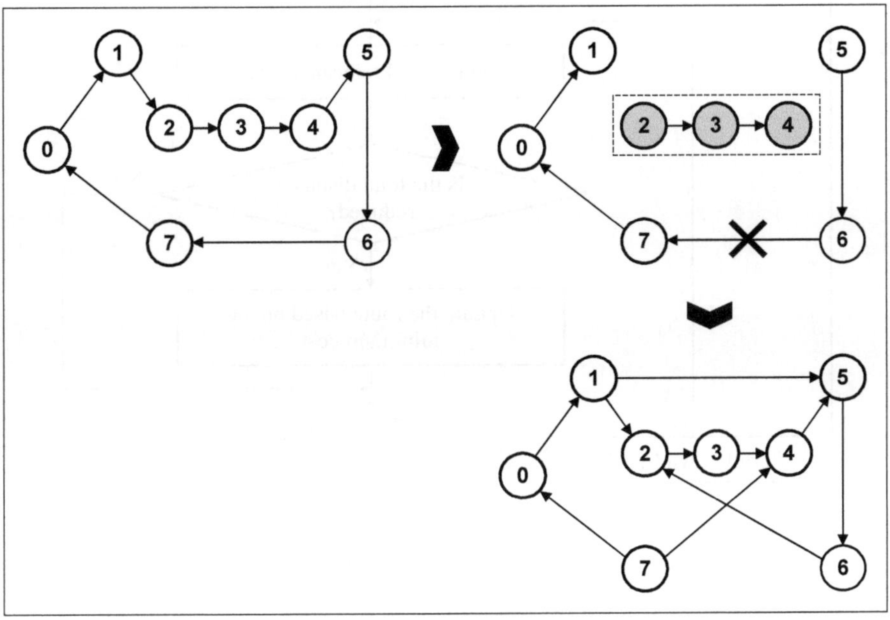

Figure 15.8. Or-opt for m-PDPTWH

15.5.2 Evaluation Function, Ranking and Temperature Assignment

This section presents the details related to the evaluation function as well as the ranking and temperature assignment of the proposed methodology as follows:

Step 1: The evaluation function is represented as the linear combination: $F(z) = C(z) + aC^d(z)$, where $C(z)$ = total route cost, $C^d(z)$ = penalty function used to indicate the non-payment in assignment of requests to the actual route $(u - u(z))*W$ (where $u(z)$ = total number of requests in the actual route), and a = positive parameter adjusted during search to lead it in different areas of solution space.

Step 2: Assign rank $R(z^{'})$ for each neighbourhood $z^{'}$ in the ascending order of the new evaluation function value $F(z^{'})$.

Step 3: Calculate the maximum temperature $t_{\max}$ for each neighbourhood $z^{'}$ as

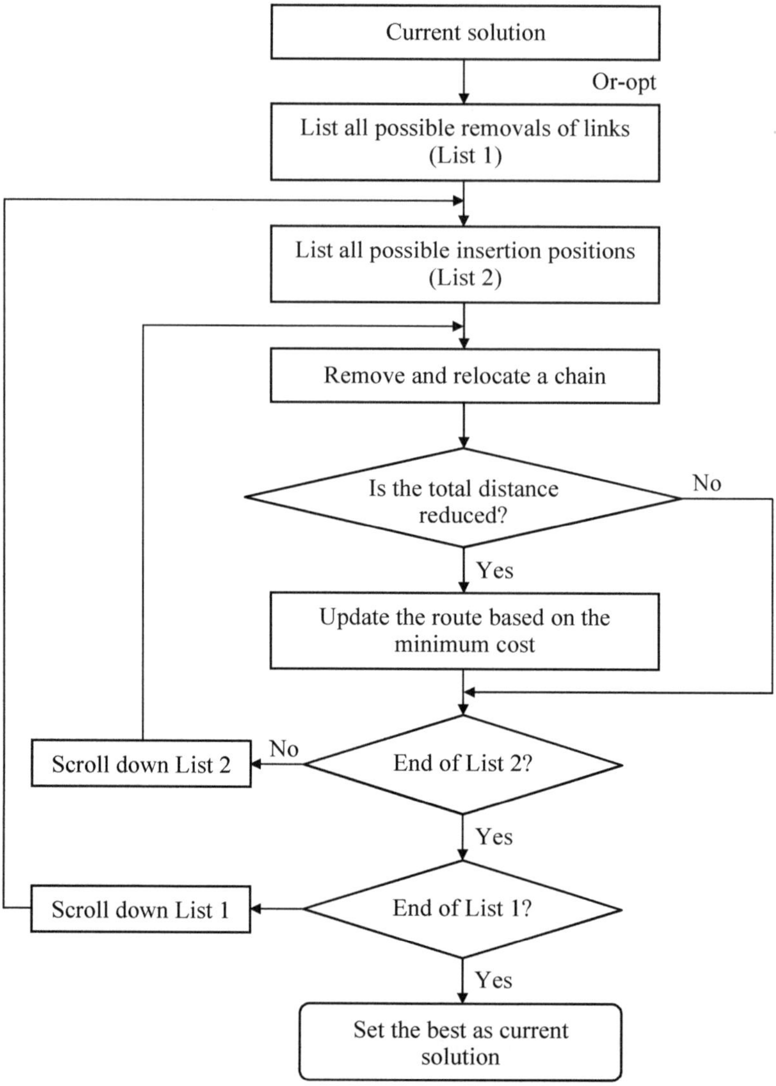

Figure 15.9. Flowchart of Or-opt exchange module for m-PDPTWH

$$t_{\max}(z') = \left(t_{\max}(z')\,\alpha^{G(z)-R(z')}\right) \tag{15.1}$$

where α ($\alpha \in [0, 1]$) is the cooling rate and $G(z)$ is the parent set.

The other steps are similar to the usual procedure of SA.

Step 4: For the post-optimisation, 2-opt* exchange, a modification of 2-opt [15.35] is used.

15.5.2.1 2-opt Exchange Modification*

For problems with multiple routes, the inter-route 2-opt* exchange, a modification of 2-opt was introduced [15.35]. It removes a pair of arcs from two different routes and divides each route into two portions. The first portion of one route and the second portion of the other route are combined to obtain two new arcs.

In Figure 15.10, (0–1–2) of the first route joins (6–7–0) of the second to form a new route (0–1–2–6–7–0). Similarly, we can obtain another route (0–4–5–3–0). 2-opt* can also link the last node of one route to the first of the other, reducing the required number of vehicles in the process. It happens when the last arc of one route and the first of the other route (arcs 3–0 and 0–4 in Figure 15.11, for example) are deleted, resulting in a single route (0–1–2–3–4–5–6–7–0) (Figure 15.11). We implemented 2-opt* with a check on vehicle capacity and maximum route length. Starting from the current solution, 2-opt* enumerates all pairs of exchanges of arcs. We rank the solutions and choose as many feasible routes as the required number of neighbourhoods, discarding those that are inferior to the current solution.

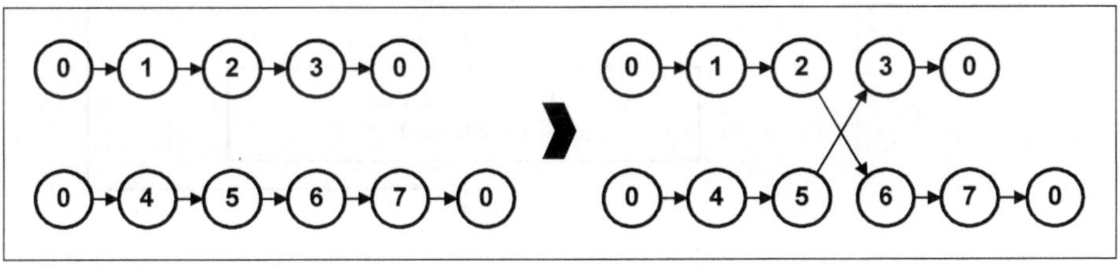

Figure 15.10. 2-opt* exchange on two routes

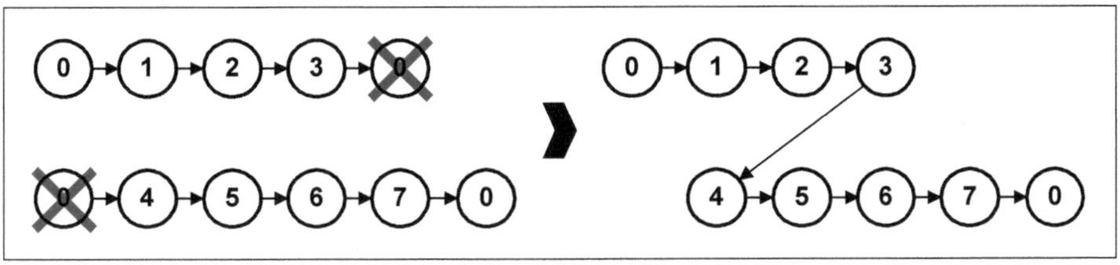

Figure 15.11. 2-opt* exchange: from two routes to one

The 2-opt* algorithm can be described as follows:

Step 1: Let T be the current route.

Step 2: For every node i ($i = 1, 2, \ldots, n$), examine all possible 2-opt* moves involving the edge between i and its successor in the route. If it is possible to decrease the route length this way, then choose the best 2-opt* move and update T.

Step 3: If no improving move can be found, then stop.

Details of the 2-opt* procedures are illustrated in Figure 15.12.

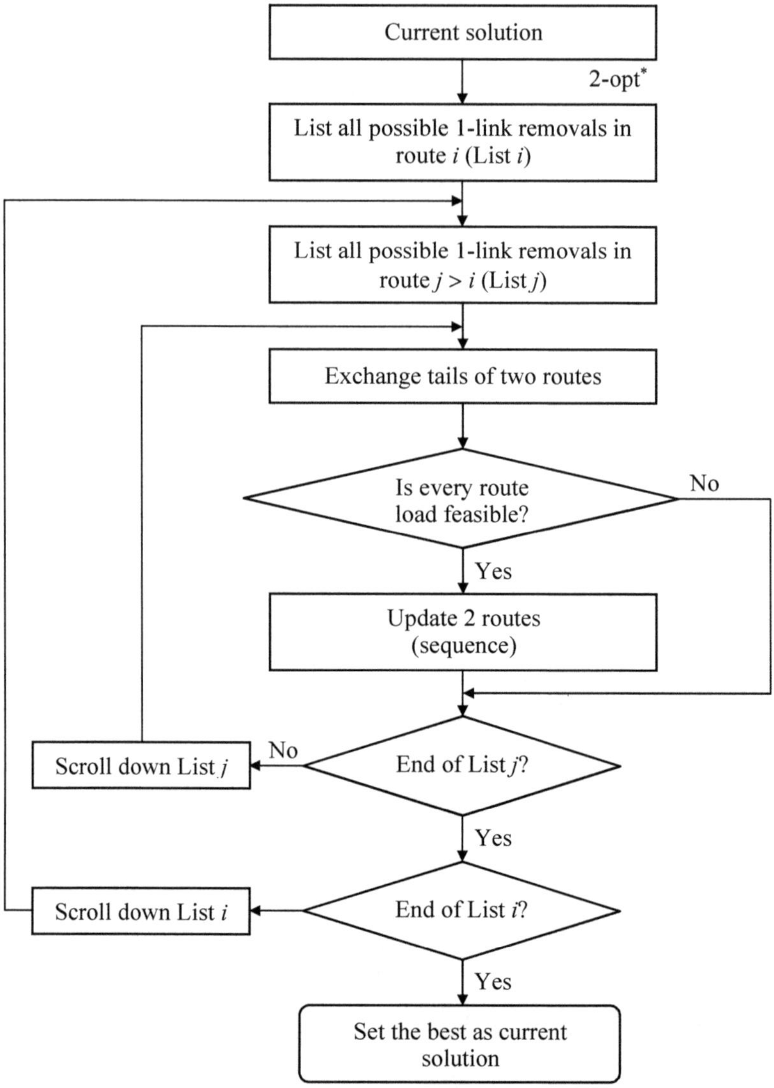

Figure 15.12. Flowchart of 2-opt* exchange module for m-PDPTWH

The pseudo-code for the basic SA is detailed below:

```
X = Generate an initial feasible solution;
C(X) = Compute initial cost of X;
best_cost = C(X);
T = Compute initial temperature;
While (stopping criterion not met)
   Repeat (pre-chosen number of times)
   Transition = Select a transition from neighbourhood (X);
   X′ = Apply Transition(X, Transition);
```

```
    ΔC = Compute change in cost (X, X', Transition);
    p = Generate random number (0, 1);
    If ((ΔC < 0) OR (e^-ΔC/T > p))
       X = X';
       C(X) = C(X) + ΔC;
    End If;
    If (C(X) < best_cost)
       best_cost = C(X);
    End If;
    End Repeat;
    T = Apply Cooing Function (T);
 End While;
 Output best_cost;
 End.
```

Note: $e^{-\Delta C/T}$ is the Boltzman function, and is used to determine whether or not to accept a poorer solution in each iteration.

15.5.2.2 Parameter Settings for ESA

The values of the parameters used in ESA are: initial (maximum) temperature (t_{max}) = 4500; cooling rate/temperature reduction coefficient (α) = 0.98; maximum number of iterations = 10000.

15.6 Computational Study

This section reports the results of the proposed SA for a number of benchmark problems for m-PDPTW. In order to test the proposed SA, 56 benchmark data sets with 100-node instances provided by [15.23] for m-PDPTW is used. The proposed SA was coded using C++ and the tests were carried out on a PC with a Pentium 4 processor. The computational results are presented in Table 15.2. From Table 15.2, it is inferred that the proposed SA proves to be competitive with the best-known solution and in minimising the number of vehicles. The best solutions are highlighted as gray-shaded rows and the relative percentage deviation from the best-known solution is also given in Table 15.2.

15.7 Conclusions

In this chapter, a heuristic to solve the blood bank logistics problem called m-PDPTWH was presented. The heuristic was tested using two publicly available sets of benchmark problem for m-PDPTW and m-PDPTWH. The experimental results demonstrate that the heuristic is able to find high-quality solutions when compared to previous methods for solving m-PDPTW. The overall findings seem to justify the employment of SA in general, as suitable techniques for solving m-PDPTW and m-PDPTWH. The critical contribution of the research is the development of a unified SA to solve both m-PDPTW and m-PDPTWH. Moreover, the SA was implemented to solve large data-sets. Future research will be dedicated to the further improvement of the proposed approach.

Table 15.2. Results for benchmark data sets [15.23]

Instance	Best-known solution		Best of proposed SA		% Deviation
	Route cost	No. of vehicles	Route cost	No. of vehicles	
lc101	828.94	10	828.94	10	0
lc102	828.94	10	828.94	10	0
lc103	827.86	10	827.86	10	0
lc104	861.95	9	861.95	9	0
lc105	828.94	10	828.94	10	0
lc106	828.94	10	828.94	10	0
lc107	828.94	10	828.94	10	0
lc108	826.44	10	826.44	10	0
lc109	827.82	10	1000.60	9	20.87169
lc201	591.56	3	591.56	3	0
lc202	591.56	3	591.56	3	0
lc203	585.56	3	585.56	3	0
lc204	591.17	3	590.59	3	–0.09659
lc205	588.88	3	588.88	3	0
lc206	588.49	3	588.49	3	0
lc207	588.29	3	588.29	3	0
lc208	588.32	3	588.32	3	0
lr101	1650.78	19	1650.78	19	0
lr102	1487.57	17	1487.57	17	0
lr103	1292.68	13	1292.68	13	0
lr104	1013.39	9	1013.39	9	0
lr105	1377.11	14	1377.11	14	0
lr106	1252.62	12	1252.62	12	0
lr107	1111.31	10	1111.31	10	0
lr108	968.97	9	968.97	9	0
lr109	1239.96	11	1208.96	11	–2.50008
lr110	1159.35	10	1159.35	10	0
lr111	1108.90	10	1108.90	10	0
lr112	1003.77	9	1003.77	9	0
lr201	1263.84	4	1253.23	4	–0.83951
lr202	1197.67	3	1197.67	3	0
lr203	949.40	3	949.40	3	0
lr204	849.05	2	860.11	2	1.302632
lr205	1054.02	3	1054.02	3	0
lr206	931.63	3	931.63	3	0
lr207	903.06	2	903.06	2	0
lr208	734.85	2	734.85	2	0

Table 15.2. Results for benchmark data sets [15.23] (continued)

Instance	Best-known solution		Best of proposed SA		% Deviation
	Route cost	No. of vehicles	Route cost	No. of vehicles	
lr209	937.05	3	937.05	3	0
lr210	964.22	3	964.22	3	0
lr211	927.80	2	927.80	2	0
lrc101	1708.80	14	1708.80	14	0
lrc102	1563.55	13	1558.07	12	–0.35048
lrc103	1258.74	11	1258.74	11	0
lrc104	1128.40	10	1128.40	10	0
lrc105	1637.62	13	1637.62	13	0
lrc106	1425.53	11	1425.53	11	0
lrc107	1230.15	11	1230.15	11	0
lrc108	1147.97	10	1147.97	10	0
lrc201	1486.96	4	1486.96	4	0
lrc202	1374.27	3	1374.27	3	0
lrc203	1089.07	3	1089.07	3	0
lrc204	827.78	3	818.66	3	–1.10174
lrc205	1302.20	4	1302.20	4	0
lrc206	1162.91	3	1162.91	3	0
lrc207	1424.60	3	1424.60	3	0
lrc208	852.76	3	852.76	3	0
Average percentage relative deviation					0.30800

References

[15.1] Spens, K., 2001, *Managing Critical Resources through Supply Network Management: a Study of the Finnish Blood Supply Network*, Swedish School of Economics and Business Administration, Helsinki.

[15.2] Lenstra, J.K. and Rinnoy Kan, A.H.G., 1981, "Complexity of vehicle routing and scheduling problems," *Networks*, **11**, pp. 221–227.

[15.3] Toth, P. and Vigo, D. (eds.), 2001, *The Vehicle Routing Problem*, SIAM Society for Industrial and Applied Mathematics, Philadelphia.

[15.4] Psaraftis, H.N., 1980, "A dynamic programming solution to the single vehicle many-to-many immediate request dial-a-ride problem," *Transportation Science*, **14**, pp. 130–154.

[15.5] Psaraftis, H.N., 1983, "An exact algorithm for the single vehicle many-to-many dial-a-ride problem with time windows," *Transportation Science*, **17**, pp. 351–357.

[15.6] Desrosiers, J., Dumas, Y. and Soumis, F., 1986, "A dynamic programming solution of the large scale single-vehicle dial-a-ride problem with time windows," *The American Journal of Mathematical and Management Sciences*, **6**, pp. 301–325.

[15.7] Dumas, Y., Desrosiers, J. and Soumis, F., 1991, "The pickup and delivery problem with time windows," *European Journal of Operational Research*, **54**, pp. 7–22.

[15.8] Savelsbergh, M.W.P. and Sol, M., 1998, "DRIVE: dynamic routing of independent vehicles," *Operations Research*, **46**, pp. 474–490.

[15.9] Savelsbergh, M.W.P. and Sol, M., 1995, "The general pickup and delivery problem," *Transportation Science*, **29**, pp. 17–29.

[15.10] Bodin, L. and Sexton, T., 1986, "The multi-vehicle subscriber dial-a-ride problem," *TIMS Studies in the Management Sciences*, **26**, pp. 73–86.

[15.11] Desrosiers, J., Dumas, Y. and Soumis, F., 1988, "The multiple vehicle dial-a-ride problem," In *Computer-aided Transit Scheduling: a Proceedings of the Fourth International Workshop on Computer-Aided Scheduling of Public Transport*, Daduna, J.R. and Wren, A. (eds.), Springer, Berlin/Heidelberg, pp. 15–27.

[15.12] Ioachim, I., Desrosiers, J., Dumas, Y. and Villeneuve, D., 1995, "A request clustering algorithm for door-to-door handicapped transportation," *Transportation Science*, **29**, pp. 63–78.

[15.13] Jaw, J.J., Odoni, A.R., Psaraftis, H.N. and Wilson, N.H.M., 1986, "A heuristic algorithm for the multi-vehicle advance request dial-a-ride problem with time windows," *Transportation Research Part B*, **20**, pp. 243–257.

[15.14] Madsen, O.B.G., Ravn, H.F. and Rygaard, J.M., 1995, "A heuristic algorithm for a dial-a-ride problem with time windows multiple capacities and multiple objectives," *Annals of Operations Research*, **60**, pp. 193–208.

[15.15] Van der Bruggen, L.J.J., Lenstra, J.K. and Schuur, P.C., 1993, "Variable depth search for the single-vehicle pickup and delivery problem with time windows," *Transportation Science*, **27**, pp. 298–311.

[15.16] Psaraftis, H.N., 1983, "k-interchange procedures for local search in a precedence constrained routing problem," *European Journal of Operational Research*, **13**, pp. 391–402.

[15.17] Toth, P. and Vigo, D., 1996, "Fast local search algorithms for the handicapped persons transportation problem," In *Meta-Heuristics: Theory and Applications*, Osman, I.C.H. and Kelly, J.P. (eds.), Kluwer, Norwell, pp. 677–690.

[15.18] Gendreau, M., Guertin, F., Potvin, J.Y. and Séguin, R., 1998, "Neighbourhood search heuristics for a dynamic vehicle dispatching problem with pick-ups and deliveries," *Technical Report CRT-98-10*, Centre de recherche sur les transports, Université de Montréal, Montréal.

[15.19] Jih, W.-R. and Hsu, Y.-J., 1999, "Dynamic vehicle routing using hybrid genetic algorithms," In *Proceedings of the 1999 IEEE International Conference on Robotics & Automation*, IEEE Computer Society, Los Alamitos, CA, pp. 453–458.

[15.20] Jung, S. and Haghani, A., 2000, "A genetic algorithm for pick-up and delivery problem with time windows," In *Transportation Research Record 1733*, Transportation Research Board, pp. 1–7.

[15.21] Nanry, W.P. and Barnes, J.W., 2000, "Solving the pickup and delivery problem with time windows using reactive tabu search," *Transportation Research Part B*, **34**, pp. 107–121.

[15.22] Lau, H.C. and Liang, Z., 2001, "Pickup and delivery with time windows – algorithms and test case generation," In *Proceedings of the 13th IEEE International Conference on Tools with Artificial Intelligence*, IEEE Computer Society, Los Alamitos, CA, pp. 333–340.

[15.23] Li, H. and Lim, A., 2001, "A meta-heuristic for solving the pickup and delivery problem," In *Proceedings of the 13th IEEE International Conference on Tools with Artificial Intelligence*, IEEE Computer Society, Los Alamitos, CA, pp. 160–167.

[15.24] Bent, R. and Van Hentenryck, P., 2006, "A two-stage hybrid algorithm for pickup and delivery vehicle routing problems with time windows," *Computers & Operations Research*, **33**, pp. 875–893.

[15.25] Lu, Q. and Dessouky, M.M., 2006, "A new insertion-based construction heuristic for solving the pickup and delivery problem with time windows," *European Journal of Operational Research*, **175**, pp. 672–687.

[15.26] Rao, S.S., 1984, "Multi-objective optimization in structural design with uncertain parameters and stochastic processes," *AIAA Journal*, **22**, pp. 1670–1678.

[15.27] Sarma, G.V., Sellami, L. and Houam, K.D., 1993, "Application of lexicographic goal programming in production planning – two case studies," *Opsearch*, **30**, pp. 141–162.

[15.28] Ganesh, K. and Narendran, T.T., 2005, "Composite heuristics for a class of vehicle routing problems," PhD Thesis, Department of Management Studies, Indian Institute of Technology Madras, Chennai.

[15.29] Ganesh, K., Sam Nallathambi, A. and Narendran, T.T., 2007, "Variants, solution approaches and applications for vehicle routing problems in supply chain: agile framework and comprehensive review," *International Journal of Agile Systems and Management*, **2**(1), pp. 50–75.

[15.30] Kirkpatrick, S., Jr., Gelatt, C.D. and Vecchi, M.P., 1983, "Optimization by simulated annealing," *Science*, **220**, pp. 671–680.

[15.31] Golden, B. and Skiscim, C., 1986, "Using simulated annealing to solve routing and location problems," *Naval Research Logistics*, **33**, pp. 261–279.

[15.32] Or, I., 1976, "Travelling salesman type combinatorial problems and their relation to the logistics of blood banking," PhD Dissertation, Department of Industrial Engineering and Management Science, Northwestern University, Evanston, IL.

[15.33] Taillard, E., Badeau, P., Gendreau, M., Guertin, F. and Potvin, J., 1997, "A tabu search heuristic for the vehicle routing problem with soft time windows," *Transportation Science*, **31**, pp. 170–186.

[15.34] Solomon, M.M. and Desrosiers, J., 1988, "Time window constrained routing and scheduling problems," *Transportation Science*, **22**, pp. 1–13.

[15.35] Potvin, J.-Y. and Rousseau, J.-M., 1995, "An exchange heuristic for routing problems with time windows," *Journal of the Operational Research Society*, **46**, pp. 1433–1446.

16

Visualisation and Verification of Communication Protocols for Networked Distributed Systems

Z.M. Bi[1] and Lihui Wang[2]

[1] Department of Engineering
Indiana University – Purdue University Fort Wayne
Fort Wayne, IN 46805-1499, USA
Email: biz@ipfw.edu

[2] Virtual Systems Research Centre, University of Skövde
PO Box 408, 541 28 Skövde, Sweden
Email: lihui.wang@his.se

Abstract

The successful design and application of a large and complex manufacturing system relies not only on the maturity of its fundamental design, but also on the technologies for seamless integration and coordination of system components, since a large manufacturing or logistic system often adopts a decentralised control architecture to manage its complexity. System components are usually distributed; their behaviours are enacted locally and autonomously. The control objective at the system-level is achieved by the executions of the sub-objectives at the component level, subjected to the condition that the controls of the sub-systems have to be coordinated via effective communication. In developing algorithms for communication and coordination of a networked distributed system, algorithm verification is complicated and trivial, due to the invisible information system. In this chapter, we propose to use the conventional simulation tool, Deneb/QUEST, for modelling and visualisation of the coordinating behaviours. Its vivid graphical environment can be a great assistance in accelerating software debugging and verification and in reducing the time for software development. General architecture of a networked distribute system is introduced, the system components are analysed, and the correspondences between these components and QUEST elements are established. A case study for the verification of ring extrema determination (RED) algorithm is used as an example to illustrate the general procedure and the feasibility of the proposed approach.

16.1 Introduction

According to Wikipedia, China has been the fastest-growing country for the past quarter century, with an average annual GDP growth rate of above 10%. The economy of China is in fact the second largest in the world after the United States [16.1].

Merchandise labelled 'Made in China' has gone beyond toys, garments and sports shoes to consumer electronics products and hi-tech gadgets. China became the world's leader in terms of its mobile phone subscriber base with 461 million users by the end of 2006. Its 368 million fixed phone lines are the largest number in the world. China is expected to surpass the United States as the world's largest PC market by 2010 [16.2]. China launched the ShenZhou VII spacecraft and performed the nation's first spacewalk in September 2008 [16.3]. China also developed the J-10 fighter as a multi-role, all-weather fighter aircraft for both air-to-air and air-to-ground missions. The J-10 fighter has comparable or even superior performance to the F-16 and Su-27 [16.4]. Despite numerous incredible achievements over these years, China has still struggled to develop some emerging complex products, *e.g.* large civil aircraft for surging demand in air services and defence systems for protecting the homeland. There is no doubt that the fundamental scientific theories and principles to design these products are well established; the critical challenge is how to deal with the complexity involved in the design, manufacturing and assembly of these products.

16.1.1 Basic Strategy to Deal with System Complexity

Complexity is defined as the measure of uncertainty in achieving the functional requirements (FRs) of a system within the specified design range. System complexity can be relevant to multiple functions or time dependent. The design and operation of some integrated systems, such as large aircrafts, defence systems or even enterprise networks, can be extremely complex and complicated. Traditionally, these systems have mostly been designed using trial-and-error processes and empiricism. To maintain the complexity of these systems to a manageable level, it is desirable to extend our capabilities to successfully synthesise and operate large systems without making them complex. The ultimate goal is to reduce complexity so as to make the system robust, guarantee its long-term stability, make it reliable and minimise the cost [16.5].

According to axiomatic design theory, complexity can be reduced by (1) minimising the number of functional requirements, (2) eliminating the time-independent real complexity, (3) eliminating the time-independent imaginary complexity, and (4) transforming a system with time-dependent combinatorial complexity into a system with time-dependent periodic complexity by introducing functional periodicity and by reinitialising the system at the beginning of each period [16.6]. The basic principle to deal with the system complexity is '*divide and conquer*'. Modularised or distributed system architecture is an effective way to deal with a complex system [16.7]. A recent review provided detailed discussion and comparison of different systems [16.8]. Nevertheless, the sub-components in these surveyed systems are all modular and/or distributed.

16.1.2 Development of a Decentralised System

The general procedure for developing a modular system has been explored in [16.8, 16.9]. This procedure applies to other type of decentralised systems; the following three issues are usually involved in the system development:

- *Architecture design* – determines the system components and their interactions. The system components are encapsulated modules, and their interactions are the options when the modules are assembled. System architecture has to be designed to produce as many system variants as possible, so that the system can deal with changes and uncertainties, cost-effectively. Architecture design is involved at the phase of system design.
- *Configuration design* – determines the system configuration under a given system architecture for a specific task. A configuration is an assembly of the selected modules; a configuration can fulfil the given task optimally. Configuration design is involved at the phase of system application.
- *Control design* – determines appropriate process variables, so that a configuration can be operated to fulfil the task satisfactorily. Control design is involved at the phase of system operation. This chapter will focus on software validation in control design.

16.1.3 Development of Decentralised Control Systems

The control of a complex decentralised system will need to meet the following requirements [16.8]:

1. The control system should be autonomous since a system-level objective can be decomposed into module-level objectives. Each module needs an encapsulated controller to fulfil its objective; the control system should be capable of integrating and coordinating the modules to implement the system-level objective.
2. The control system should be distributed and modularised, since system components are decentralised and geographically distributed.
3. The control system should be open so that it can update controlling components. These controlling components might be developed on heterogeneous operation systems, languages, networks, databases and protocols, and supplied by different vendors.
4. The control system should be scalable and upgradeable because adding/removing/upgrading system components are needed when the functionality, capability or enabling technologies have been changed.
5. The control system should be self-reconfigurable. Since the configuration of a system can be shifted from one configuration to another frequently, the corresponding control system should also be quickly self-reconfigurable.
6. The control system should be capable of identifying any changes to the task specifications. These changes can cause system reconfiguration.

Many concepts of the control paradigms, such as holonic manufacturing [16.10], bionic manufacturing [16.11], fractal companies [16.12], interactive manufacturing [16.13] and random manufacturing [16.14], have been proposed over the last 15 years for next-generation manufacturing systems. Among them, Bussmann and McFarlane [16.15] analysed the rationales to apply agent technologies in manufacturing; it seems that agent-based technologies are feasible to implement these concepts because of their capability to deal with autonomy, distribution, scalability and disturbance. Recent achievements on agent-based technologies in manufacturing were surveyed and documented in [16.16–16.21]. Various control architectures for manufacturing systems control were also proposed [16.22–16.27]. In 1999, Brussel *et al.* [16.28] presented a fundamental work to identify various 'holons' for holonic manufacturing systems.

However, efficient methodologies are still needed to support collaborations in a large-scale multi-agent system. Most of the prototype systems are developed for simple or simplified systems with fewer components [16.29–16.35].

Open architecture control (OAC) provides the infrastructure to implement decentralised system control. Advances in OAC development have been reviewed extensively [16.36–16.40]. The hierarchical structures, which are used widely in mass production and computer integrated manufacturing (CIM), could also be utilised with the consideration of time and changes. Monfared and Weston [16.41] and Harrison *et al.* [16.42] proposed a model-driven approach based on CIM-OSA (open system architecture); Park *et al.* [16.43] developed a generic control framework for modular flexible manufacturing systems; Kalita and Khargonekar [16.44] introduced a formal verification approach for the design of logic controllers for reconfigurable manufacturing systems.

16.1.4 Life Cycle of Control Systems Development

Researchers have observed the repeatable, predictable processes that can improve productivity and quality. The international standard for describing the method of selecting, implementing and monitoring the life cycle for system development is ISO 12207. One of the most popular models for standardising the process of system development is the *waterfall model*. As shown in Figure 16.1, it can be divided into five stages [16.45]:

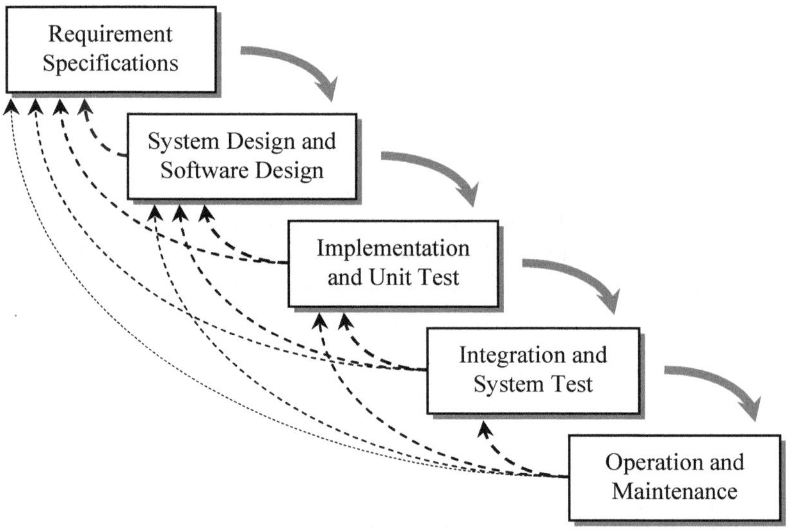

Figure 16.1. The stages of control systems development – the waterfall model

- requirements specifications;
- system design and software design;
- implementation and unit testing;
- integration and system testing;
- operation and maintenance.

At the stage of implementation and unit testing, the design is converted into code. The system is divided into modules and each module is further divided into units. A unit can be defined as a logically separable part of the program. Each unit is tested separately to ensure that it works without any defects. At the stage of integration and system testing, all the units are combined together and the system is built. The complete system is then tested against its functionality and performance requirements. System verification and validation are the essential activities to pass through these two stages successfully.

Developing the control sub-system for a decentralised system is a complicated task. Testing the control system prior to putting it into use is crucial for the system development. A graphic simulation tool is used as an assistant for the visualisation and verification of the control algorithms and protocols for communications.

16.1.5 Overview of the Presented Work

With the rapid development of information technology (IT) and sensing techniques, more and more distributed sensor-based information systems (DSBIS) have been developed for applications in manufacturing, military, anti-terrorism and utility management. Agent-based distributed systems are becoming more and more attractive because of their flexibility, robustness and efficiency.

A DSBIS is capable of making quick decisions based on massive real-time information collected from geographically distributed sensors. The system includes a *data acquisition* system and a *decision-making* system. The data acquisition system has a large number of equitable intelligent sensors with a similar function. Here, 'intelligent' means that sensors not only acquire real-time data from environment, but also make local decisions in regards to their communication behaviours and the type of data received. Intelligent sensors communicate via a wireless communications network. The data acquisition system can be easily extended or dynamically reconfigured. The decision-making system can make decisions for the system-level optimisation over an entire network. Note that the decision-making system is not responsible for making decisions about communications among intelligent sensors. The communication-related decisions are made by intelligent sensors themselves via mutual negotiation and coordination.

Great challenges may arise when developing a software tool for the negotiation and coordination of intelligent sensors, since negotiation and coordination are information exchanges that are invisible and hard to follow from a human perspective. A large number of communication patterns have to be dealt with when a system is used in an unanticipated environment. The system development becomes much more complicated and trivial. In this chapter, efforts are made to facilitate the system development by visualising the procedures of information exchanges via commercial manufacturing simulation tool – Deneb/QUEST. QUEST (Queuing Event Simulation Tool) is a simulation package produced by Delmia Solutions.

For DSBIS development, a graphical visualisation and simulation package can contribute to: (1) validate the functions of wireless sensors; (2) visualise the communication behaviours; (3) analyse the negotiation and coordination algorithms; (4) investigate system responses to unanticipated events; (5) evaluate the system performance; and (6) serve as a real-time monitoring system during the system operation.

The remainder of the chapter is organised as follows. In Section 16.2, the application scenario of a DSBIS is described. System hardware and software architecture is discussed. Unified modelling language (UML) is used to describe its components. An example of a coordinating algorithm is introduced. In Section 16.3, a QUEST modelling approach for system verification is introduced. In Section 16.4, the correspondence between the components of the DSBIS and QUEST elements are established, so that the system can be represented and simulated via the QUEST software. In Section 16.5, an application example is introduced to demonstrate the approach developed in this research. Finally in Section 16.6, a summary is provided and some challenges are discussed.

16.2 Distributed Sensor-based Information System

16.2.1 Application Scenarios

In many networked systems, *e.g.* distributed manufacturing systems, transportation systems, power transmission networks and military surveillance systems, their system components are distributed in a complex geographical environment. The benefits provided by a distributed sensor-based control and monitoring system have drawn great attention in modern manufacturing and logistics systems.

Figure 16.2 shows how wireless sensors are used for tracking, locating and monitoring progress along the supply chain of a car manufacturing plant. In this application, the supply of the sub-components for different models on different production lines involves dozens of variations in sub-components, and the task of ensuring continuity of supply is complex and critical to the business. A wireless tracking system is applied to (i) provide real-time visibility of assets and stock, (ii) improve supply chain management, (iii) monitor the progress of assembly lines, (iv) identify operational and logistic pinch points, (v) improve production planning, and (vi) reduce stock level. A ZigBee network is developed to support the communications among sensor nodes. ZigBee works in the licence-free and globally available 2.4 GHz bandwidth, based on the IEEE 802.15.4 Private Area Network (PAN) standard [16.46].

Figure 16.3 illustrates another application example of DSBIS in the power transmission network. Electrical energy is generated from other energy sources, and the power is delivered via high-voltage lines from power plants to distribution sites. The high-voltage electricity is then transferred into low-voltage power and delivered to consumers [16.47]. When abnormal situations occur, the manual monitoring and inspection of such a distributed system may incur high cost, low reliability, long delay and difficulty in predictable maintenance. This is one of the reasons why we occasionally encounter blackouts and pay a high price for utilities.

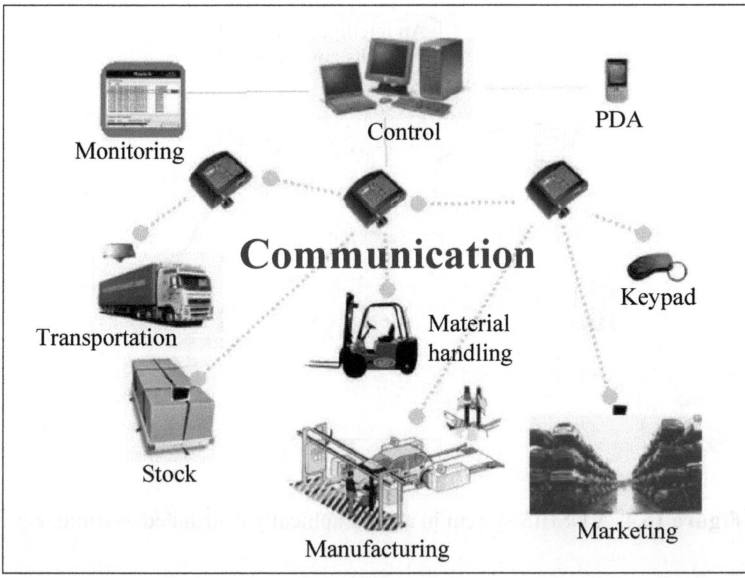

Figure 16.2. Tracking, locating and monitoring with ZigBee networks [16.46]

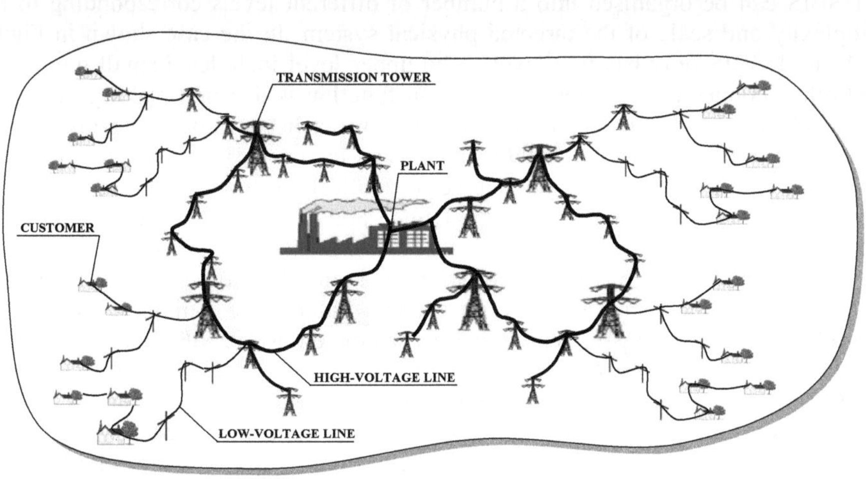

Figure 16.3. A transmission network and its geographically distributed environment

Figure 16.4 shows a scenario that a DSBIS is built upon power transmission network. Intelligent sensors are installed on the monitored objects over the geographically distributed environment. Each senor is capable of collecting real-time data, sharing data with others in a given area through wireless communication, and making local decisions of its communication behaviours. As a result, the sensor controllers can obtain all essential information dynamically. Based on real-time collected information, the system can respond to any abnormal phenomena quickly and efficiently.

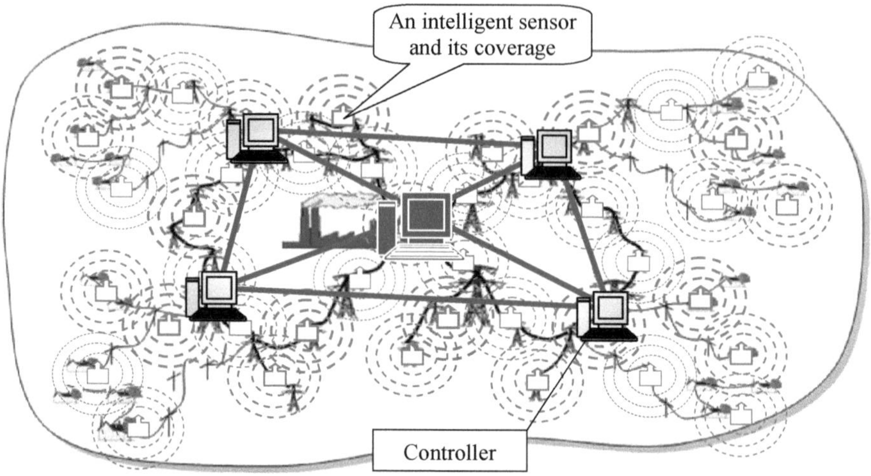

Figure 16.4. A DSBIS system in a geographically distributed environment

16.2.2 Classes of Components in a DSBIS

A DSBIS can be organised into a number of different levels corresponding to the complexity and scale of the targeted physical system. In the case shown in Figure 16.4, the DSBIS only has two levels. The upper level includes a small number of controllers connected by wires, whose main function is decision making. Intensive data exchanges and calculations are involved in each controller. The lower level consists of a large number of intelligent sensors connected via a wireless protocol (such as IEEE 802.11b [16.48]); each sensor collects data and communicates with other sensors locally. In this section, the DSBIS architecture is described using UML [16.49].

This section concerns the negotiation and coordination of communications among the controllers and sensors. From the viewpoint of the communication, both a controller and an intelligent sensor can be abstracted as the inheritance of a more general class, the *MessageProcessor*.

As shown in Figure 16.5(a), a MessageProcessor has common attributes, such as type, name, location and priority, and common functions including receiving/ sending messages and local decision making. In Figure 16.5(b), a MessageProcessor also defines some special attributes and functions related to an instance of the class; these attributes and functions are able to tell the difference between a controller and an intelligent sensor.

The MessageProcessor class could be further modelled as an aggregate of a set of components. As shown in Figure 16.6(a), these components include *receiver*, *database*, *sensor*, *personal digital assistant* (PDA), *storage*, and *sender*. Their relationships are also illustrated in Figure 16.6(b). Note that a MessageProcessor can have multiple objects of the same component class. For example, n_s indicates that the MessageProcessor has n sensors, which can be used to collect various data including environmental temperature, the voltage of the power transmission line, and so on.

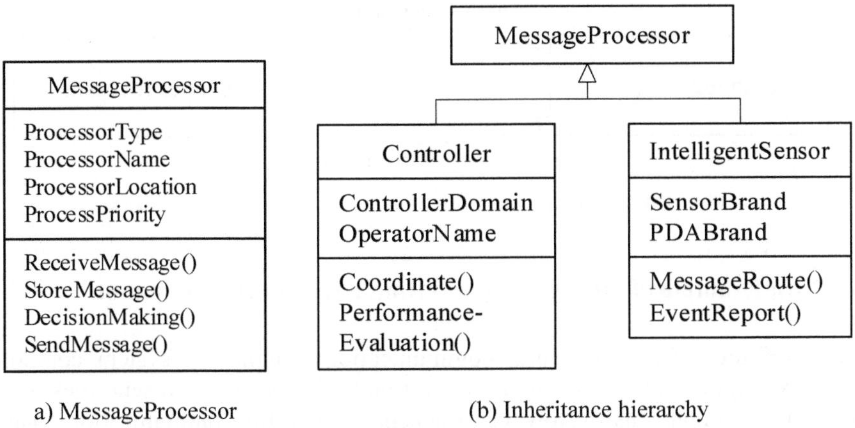

Figure 16.5. MessageProcessor and its inheritances

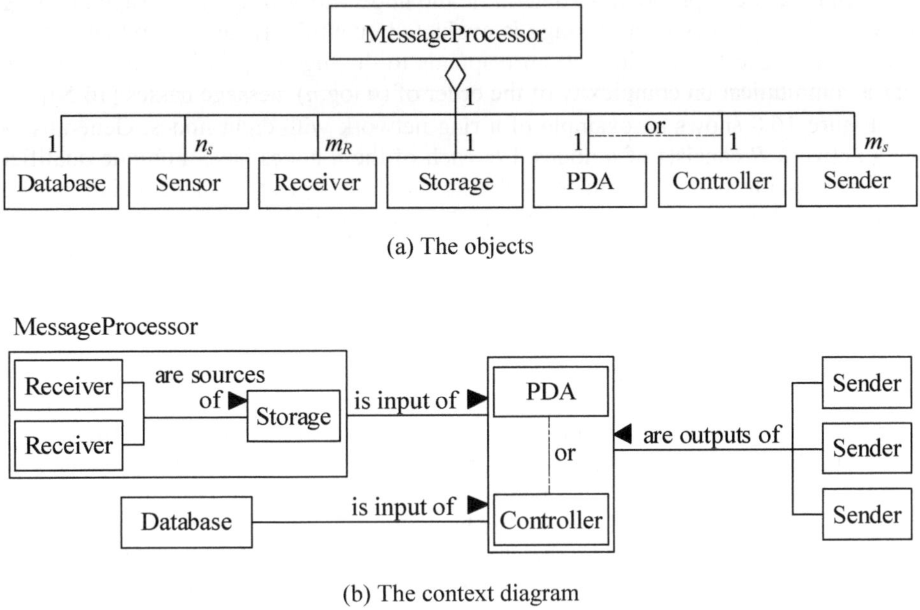

Figure 16.6. Objects and their context diagram in a MessageProcessor

Communication between two MessageProcessors can be classified by the directions of message flow, *i.e.* 'forward' and 'backward'. As shown in Figure 16.7, the meaning of 'forward' and 'backward' are defined with respect to an individual object. If the communication comes from its 'sender', it is 'forward'; if the communication goes to its 'receiver', it is 'backward'. A specific protocol will be applied to make these mutual communications understandable and manageable within a system.

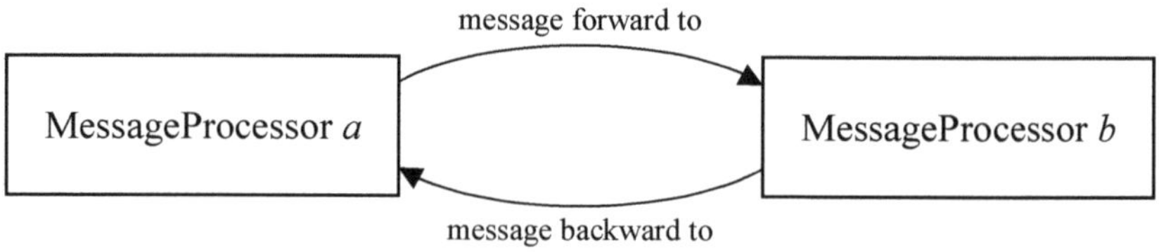

Figure 16.7. Communications between two MessageProcessors

16.2.3 An Example of the Algorithms – Ring Extrema Determination

In a distributed system network, communications among the system component nodes are required. It is very often that a group of nodes have to send messages to each other simultaneously. However, the capacities of the communication channels are physically limited. The communications among these nodes have to be carefully coordinated so that emerging messages can be delivered first and the rest of the messages can be broadcast effectively. The *ring extrema determination* (RED) algorithm has been proposed to determine the node with a priority message in a ring network. The priority of a message is evaluated quantifiably, and an extrema value (either the highest or the lowest) corresponds to the highest priority. The algorithm has a communication complexity of the order of (n log n) message passes [16.50].

Figure 16.8 shows an example of a ring network with eight nodes. Generally, a ring network R consists of n nodes. Let each of the n nodes have a unique qualifier associated with them. Each node is able to pass messages to either a neighbouring node located immediately clockwise and counter-clockwise to it. Each of these dispatched messages originally consists of the node's unique qualifier, indicating the latest known extrema value, a count of the number of hops the message has experienced, the maximum number of hops the message allowed, and a type field indicating whether the message is an outbound or inbound. Each node has the dual responsibility of managing the messages that it has sent out and taking action on messages that belong to other nodes in the ring network. Consider a 'round' for a node to be the total activity of the messages belonging to that node, from the start of outbound messages at the time of dispatch until the return of those messages as either inbound (returned by other nodes) or outbound (ones that have made it all the way around the network and are coming back to their starting point) messages. Each round of a node is numbered by the kth power of 2, where 2^k indicates the maximum number of hops allowed on that round, thus round 0, round 1, round 2, ... means that $2^0, 2^1, 2^2, \ldots$ maximum hops are allowed for those rounds.

Every round is started by a node dispatching outbound messages. A round for each node continues until it has received back its messages, irrespective of whether or not the node has determined itself as an extrema or not. No messages are to be left in the ring at the end of the algorithm, and each node is responsible for taking off the ring the messages that it placed on it.

All nodes commence with round 0. That is, the round where messages with a range of one hop can travel. As each node finishes its round, it continues to the next round. Nodes do not need to be on the same round. Nodes either continue rounds or assist other nodes in their rounds until all messages on the ring are cleared and an extrema is determined. Once a round has commenced and a node has dispatched two outbound messages, one of two events can occur during the round: a node may receive an outbound message from a neighbouring node, or a node may receive an inbound message from a neighbouring node. If a node receives an outbound message, the operations outlined in the following sub-sections can occur together with consequent actions.

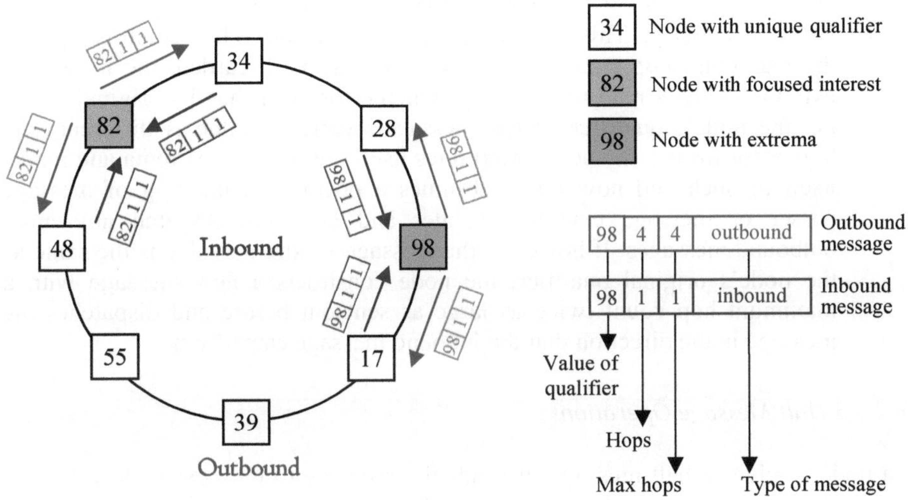

Figure 16.8. Illustration of the RED algorithm

16.2.3.1 Outbound Message Initiated Operations

(a) Send the message forward to the next node. This is required when the message's hop count is less than the message's maximum hop count and the message's extrema value is greater than the node's qualifier. Before the message is forwarded, the hop count is incremented by 1.

(b) Send the message back to its home node. This is required when (i) the message's hop count is equal to the maximum hop count, and/or (ii) the message's extrema value is less (or greater) than the node's qualifier. Before the message is sent back, the node resets the hop count to 1 and the type field in the message is reassigned from outbound to inbound. Node also replaces the message's extrema value with its own qualifier and sets the maximum hop count to the hop count of the message.

(c) Keep the message. This is required when the message's extrema value is equal to the node's qualifier. The message is discarded. This node is now marked as the largest (or smallest) qualifier; since its outbound message has travelled the entire ring and has come back to its origin. The node now has the responsibility to inform all other nodes that an extrema has been determined and that all other nodes can now halt their RED algorithms.

16.2.3.2 Inbound Message Initiated Operations

If a node receives an inbound message, the following operations can occur with consequent actions:

(a) Send the message forward to the next node. This is required when the message's hop count is less than message's maximum hop count. Before the message is forwarded, the hop count is increment by 1.
(b) Keep the message. This is required when the message's hop count is equal to the maximum hop count. Before the message is discarded, the message's extrema value is read to determine whether or not it has its original value, *i.e.* the node's qualifier. If the message's extrema value is different, *e.g.* higher (or lower), the node determines itself not to be an extrema and marks itself as such and now only subsumes a position in the ring of message passer or message extrema changer and does not dispatch any more outbound messages. If however, the message's extrema value is the same as the node's original qualifier, the node constructs a new message with a maximum hop count twice as large as sent out before and dispatches the message in the direction that the inbound message came from.

16.2.3.3 Halt Message Operations

If a node receives a halt message, it sends the message to the next node when the message's hop count is less than the message's maximum hop count. The message is forwarded; the hop count is incremented by 1. As the rounds pass, more and more nodes determine that they are not the extrema being searched for, with message extrema values being compared between nodes. As these nodes are determined, they are left to play a secondary role of either just passing messages or helping other nodes determine that they are not the desired extrema. This continues until the last round where there is only one node left, which determines it to be the extrema node. This round is uniquely characterised as the only round where there are no inbound messages, only the outbound messages of the extrema node exist for the last round. Once explicitly determines itself as the extrema node, the node has one last duty to perform, *i.e.* to notify all other nodes that the RED algorithm is now finished and therefore to stop their algorithms (this is also known as the halt condition).

16.2.3.4 Algorithm

Let i denote an integer reference to a node in a ring R populated with n nodes. For the purposes of simplicity, we assume a sequential numbering of the nodes. It is not necessary to know R's population n; the RED algorithm does not require it. In the pseudo-code listed on the next pages, three abstract classes are used. These classes are *Node*, *MessageSystem* and *Message*.

(a) An object from the *Node* class contains data pertaining to the node that the algorithm is running. It includes two variables, the node's qualifier, *Node.Qualifier*, and the node's status, *Node.Status*. The different types of statuses for the Node.Status variable are held in an internal structure, *Node.STATUSES*, where the constant values {UNKNOWN, EXTREMA, NONEXTREMA} are stored. As the default status when the Node is created, set Node.Status = UNKNOWN.

(b) An object from the *MessageSystem* class has all the requirements needed for handling messages. It contains methods to *Create* and *Destroy*, to *Send* and *Read* messages. Physically, the network ring can be connected with one communication path between nodes as shown in Figure 16.7. Logically, we use an inter-node connection as illustrated in Figure 16.8, where each node internally represents sending messages to their neighbours via *out* queues and receiving messages via *in* queues. When outbound messages are required to be sent, only the out queue number is required; in the case of sending messages clockwise (node IDs getting bigger), out queue = 1; otherwise when sending messages counter-clockwise (node IDs getting smaller), out queue = 2. A similar argument applies to the in queues for the inbound messages. This way when the messages only need to be relayed forward, a simple expression, like *Message.OutQueue* = (*Message.InQueue Mod* 2) + 1, is required for either clockwise or counter-clockwise message forwarding. For those messages to be returned, *Message.OutQueue* = *Message.InQueue* suffices to have the message send back in the direction that it came from. The MessageSystem object also contains a data structure containing the various kinds of messages that may exist. This structure is named *MessageSystem.TYPES*, and contains the constant values {NULL, INBOUND, OUTBOUND, HALT, QUIT}.

(c) An object from the *Message* class contains all the necessary information that a message needs to be delivered and the relevant queue that the message needs to be placed in. *ExtremaValue*, *HopCount*, *MaximumHopCount* and *MessageType* are the data needed for the message to be transmitted between nodes. ExtremaValue is the Node.Qualifier value that is used in the RED algorithm to determine the extrema of a ring network. HopCount is the count of hops that the message is currently away from its node of origin. The MaximumHopCount is the maximum number of hops that a message is allowed to proceed before it is returned, and the MessageType is a value of the MessageSystem.TYPES indicating the kind of message that we have. In addition, there are internal variables, *OutQueue* and *InQueue*, used by a MessageSystem object to understand which out queue to use and which in queue were used, for sending and reading methods, respectively.

List of pseudo-code

```
Extrema = Node.Qualifier
HopCount = 1
MaximumHopCount = 1
MessageType = MessageSystem.TYPES[NULL]
HaltActivated = 0
Message = MessageSystem.Create( Extrema, HopCount,
MaximumHopCount, MessageType )

While MessageType <> QUIT Do
Case MessageType
```

```
       MessageSystem.TYPES[NULL]:
        MessageType = MessageSystem.TYPES[OUTBOUND]
        Message.OutQueue = 1
        MessageSystem.Send( Message )
        Message.OutQueue = 2

      MessageSystem.TYPES[OUTBOUND]:
        If Message.HopCount < Message.MaximumHopCount AND \
           Message.Extrema > Node.Qualifier Then
           Message.OutQueue = (Message.InQueue Mod 2) + 1
           Message.HopCount = Message.HopCount + 1
        Else
        If Message.Extrema < Node.Qualifier Then
           Message.Extrema = Node.Qualifier
           Message.MaximumHopCount = Message.HopCount
           MessageType = MessageSystem.TYPES[INBOUND]
        Else
        If Message.Extrema = Node.Qualifier Then
           Node.Status = Node.STATUSES.EXTREMA
           MessageType = MessageSystem.TYPES[HALT]
           HaltActivated = HaltActivated + 1
        Else
        If Message.HopCount = Message.MaximumHopCount Then
           MessageType = MessageSystem.TYPES[INBOUND]
        End If
        Message.HopCount = 1
        Message.OutQueue = Message.InQueue

      MessageSystem.TYPES[INBOUND]:
        If Message.HopCount < Message.MaximumHopCount Then
           Message.OutQueue = (Message.InQueue Mod 2) + 1
           Message.HopCount = Message.HopCount + 1
        If Message.HopCount = Message.MaximumHopCount Then
           If Message.Extrema > Node.Qualifier Then
              Node.Status = Node.STATUSES[NONEXTREMA]
           End If
           If Message.Extrema = Node.Qualifier Then
              Message.OutQueue = Message.InQueue
              Message.MaximumHopCount = 2 * \
              Message.MaximumHopCount
              MessageType = MessageSystem.TYPES[OUTBOUND]
           End If
        End If

      MessageSystem.TYPES[HALT]:
        If Message.HopCount < Message.MaximumHopCount And \
           HaltActivated <= 1 Then
           Message.OutQueue = Message.InQueue
           MessageSystem.Send( Message )
           MessageType = QUIT
        Else
           MessageType = QUIT
        End If
      End Case

   If MessageType <> QUIT AND HaltActivated <= 1 Then
      MessageSystem.Send( Message )
      Message = MessageSystem.Read()
   End If
   End While
   MessageSystem.Destroy( Message )
```

The algorithm commences by first creating and initialising a Message object. Lines 1–6 illustrate this process in the pseudo-code. Lines 1–5 are the initial values of the Message object that is created in Line 6. The Message object has an initial message type of NULL, indicating that it is the first message of the algorithm. Upon entering into the message-handling loop, the first one is the NULL message-type handler. Here, the algorithm sets up messages for its neighbours, via out queues 1 and 2, and sends messages to them. Note that the second message is sent at the bottom of the message handling routine near the end of the *while* loop. Now when the messages are sent, the *Read* method of the MessageSystem waits until a message is returned (we assume a synchronous behaviour).

Upon receiving a message, the message-handling loop is recommenced again. Three other possible messages, OUTBOUND, INBOUND and HALT, are expected to arrive other than NULL. As explained earlier, the OUTBOUND, INBOUND and HALT messages trigger specific actions when certain conditions are met.

As the RED algorithm proceeds to the end, a HALT message is sent out to all nodes, notifying them that it is the time to end. The last node, the node determined to be the extrema, has this responsibility. It sends a HALT message throughout the ring network where it is propagated from node to node until it comes back. Note that only one HALT and hence one QUIT message is sent throughout the ring network – the determined extrema node can send two notifications one per direction around the ring network – the variable *HaltActivated* counts the number of times that the HALT message is on the ring (Line 27). The condition in Line 49 is to make sure that every node other than the extrema node has the chance to receive the notification, while Line 54 comes into operation for the extrema node itself, so that when the HALT message circumnavigates throughout the ring network and returns, the extrema node can also quit. With the major job done, the message-handling loop terminates. The only job left to do is to destroy the Message object that was created at the beginning, as described in Line 62.

16.3 Modelling Methodologies

The core of the QUEST simulation is its logical model. The logical model has two parts in the context of 'logical components' and 'logic'. The logical components of a QUEST model comprise *elements* and *parts*. As shown in Figure 16.9, the elements can be further classified into different classes. Some elements, such as *sources*, *sinks*, *buffers* and *machines*, are machines or tools to operate *processes* for the parts, and the processes are classified into *setup*, *load*, *unload*, *cycle* and *repair*. Other elements, such as *conveyors*, *automated guided vehicles* (AGV) and *labour*, are material-handling elements, which are tools or paths to move the parts between the elements. *Decision points* are special elements acting as sensors on material handling equipment. Parts are objects that a system makes.

The other part of the QUEST context is the logic. When an interaction occurs between a part and an element, a decision that governs the system behaviour at a specific time must be made. Logic is used to describe decision-making activities to assign tasks to manufacturing resources. Logic acts on, or in response to, a part arriving at a particular element. Logic in QUEST is classified into *process logic*, *route logic*, *reQUEST logic*, *part input logic*, *queuing logic*, *initial logic* and *simulation logic*. Logics could be used-defined or default provided by the QUEST. More details of the QUEST modelling methodology are provided in the reference [16.51].

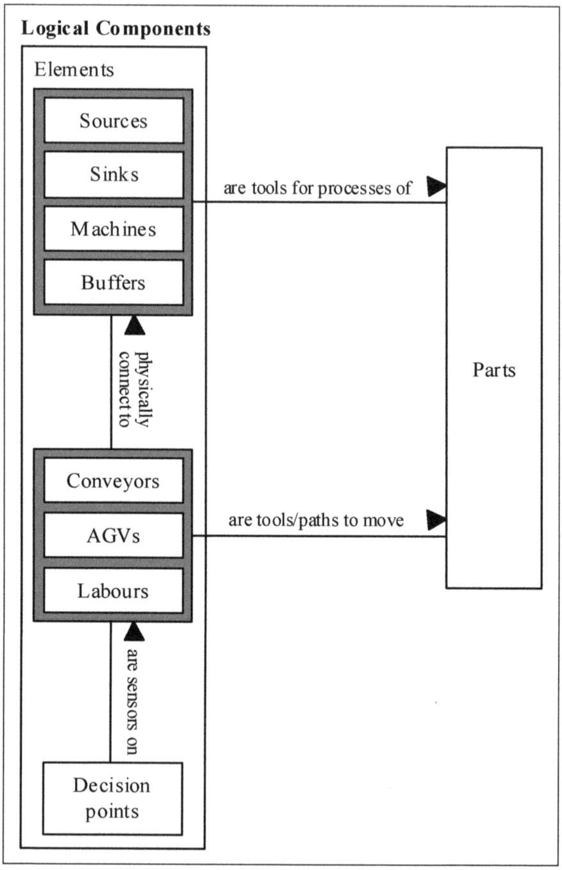

Figure 16.9. Logical components in QUEST

16.4 DSBIS Modelling in QUEST

Although a DSBIS is an information system, it possesses the common characteristics of a manufacturing system: discrete events, synchronous operation and distributed decision making. In this section, we illustrate how the QUEST is applied to simulate a DSBIS. For this purpose, the physical components have to be mapped into the logical components of a QUEST model, where the mapping is established in terms of the functions of components.

As shown in Figure 16.10, the mapping from a DSBIS model to a QUEST model is straightforward. Sensors are mapped to *sources* as the creation of information is similar to the creation of a *part* when a piece of information (message) is mapped to a part in the QUEST model. The PDA/controller could be mapped to a source, a sink or a machine, since a local decision may create the message of a command, ignore unnecessary information, or condense detailed information into an abstracted one. Receivers, storage and senders are modelled as *buffers*, since their functions are in common to keep data temporarily. Communication ports are modelled as *decision points* where some decisions are required to determine the routes of parts, while wireless channels are modelled as *labours*, since both of them provide the paths of transmission for parts and have loading capacities. Finally, control algorithms such as coordinating algorithm could be treated as *logics* in the QUEST model. Note that the mapping is not of one-to-one correspondence.

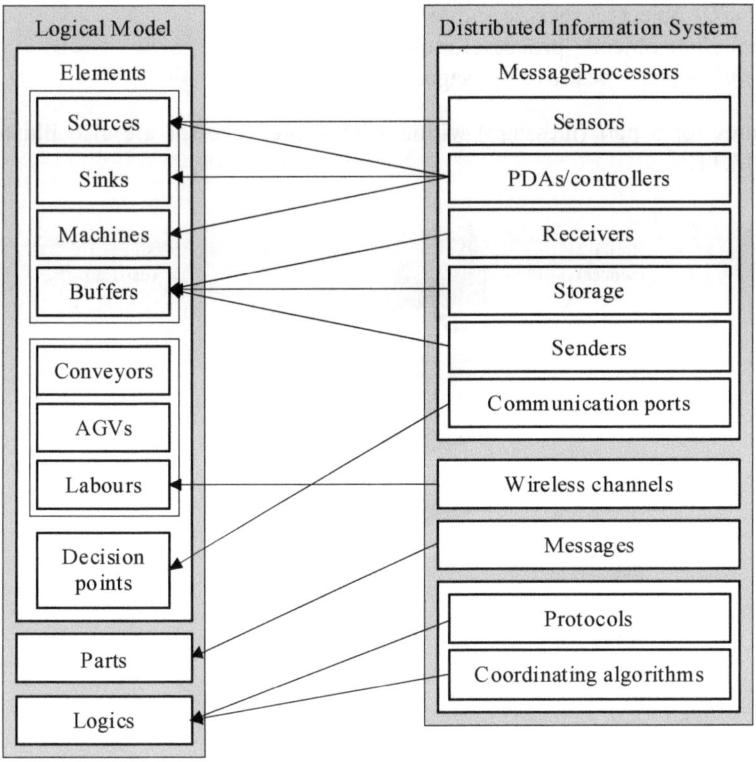

Figure 16.10. Mapping from a DSBIS model to a QUEST model

16.5 Case Study

This section reports a hypothetical wireless sensor network, commonly found in an enterprise network in a decentralise environment. As the case study, its DSBIS is modelled in the QUEST environment, and the RED algorithm introduced in Section 16.2.3 is then simulated and verified to demonstrate its effectiveness.

16.5.1 Basic Components and Communications

The following are considered for modelling: (i) the functions of an intelligent sensor; (ii) the network formed by a set of intelligent sensors; (iii) communication among the intelligent sensors under a *priority rule*, meaning that the message with the highest priority is delivered first. The modelling procedure is as follows.

16.5.1.1 Modelling of MessageProcessor

In Figure 16.11, the components of a MessageProcessor are modelled as a set of the following elements: source, sink, buffer and machine:

- *source* object – produces messages;
- *buffer* objects – receive and send messages;
- *machine* object – processes messages;
- *sink* object – destroys messages.

The routes for a part (message) within a MessageProcessor are also illustrated in Figure 16.11.

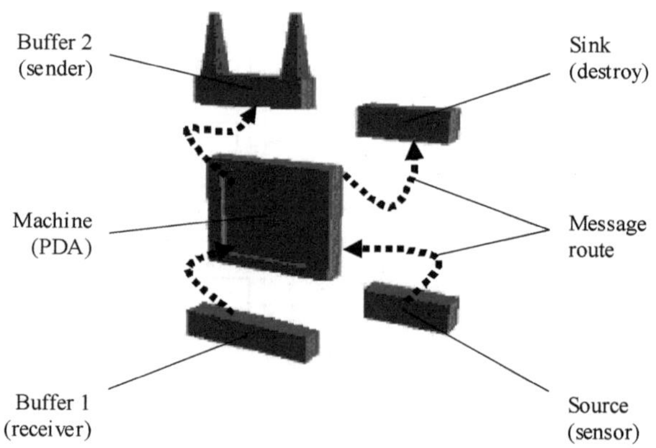

Figure 16.11. MessageProcessor model in QUEST

16.5.1.2 Modelling of Wireless Communication

Communication is implemented by wireless channels. There are some constraints in the communication, *e.g.* the constraints of the maximum capacity and one-direction delivery at any instant of time. To model these constraints, the labour system of the QUEST is considered to be appropriate. As showed in Figure 16.12, a labour system has the following elements:

- labour controller;
- labour;
- labour path;
- decision points on a labour path.

Labour controllers can control the communication speed based on the channel load; labour elements allow it possible to model a wireless channel to transfer a one-directional message at an instant of time; a labour path models the geographic shape between two physical MessageProcessors; and decision points are the ports to connect the MessageProcessors.

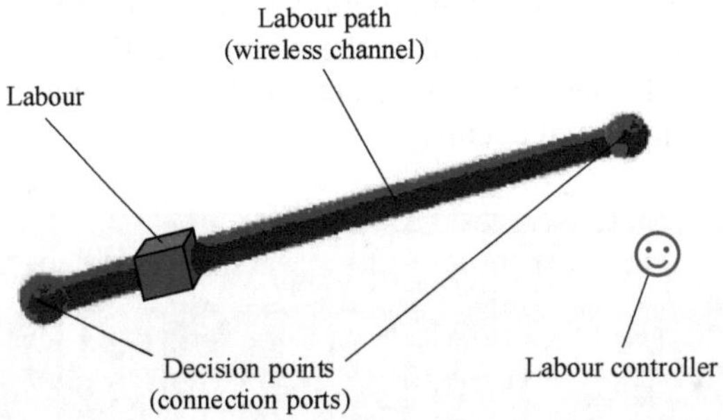

Figure 16.12. Wireless communication model in QUEST

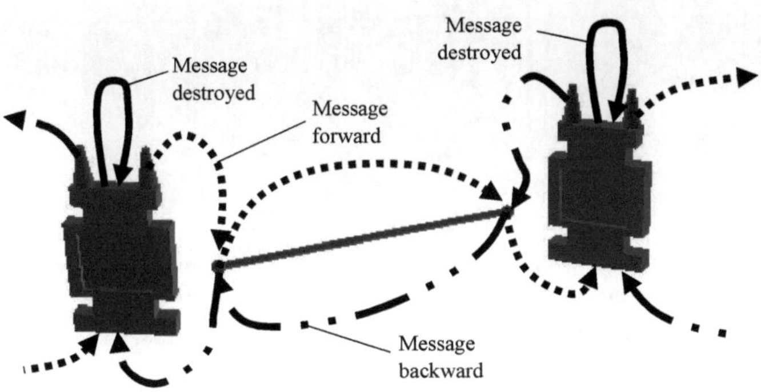

Figure 16.13. Connection between MessageProcessors

16.5.1.3 Modelling of Connection

Two MessageProcessor objects are logically connected by a wireless channel, which supports bi-directional communications. Corresponding to the QUEST model, the connections among objects are defined in Figure 16.13.

16.5.1.4 Modelling of Message

Messages correspond to *parts* in QUEST. Messages can be created and destroyed by a MessageProcessor. Contexts of messages, such as the destination and the source of the message, can be used for the control logic; in addition, they have to be defined as user-defined attributes of *parts* components.

16.5.1.5 Coordinating Algorithms Modelling

Negotiating and coordinating algorithms are modelled as *logic* of a *process* in different elements. For example, if there are multiple outputs for an element, one control logic is needed to determine the appropriate output to which a message is forwarded. This logic can be defined in *route logic* of the element.

As shown in Figure 16.14, a sample model with five MessageProcessor objects is given. This model is used to simulate the communication under the conditions of varying frequency of messages, different decision-making logics, and various combinations of the abnormal events.

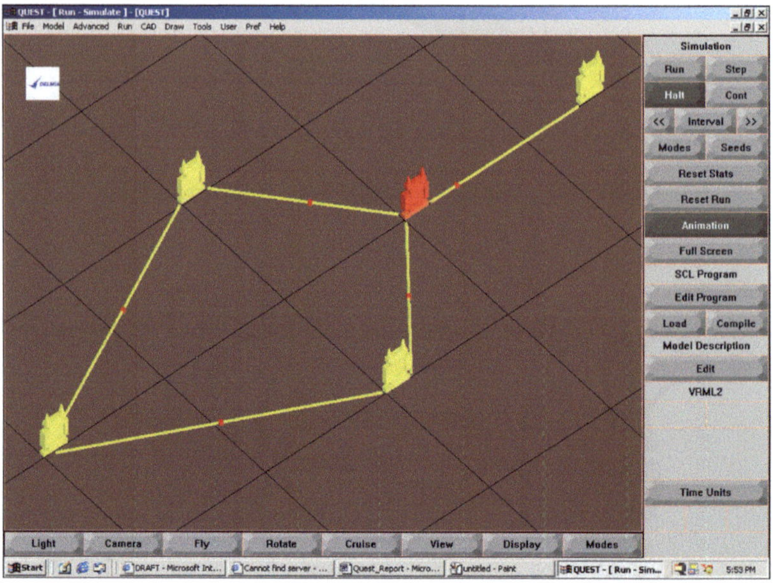

Figure 16.14. Example 1: a QUEST model with five objects

16.5.2 Coordinating Algorithm

This example shows how a QUEST model can be used to verify the coordination algorithms for a DSBIS. Note that the system is totally distributed; the network communication has to be self-disciplined by distributed intelligent sensors in the system. In this case, a ring of *n* nodes are required to determine by themselves which of the nodes will take priority in delivering data. The RED algorithm is adopted to determine the priority message to be delivered.

In this example, we assume that the MessageProcessor objects require a large number of messages to be delivered and their message transmission must be accomplished individually. RED helps to find the most urgent message (extrema) among a set of MessageProcessor objects based on the attribute of Qualifier of a MessageProcessor. The requirement of the message transmission from the extrema object is the first priority to be satisfied.

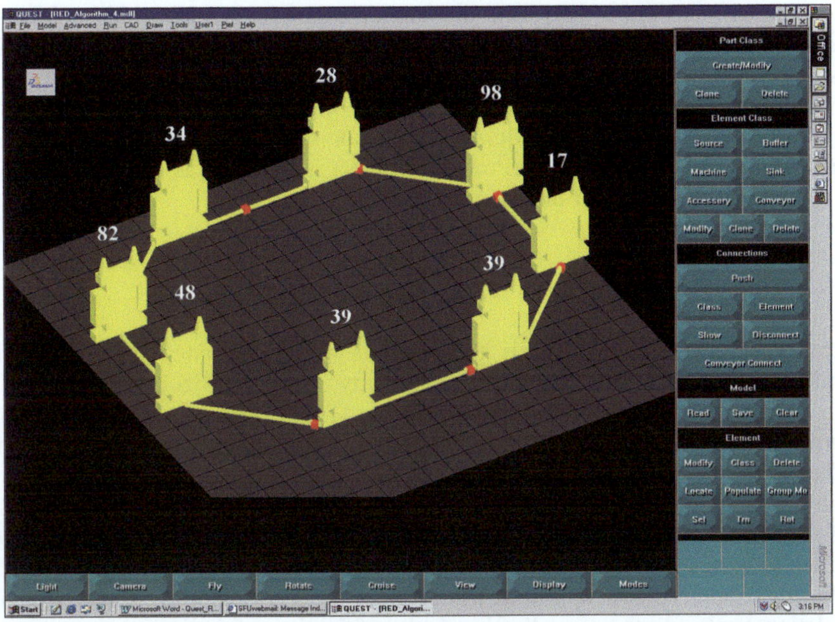

(a) Before simulation

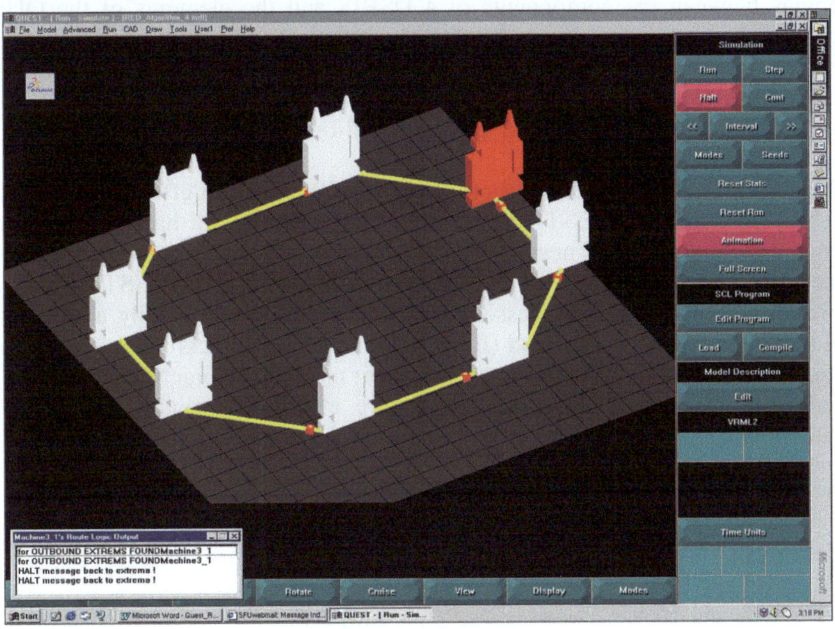

(b) After simulation

Figure 16.15. Example 2: a QUEST model with eight objects

As shown in Figure 16.15(a), the model example includes eight objects of MessageProcessor. The RED algorithm is defined as the route logic for all of the MessageProcessor objects. The initial values of Qualifier are assigned to the corresponding objects. The simulation result is illustrated in Figure 16.15(b). The original colour for all of the MessageProcessor objects is yellow; the colour is changed to white if the MessageProcessor object is detected as 'not extrema', and red if the MessageProcessor object is detected as 'extrema'.

16.6 Conclusions

A comprehensive review reveals that modular and distributed architecture is a premier strategy to deal with system complexity, when DSBIS becomes more and more important in daily life and manufacturing. This is also true when an extended enterprise network is formed in a decentralised environment. One of the challenges in the design of a DSBIS is to reduce the development time and increase the reliability of its control software. The software test and verification is critical at the phases of unit test; integrated graphical simulation tools have barely been applied to accelerate the system testing and verification.

In this chapter, a new approach has been proposed using conventional graphical simulation tools for the visualisation and verification of information systems. A DSBIS for the control and monitoring of a large complex system has been analysed and its similarities to the QUEST model are discussed. It is concluded that the mappings between the components of a DSBIS and the elements of the QUEST model can be established, in order that the information system is visualised and simulated in the Deneb/QUEST environment. Visualisation and simulation can accelerate the development of control system tools (software control modules) of an information system. It can also be used as a real-time monitoring system during the course of the system operation.

Despite the superior advantages of vivid visualisation and effective verification, we observed that the QUEST-based graphical simulation still has two limitations: (i) the modelling process needs extensive time to create the elements and their connections since the graphical user interface (GUI) of the QUEST is relatively restricted, *e.g.* all MessageProcessor objects are created individually and manually even though they are similar; and (ii) the number of the objects in the simulation is confined, and modelling of a large-scale system may not be practical because of the complexity of the modelling process. Future efforts are required to solve the aforementioned shortcomings by developing an automated modelling program that is compatible with the QUEST model.

References

[16.1] Wikipedia, 2009, *Economy of the People's Republic of China*, online access available at http://en.wikipedia.org/wiki/Economy_of_the_People's_Republic_of_China.

[16.2] Cong, C., 2008, "Technological development challenges in Chinese industry," In *China's Science and Technology Sector and the Forces of Globalisation*, Thomson, E. and Sigurdson, J. (eds.), World Scientific, Singapore, pp. 14–44.

[16.3] Sinodefense, 2009, *ShenZhou Manned Spacecraft*, online access available at http://www.sinodefence.com/space/spacecraft/shenzhou.asp.

[16.4] Sinodefense, 2009, *Jian-10 Multirole Fighter Aircraft*, online access available at http://www.sinodefence.com/airforce/fighter/j10.asp.

[16.5] Suh, N.P., 2005, *Complexity: Theory and Applications*, Oxford University Press, Oxford.

[16.6] Suh, N.P., 2005, "Complexity in engineering," *Annals of CIRP – Manufacturing Technology*, **54**(2), pp. 46–63.

[16.7] Bi, Z.M. and Zhang, W.J., 2001, "Modularity technology in manufacturing: taxonomy and issues," *International Journal of Advanced Manufacturing Technology*, **18**(5), pp. 381–390.

[16.8] Bi, Z.M., Lang, S.Y.T., Shen, W. and Wang, L., 2008, "Reconfigurable manufacturing systems: the state of the art," *International Journal of Production Research*, **46**(4), pp. 967–992.

[16.9] Bi, Z.M. and Zhang, W.J., 2000, "Concurrent optimal design of modular robotic configuration," *Journal of Robotic Systems*, **18**(2), pp. 77–87.

[16.10] Markus, A., Vancza, T.K. and Monostori, L., 1996, "A market approach to holonic manufacturing," *Annals of CIRP*, **45**(1), pp. 433–436.

[16.11] Okino, N., 1993, "Bionic manufacturing system, flexible manufacturing systems past-present-future," *Annals of CIRP*, **42**(2), pp. 73–95.

[16.12] Sihn, W. and Rist, T., 1998, "Experiences with the fractal company value shift," *CIRP Journal of Manufacturing Systems*, **27**(1), pp. 23–30.

[16.13] Ueda, K., Markus, A., Monostori, L., Kals, H.J.J. and Arai, T., 2001, "Emergent synthesis methodologies for manufacturing," *Annals of CIRP*, **50**(2), pp. 535–551.

[16.14] Uata, K. and Onosato, M., 1994, "Random manufacturing system: a new concept of manufacturing systems for production to order," *Annals of CIRP*, **43**(1), pp. 379–383.

[16.15] Bussmann, S. and McFarlane, D.C., 1999, "Rationales for holonic manufacturing control," In *Proceedings of the 2nd International Workshop on Intelligent Manufacturing Systems*, Leuven, Belgium, pp. 177–184.

[16.16] Shen, W., Norrie, D.H. and Batthes, J.-P., 2001, *Multi-Agent Systems for Concurrent Intelligent Design and Manufacturing*, Taylor & Francis, New York.

[16.17] Caplinskas, A., 1998, "AI paradigms," *Journal of Intelligent Manufacturing*, **9**, pp. 493–502.

[16.18] Tommila, T., Venta, O. and Koskinen, K., 2001, "Next generation industrial automation – needs and opportunities," *Automation Technology Review*, pp. 34–41.

[16.19] Kim, B., 2002, *Intelligent Agent Based Planning, Scheduling, and Control: Warehouse Management Application*, PhD Thesis, Rensselaer Polytechnic Institute, Troy, NY.

[16.20] Marik, V., Fletcher, M. and Pechoucek, M., 2002, "Holons & agents: recent developments and mutual impacts," In *Multi-Agent Systems and Applications II*, Marik, V., Stepankova, O., Krautwurmova, H. and Luck, M. (eds.), Springer, Heidelberg, pp. 233–267.

[16.21] Tian, G., Yin, G. and Tayloy, D., 2002, "Internet-based manufacturing: a review and a new infrastructure for distributed intelligent manufacturing," *Journal of Intelligent Manufacturing*, **13**, pp. 323–338.

[16.22] Bongaerts, L., Monostori, L., McFarlane, D. and Kadar, B., 2000, "Hierarchy in distributed shop floor control," *Computers in Industry*, **43**, pp. 123–137.

[16.23] Langer, G. and Alting, L., 2000, "An architecture for agile shop floor control systems," *Journal of Manufacturing Systems*, **19**(4), pp. 267–280.

[16.24] Zhang, X., Balasubramanian, S., Brennan, R.W. and Norrie, D., 2000, "Design and implementation of a real-time holonic control system for manufacturing," *Information Science*, **127**, pp. 23–44.

[16.25] Balasubramanian, S., Brennan, R.W. and Norrie, D., 2001, "An architecture for metamorphic control of holonic manufacturing systems," *Computers in Industry*, **46**, pp. 13–31.

[16.26] Heikkila, T., Kollingbaum, M., Valckenaer, P. and Bluemink, G.-J., 2001, "An agent architecture for manufacturing control: ManAge," *Computers in Industry*, **46**, pp. 315–331.

[16.27] Zaremba, M.B. and Morel, G., 2003, "Integration and control of intelligence in distributed manufacturing," *Journal of Intelligent Manufacturing*, **14**, pp. 25–42.

[16.28] Brussel, H., Bongaerts, L., Wyns, J., Valckenaers, P. and van Gingerachter, T., 1999, "A conceptual framework for holonic manufacturing: identification of manufacturing holons," *Journal of Manufacturing Systems*, **18**(1), pp. 35–52.

[16.29] Tatra, M., Soparkar, N., Yook, J. and Tilbury, D., 1997, "Real-time data and coordination control for reconfigurable manufacturing systems," In *Real Time Database and Information Systems*, Bestavros, A. and Fay-Wolfe, V. (eds.), Springer, London, pp. 23–48.

[16.30] Lucas, M.R., Endsley, E.W. and Tilbury, D.M., 1999, "Coordinate logic control for reconfigurable machining systems," In *Proceedings of the American Control Conference*, San Diego, CA, pp. 2107–213.

[16.31] Tilbury, D.M. and Kota, S., 1999, "Integrated machine and control design for reconfigurable machine tools," In *Proceedings of the 1999 ASME/IEEE International Conference on Advanced Intelligent Mechatronics*, 19–23 September, pp. 629–633.

[16.32] Shu, S., Wilkes, M. and Kawamura, K., 2000, "Development of reusable, configurable, extensible holonic manufacturing system," In *Proceedings of the IEEE International Conference on Systems, Man, and Cybernetics*, Piscataway, pp. 1679–1684.

[16.33] Wang, S. and Shin, K.G., 2002, "Constructing reconfigurable software for machine control systems," *IEEE Transactions on Robotics and Automation*, **18**(4), pp. 475–486.

[16.34] Arbib, V. and Rossi, F., 2000, "Optimal resource assignment through negotiation in a multi-agent manufacturing system," *IIE Transactions*, **32**, pp. 963–974.

[16.35] Ottaway, T.A. and Burns, J.R., 2000, "An adaptive production control system utilizing agent technology," *International Journal of Production Research*, **38**(4), pp. 721–737.

[16.36] Albus, J.S., 1993, *A Reference Model Architecture for Intelligent Systems Design*, http://www.isd.mel.nist.gov/documents/albus/Ref_Model_Arch345.pdf.

[16.37] Faulkner, D., Levy, B. and Garner, T., 1999, "Open-architecture platforms," *Circuits Assembly*, January, http://www.cassembly.com.

[16.38] Erol, N.A., Altintas, Y. and Ito, M.R., 2000, "Open system architecture modular tool kit for motion and machining process control," *IEEE/ASME Transactions on Mechatronics*, **5**(3), pp. 281–291.

[16.39] Pritschow, G., Altintas, Y., Jovane, F., Koren, Y., Mitsuishi, M., Takata, S., van Brussel, H., Weck, M. and Yamazaki, K., 2001, "Open controller architecture – past, present and future," *Annals of CIRP*, **50**(2), pp. 463–470.

[16.40] Katz, R., Li, Z. and Pierrot, F., 2002, *Engineering Research Center for Reconfigurable Machining Systems: Conceptual Design of a High-Speed Drilling Machine (HSDM) Based on PKM Module*, ERC/RMS Report #37, College of Engineering, The University of Michigan, Ann Arbor, MI.

[16.41] Monfared, R.P. and Weston, R.H., 1997, "Reconfiguration of manufacturing cell control systems and reuse of their components," *Proceedings of the Institution of Mechanical Engineers*, **211**(B), pp. 495–508.

[16.42] Harrison, R., Weston, R.H. and Monfared, R.P., 2001, "Distributed engineering of manufacturing machines," *Proceedings of the Institution of Mechanical Engineers*, **215**(B), pp. 217–231.

[16.43] Park, J.H., Park, J.W. and Kim, J., 2000, "A generic event control framework for modular flexible manufacturing systems," *Computers & Industrial Engineering*, **38**, pp. 107–123.

[16.44] Kalita, D. and Khargonekar, P.P., 2002, "Formal verification for analysis and design of logic controllers for reconfigurable machining systems," *IEEE Transactions on Robotics and Automation*, **18**(4), pp. 463–474.

[16.45] Pressman, R.S., 2002, *Software Engineering – A Practitioner's Approach*, 5th edn., McGraw-Hill, New York.

[16.46] Intelligent Distributed Controls Ltd., 2009, *Track, Locate and Monitor Progress in Real Time with ZigBee Networks*, http://www.zig-bee.co.uk/.

[16.47] Energy Information Administration, 2009, *Chapter 3: The U.S. Electric Power Industry Infrastructure: Functions and Components*, online access available at http://www.eia.doe.gov/cneaf/electricity/chg_stru_update/chapter3.html.

[16.48] IEEE Standards, 2009, *IEEE 802.11™ Wireless Local Area Networks*, online access available at http://grouper.ieee.org/groups/802/11/.

[16.49] Booch, G., Rumbaugh, J. and Jacobson, I., 2005, *The Unified Modeling Language User Guide*, 2nd edn., Addison-Wesley.

[16.50] Hirschberg, D.S. and Sinclair, J.B., 1980, "Decentralized extrema-finding in circular configurations of processes," *Communications of the ACM*, **23**(11), pp. 627–628.

[16.51] Delmia Solutions, 2000, *QUEST User Manual*, Version 5.3, Delmia Corporation.

17

Robustness and Capability Indices in the Optimisation of an Airline's Fleet – Bridging Contradicting Outcomes

Leo D. Kounis

Department of Aviation Technology, Halkis Polytechnic, 34400 Psachna, Evia
State Aircraft Factory, KEA, Research Department, 300 Vouliagmenis Ave.
Hellinikon, Athens, Greece
Email: leo_kounis@hotmail.com

Abstract
This chapter discusses the implication of robustness and capability indices in the optimisation process of an airline's fleet. The introduction of European Commission Regulation No 1617/93 of June 25th, 1993 aimed at minimising observed bottlenecks in air traffic. The importance of maintaining specific hub-and-spoke systems, as well as optimising slot controls have been raised by many authors and field experts to determine optimum market-related performance and shares. The introduction of Taguchi's design of experiments (DoE) is applied as a means of improving overall performance levels. Conducted research reveals that the application of robustness and capability indices may lead to contradicting outcomes regarding the most appropriate courses of action. The aim of this work is to introduce a technique capable of effectively addressing contradicting outcomes by optimising processes and procedures-alike and thus minimising potential losses to the company.

17.1 Introduction

The introduction of the European Commission Regulation No 1617/93 of June 25th, 1993 with effect as of 01/01/1993 [17.1] places an emphasis on sound competition among airliners. Airline companies can achieve this by offering services within popular times between popular airports. Time allowances are referred to and therein as slots, whereas major airports making up the bulk of connecting and inter-connecting flights are characterised as hubs.

In order to enable a smooth and optimum number of aircraft movements in terms of runway(s) utilisation, taxiing, landing and take-off, each and every airport has formed so-called scheduling committees. The latter act in accordance to IATA-(International Air Transport Association) implied regulations pertaining to the maximum number of scheduled landings and take-offs. The liberalisation of the world's air space has led to a gradual congestion, particularly between major European hubs. In order to address the bottleneck of continuous delays, the Federal Aviation Authority (FAA) has proposed the materialisation of a slot-specific re-design process according to a Ground Delay Program (GDP).

Performed literature review indicates a number of papers introducing new management techniques in order to handle the increasing number of passengers. It is worth mentioning that emphasis is placed on intermodal transport means, as well as the grouping of unified cargo transports. The latter poses a constraint at major hub terminals as well as other non-hub airports. A potential loss of a hub-and-spoke system is regarded as potentially detrimental in terms of financial, operational, ethical, and strategic planning of the associated airline companies.

Literature review reveals that an emphasis be placed on the strategic importance of maintaining a hub-and-spoke system in relation to ground handling schemes. An in-depth comparative analysis highlights the capabilities and limitations between full-service carriers and low-cost airliners, taking into consideration the effect of slot controls in determining best market practices. The proposed model is to establish a platform for further research and hub-and-spoke model optimisation.

Further work focuses on analysing, evaluating and optimising parameters that may inadvertently affect the performance-oriented effectiveness of an airport-airline companies system. In view of the above, the development of a dynamic model incorporating a clear definition, robust design and a proper and optimised development of a hub-and-spoke network system is regarded by the author as being of paramount importance.

For effectively addressing observed bottlenecks in a hub-and-spoke network, Taguchi's DoE is applied. The latter is based upon already existing outcomes – albeit of an experimental nature – and therefore, takes into consideration influencing parameters and interrelating factors.

The outcomes of this study will assist in the optimisation of

- an airline's operational effectiveness
- improving overall financial performance
- reducing an airline's overall environmental footprint

Based on observed limitations of proposed techniques, this work aims to

- emphasise the criticality of maintaining and utilising specific hubs
- evaluate current operating practices
- introduce a method for effectively meeting market-driven demands
- develop a statistical model for analysing contradicting parameters
- apply obtained outcomes and a statistical model in selected cases

17.2 Literature Review

The advent of Commission Regulation EEC 95/93 concerning 'common rules for the allocation of slots at Community airports' [17.2.] emphasises slot allocation and introduces a two-phase approach to allowing for greater flexibility pertaining to 'market mechanisms to improve market accessibility.' Indeed, phase two of the Directive proposes a review of the regulation taking into consideration European Union's white paper that forecasts double aircraft movements until 2010 [17.3].

As such, the objectives of the aforementioned regulations are to

- ensure mobility of slots and efficient use of airport capacity
- maintain effective competition at EU airports

- ensure the compatibility of proposed and implied scheme with worldwide practices and procedures

To effectively address the continuous growth in air traffic, the aforementioned Directive stresses the fact to allow for '…the development of a common platform pertaining to airport-related capacity in congested EU-airports, where slot allocation has been allocated on the basis of clear, objective and unbiased rules.'

The regulation has been enacted on January 18th, 1993, so as to serve as a legal platform on already existing practices concerning slot-allocation regime in Europe. It is worth mentioning that the Directive's main points originated from IATA.

In view of the above, the long-term goal of the EU is to develop, introduce and imply a drastically amended regulation, addressing existing operating mechanisms and practices pertaining to slot-allocation.

Furthermore, the new Directive EEC 95/93 suggests the introduction and the development of a Code of Conduct, including all airport-related activities, such as

- financial management
- hybrid demand management approaches
- optimisation of airport infrastructure
- improved and sound approach of new airline entrants/alliances, and
- the enactment of market values relating to an airport's capacity

Continuous growth of passenger traffic has led to a five-fold increase of airport handling capability, as documented in Table 17.1 referring to Athens International Airport (AIA).

Table 17.1. Domestic and international air traffic, 1997–2007

Year	Aircraft traffic	Passenger (*thousands*)		Freight and mail (*tons*)	
		Embarked	Disembarked	Loaded	Unloaded
1997	332,491	13,765	14,276	70,714	97,684
1998	343,414	13,932	14,524	62,986	87,383
1999	396,624	16,346	16,459	58,765	89,600
2000	427,309	18,314	17,918	65,933	90,394
2001	403,170	17,856	17,591	–	136,534*
2002	366,524	16,750	16,672	–	127,106*
2003	395,807	17,054	16,973	–	127,856*
2004	408,561	17,805	17,582	–	133,496*
2005	403,129	18,319	18,131	–	133,652*
2006	429,488	19,454	19,265	–	143,958*
2007	442,269	20,789	20,523	–	136,430*

* Since 2001, the loaded and unloaded freights are presented in total. (Source: Hellenic National Statistical Society)

It is worth mentioning that during the years 1997–2000, embarking passengers show an increase of 132%, whereas the corresponding aircraft movements follow the pace with an increase of 128%. Table 17.1 also indicates that world-wide air transport was marginally hit and affected by the event of September 11th, 2001. During the years 2002–2007, passenger growth increased by 128%, whereas aircraft traffic rose by 120%, respectively.

The observed outcomes of Table 17.1 are indicative of the trend concerning annual air-traffic growth. The latter may influence the following areas

- airport capacity,
- runway capacity, and
- safe time and distance between taxiing aircrafts.

Rassenti *et al.* [17.4] are the initiators to propose the method of slot auctioning as a means of fair, just and sustainable policy in the field of air transport. Among others, Kounis *et al.* [17.5] claim that auctioning access, on one hand to existing airport capacity and to new capacity on the other hand, may result in 'perverse incentives' and limit their practical application. They proposed

- accessibility auctioning on already existing airport capacity; and
- accessibility auctioning regarding new capacity.

In view of the above, the European Union suggested during the 1990s the liberalisation of the air transport market. Performed research revealed that the last part of EEC 95/93 was enacted approximately 3 years ago, when the former Eastern European countries gradually joined EU. For the complete implementation of the Directive, the normalisation of the legal framework of the member states ought to materialise by incorporating European law into their national law.

However, the strive for continuous improvement in passenger services in conjunction with financial issues affecting flight scheduling is regarded by the author as forming an integral part of the liberalisation of European skies. For this to be facilitated, the aforementioned factors ought to take into consideration the following parameters.

- Equipment and infrastructure maintenance. This includes human resources and technical maintenance, including servicing areas;
- Continuous training of staff, including flight personnel and ground handling staff;
- Adequate capacity at airports that effectively deal with passengers, luggage and cargo handling, check-in facilities and waiting lounges;
- Marketing parameters, such as market-size, duration of travel, including time zones, distance of travel and distance between airport and city;
- Other parameters, including weather-related conditions (winds, *etc.*)

In order for an airline company to be flexible and to meet market demands effectively, most air carriers have introduced the following four scheduling types.

- skip stop
- local services
- hub-and-spoke networks
- non-stop flights

As shown in Figure 17.1, skip-stop flights may serve points A, B, C, D, E, F, G, *etc.* by scheduling the flight in a pattern A, C, E, G, or A, D, G, or by any other possible combinations.

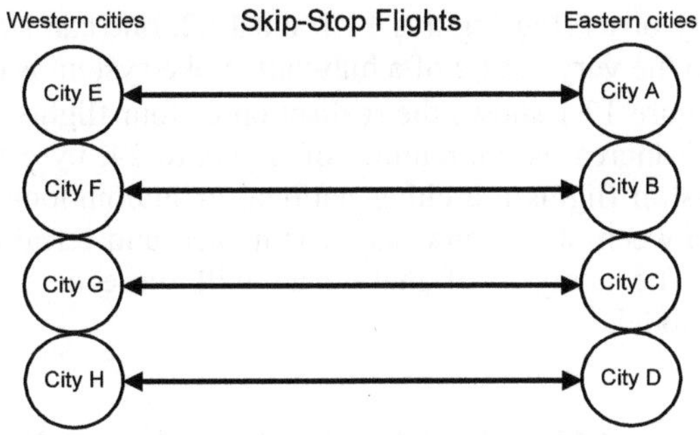

Figure 17.1. Skip-stop flights

The advantages of such a system are summarised below:

- quick call at intermediate airports;
- servicing of major routes by other flights/operators.

Connecting flights offer the following advantages:

- quick servicing of in-between major airports;
- emphasis is placed on local services and non-stop flights;
- parallel running of skip-stop and hub-and-spoke networks.

Local services connect cities with each other and have a flight duration of not more than one hour. Their services normally utilise small-range aircrafts. The latter call at bigger airports, so that passengers may embark on extended range aircrafts for their final destinations. The advantages of this policy are associated with a frequent and fast service between smaller and bigger airports. A drawback of this strategy is the increased number of aircrafts used to materialise such a diagram.

Hub-and-spoke networks form the bulkhead of airline companies and they are detailed in Figure 17.2.

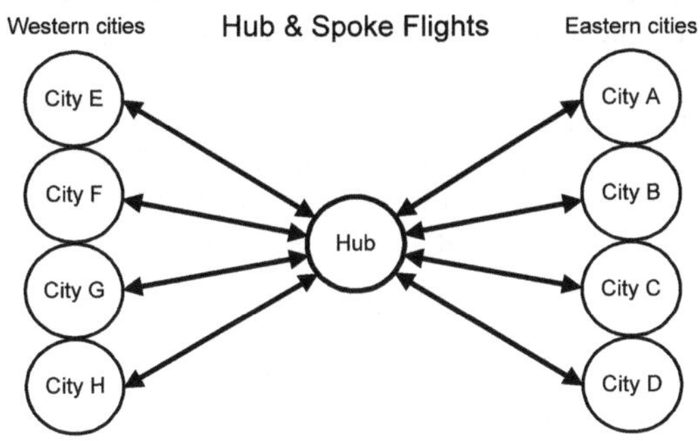

Figure 17.2. Hub-and-spoke flights

In such a system, the incoming flights allow passengers to change over and use a connecting flight at a hub. The connecting flight then serves the passengers' final destination point. By observing Figures 17.1 and 17.2, one can state that the growth of services owing to the very nature of a hub-and-spoke system is evident.

Furthermore, Figure 17.1 shows the realisation of four flights between the cities, whereas Figure 17.2 increases the number of flights to 24, by establishing one city as a hub. The non-stop flights are either national or international flights that serve cities without in-between stops, and may commence and terminate their trip at a secondary airport. The analysis of these data will serve as input factors to the proposed Taguchi model.

17.3 Contribution of Quality Standards in the Airline Industry

In 1987, the EC Commission requested CEN/CENELEC to adopt the International Standards ISO 9000 as the appropriate European Standards known as EN 29000. This is regarded as a decision of momentous importance, as it set the fundamentals for the creation of an EU-internal market, but also served as a milestone for the industry worldwide. By 1977, a number of countries had already produced national standards for the operation in manufacture of quality control systems. In 1979, British Standards Institution (BSI) published the BS 5750 series of standards [17.6]. At that time, ISO formed a technical committee, TC 176, with the task to develop a single standard for the operation and the management of quality assurance. The scope of this team was to bring together expertise from the standard bodies of the various countries, where similar work was in hand at national level.

Meanwhile, the ISO processes for transforming draft standards into voting documents and subsequently final published standards proceeded through the system of worldwide consultations. By 1987, ISO has published the ISO 9000 series of standards. Equally at that time, several countries took the opportunity to bring their own national standards in line with the final ISO-specific requirements. Indeed, Germany with the Deutsches Institut fur Normung (DIN) developed the Deutsche Industrie Normen (German Industrial Norms), which is common to BS 5750 and now referred to as BS EN ISO 9000.

The Single European Act declared the end of 1992 to be the beginning of the single market. The EC accepted this date as the formal adoption of ISO 9000, but some member states were advanced in promoting the aforementioned standard.

The contemporary interest in quality audits and the concomitant development of fundamental quality systems, especially in the wake of ISO 9000, has accelerated over the last decade [17.7]. Owing to the advent of QS 9000 resulted as the merger of the quality standards of GM, Ford, Chrysler and Mack Trucks, the motor industry sector has built upon British Standards / European Norm / International Standards Organisation – BS EN ISO 9001, and further developed it to include continuous improvement techniques and methods [17.8].

The above summarises the result of worldwide competition practices and market globalisation in conjunction with increasing customer expectations for high-quality items or services that set the pace. Many industries and specialist standards institutions adopt accepted techniques, such as statistical process control (SPC), failure mode and effects analysis (FMEA) and signal-to-noise ratios (SNR) to improve their processes [17.9].

Design of experiments includes the use of SNR, which are widely applied in the automotive manufacturing industry. In order to examine the percentage of rejects, Capability Index (Cp_k) is used. However, performed research reveals that SNR and Cp_k indices are used independently from each other.

Literature review indicates that there are many different types of SNRs. Three principal ones are generally used, *i.e.* minimum-the-best, maximum-the-best, and nominal-the-best.

Each of the aforementioned SNRs requires individual treatment. The implication of SNRs are normally applied in industry to show a form of robustness, but seldom as long-term commitment to achieving optimum settings. It has been observed in a variety of cases that SNR ratios do meet their targets, but showing considerable variations [17.10]. Figure 17.3 depicts the SNR vs. Cp_k.

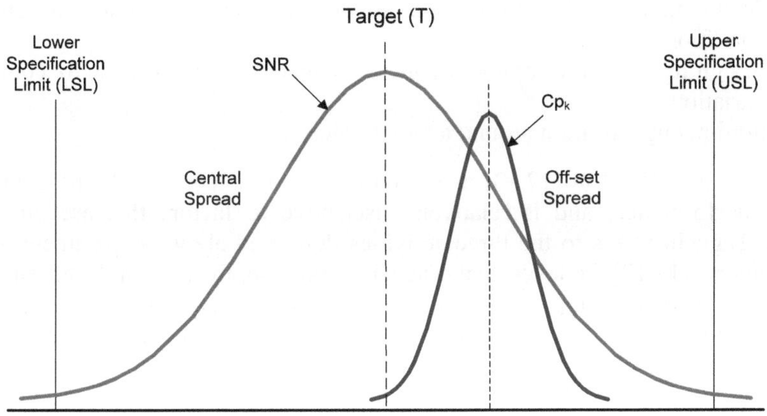

Figure 17.3. Signal-to-noise ratio vs. capability index

Another quality technique tool used by companies is Cp_k, which is a method of describing the likelihood of rejects [17.11]. Cp_k focuses on the target value and the variations that exist; however, the ends of the variation are related to specification limits. By observing the behaviour of the bell-shaped curve at the specification limits, the probability of failure can be determined.

The imposition of ever stricter standards direct companies not only to maintain and ensure quality standards are met, but also to strive for continuous improvements. To this end, SNRs are used to determine the robustness of a process. For evaluating a process's behaviour regarding the likelihood of rejects, Cp_k is used. Only when both quality technique tools are used may contradicting outcomes occur.

17.3.1 Design of Experiments: Industrial Application of SNRs

In 1979, Dr Genichi Taguchi's book 'Off-line Quality Control' was published in English and attracted a number of interesting western industries. Indeed, Ford Motor Company, among many others, arranged for a series of conferences and seminars to widespread Taguchi's ideas concerning quality and the associated improvements. As such, a number of industrial applications based on Taguchi's methods have been published [17.12]. The latter resulted in further evaluations of Taguchi's ideas and subsequent discussions and elaborations of techniques for further improvement.

The accumulation and the systematic treatment of data pertaining to processes and procedures alike is the input variable to sound managerial and engineering decisions. For the latter to have a direct effect on quality, knowledge build-up has to commence earlier and be planned in advance in the product and/or process development. Well-planned experiments provide rapid knowledge of the values that have to be chosen for design and process parameters to achieve the best possible products or processes at the lowest cost level. DoE, therefore, is considered to form an important stage in quality improvement, and is related to robust design.

Literature review shows that Dr Taguchi's work calls for the following three distinctive experimental design approaches, namely:

- designing products or processes so that they are robust to environmental conditions;
- designing and developing products so that they are robust to component variation;
- minimising variation around a target value.

Robustness is defined in [17.13] as the on-target consistency of the product or the process performance, and is relatively insensitive to factors that are difficult to control. Taguchi refers to the three activities described above as parameter design. Montgomery [17.12] believes that Taguchi's philosophy is sound and should be included in the quality improvement process of any organisation.

In his book 'Quality Engineering in Production Systems' [17.13], Dr Taguchi introduced design of experiments and the associated signal-to-noise ratios that are applied on the analysis of selected case studies. The use and the mix of complicated mathematics with economics make it indeed difficult for an inexperienced person to deal with it effectively.

A key component of Dr Taguchi's philosophy is reduction of variation. It is often required that each quality characteristic has a target or nominal value. The objective is to reduce variability around this target. Taguchi models the departures that may occur from this target value with a loss function. The loss refers to the cost that is incurred by society when a consumer uses a product whose quality characteristics differ from the nominal. In summary, Taguchi's philosophy involves three central ideas:

- products and processes should be designed so that they are robust to external sources of variability;
- experimental design methods are an engineering tool to help accomplish this objective;
- operation on target is more important than conformance to specifications.

Taguchi furthermore separates variables into two types:

- control factors are those variables that can be practically and economically controlled, such as a controllable dimensional or electrical parameter; and
- noise factors are those variables that are difficult or expensive to control in practice. These, however, may be controlled in a laboratory by experimental

means, *e.g.* the ambient temperature or parameter variation within a tolerance range. The objective is then to determine the combination of control factor settings (design and process variables) that will make the product have the maximum 'robustness' to the expected variation in the noise factors. The measure of robustness is the signal-to-noise ratio [17.14].

With SNR analysis, the calculations take into consideration both the mean and the variation from the result to the next. SNR analysis can therefore be regarded as being two-dimensional as opposed to regular analysis being one-dimensional. The signal-to-noise ratio is characterised as S/N or represented by the Greek letter η.

Noise can denote, according to Dr Taguchi, environmental conditions (outer noise), internal deterioration (inner noise), or variation from unit to unit (between product noise). The relevance of the SNR equation is tied to interpreting the signal or numerator of the ratio as the ability of the process to build good product, or of the product to perform correctly [17.15].

By including the impact of the noise factors on the process or product as the denominator, the signal-to-noise ratio can be adopted as a barometer of the ability of the system (process or product) to perform well in relation to the effect of the noise. The successful application of this concept to experimentation is capable of delivering the control factor settings that can produce the following results:

- they can either result in the best performance (high signal); or
- they can minimise the effect of those influences, which cannot be controlled (low noise) [17.16].

Signal-to-noise ratios are problem-type-specific; however, as already mentioned, the three principal ones are predominantly used in industry. Indeed, Peace [17.17] discussed the concerns of each of them and showed that the method should centralise the process and then reduce the variability. This in its own right is practical, but the application falls short due to commercial constraints.

McAndrew and O'Sullivan [17.18] found that SNRs were normally used by industry to show a form of robustness in design and process but seldom as long-term commitment to achieving optimum settings. Thus, robustness is more likely to produce situations where target values are met but no real reduction in variation is achieved. This statement is shared with Garvin [17.19] who criticised Taguchi's approach regarding the loss function and the different definitions of target values.

Taguchi introduced the concept of a 'loss function'. The latter measures the losses that a product imparts to society from the time it is shipped and/or forwarded. Such losses include warranty costs, dissatisfied customers, and other problems related to performance failures. Taguchi continues by comparing the resulting losses by applying two alternative approaches to quality. These are:

- conventional approach entailing the fundamental concept of conformance meeting specifications; and
- the second approach equating conformance with the degree of variability (inversely, the degree of uniformity) around the target dimension or centre-line. Variation within specification limits is thus, explicitly acknowledged by this approach.

Figures 17.4 and 17.5 illustrate the differences between the two definitions, respectively. Advocates of the traditional approach to conformance normally prefer the second approach. Even though it is poorly centred, all items fall within specification limits and none are rejected for failure to conform. Followers of the Taguchi idea, however, favour the first approach instead, as according to Taguchi's loss function it will result in lower long-term costs.

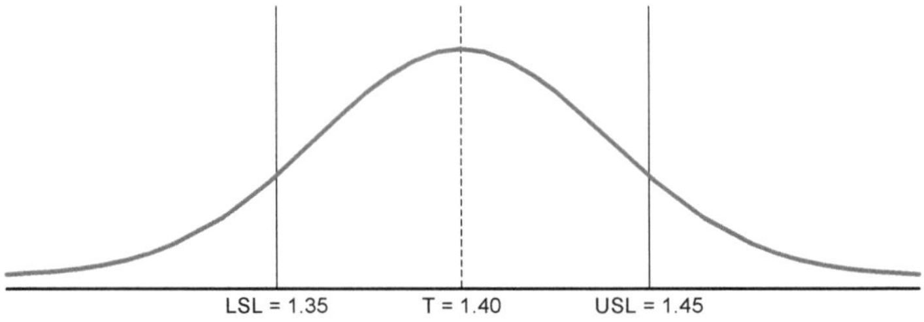

Figure 17.4. Taguchi's approach to robustness

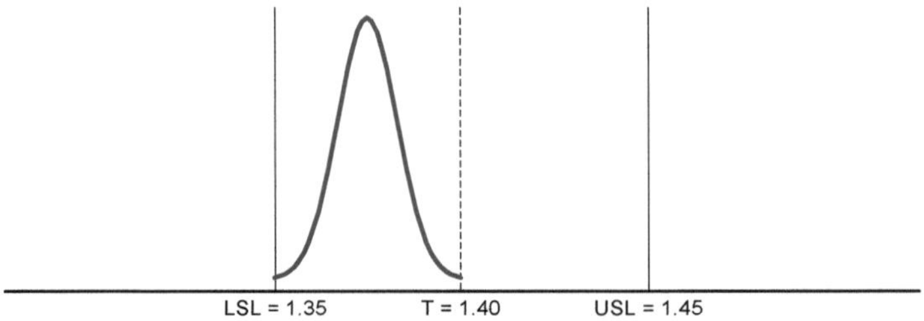

Figure 17.5. Position of capability indices as opposed to robustness

Even though some items fail to meet specifications and will eventually be rejected by inspection and/or consumers, the vast majority is clustered tightly around the target, suggesting fewer problems due to tolerance stack-up. The link between conformance and reliability results in smaller losses. This statement is justified, because conformance to specifications is defined as meeting as close as possible the target value. By achieving a high target value, in addition to a high degree of a system's reliability, the overall losses of the system's performance will be smaller. While hard evidence supporting this claim is still scarce, the approach is regarded as especially desirable for products that either require the matting of large numbers of parts (*i.e.* automobiles, aircrafts), or involve an increasing number of processes and procedures, and whose reliability and robustness is of paramount importance.

Despite their differences, both approaches to conformance rely on similar data for monitoring production. Within the factory, measures of the incidence of defects (the proportion of all units that fails to meet specifications and so require rework and repair) form the basic building blocks. If the traditional approach to conformance is being used, simple counts of defect percentages are usually enough. If Taguchi's approach is being followed, more elaborate measures of the distribution of output are required. These measures will aim to:

- centralise the process output around a target value; and
- maintain the process performance within the desired specification limits.

The aforementioned measures include the process capability ratio that measures the width of specification limits relative to the spread of the process, and the process location ratio that locates the centre of the process relative to the target dimension or centre line. Without such measures, improvements in conformance within the specification limits are impossible to track [17.20].

Figure 17.6 visualises the core point of all items meeting stated specifications, as their graphical representation (in terms of the equivalent 'bell-shaped' curves) falls within the range of the upper and lower specification limits, USL and LSL, respectively. As such, the quality measures based on the simple counts of the number of defects treat the three processes as identical, as each is responsible for and results in zero defects.

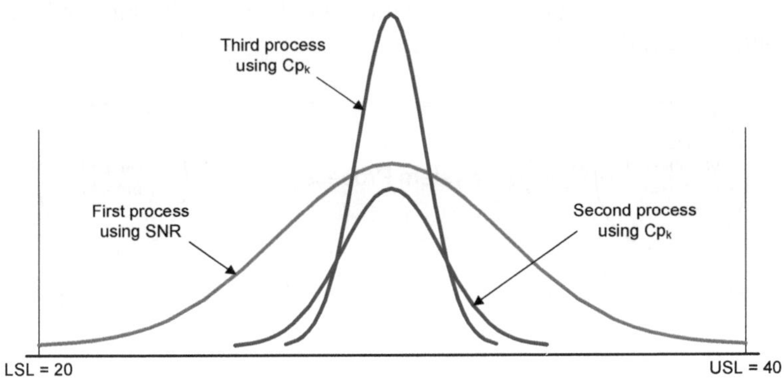

Figure 17.6. Differences between process capability ratios and robustness

Process capability ratios defined as the ratio of specification width vs. process width, however, contradict robustness. These ratios vary significantly and quickly demonstrate the superiority of the second and third processes, indicated as the red and blue coloured bell-shaped curves, respectively.

For SNRs to be used successfully, it is important that they take the form of an iterative stage; otherwise, the improvements in robustness are not at their optimum.

17.3.2 Implications of Capability Indices

For companies to meet the needs of quality-related standards and to satisfy customer expectations, a variety of quality technique tools are, therefore, adopted. Indeed, companies now use techniques to ensure that the predicted quality levels not only can be met, but also can be further improved, *e.g.* statistical process control, failure mode and effects analysis, quality function deployment and Taguchi. It is worth mentioning that although statistics is the branch of mathematics that enables the interested bodies to handle variable measurements, statistical methods like the ones described overleaf, are ever so important in controlling quality. Since their first introduction, the quality technique tools have been applied in various markets and branches [17.21].

Before addressing the use of capability indices and signal-to-noise ratios in various areas of modern manufacturing industry, it is noteworthy to mention how variation occurs and how it is defined.

No two parts manufactured to the same specification under the same process conditions will be identical. Thus, observations or measurements of these items will also differ. This difference is known as *variational noise* that may arise due to either assignable (special) causes or random (common) causes.

Montgomery [17.12] stated that there is a natural (inherent) variability in a process when it is stable and in a state of statistical control. This condition is referred to as natural variability and is also known as *background noise*. The latter is the result of 'the cumulative effect of many small, essentially unavoidable causes'. This inherent variability is called *process capability* and is normally reflected in the variations pertaining to observations concerning a number of samples drawn from a process. Figure 17.7 illustrates the relationship between background noise and desired outputs at end.

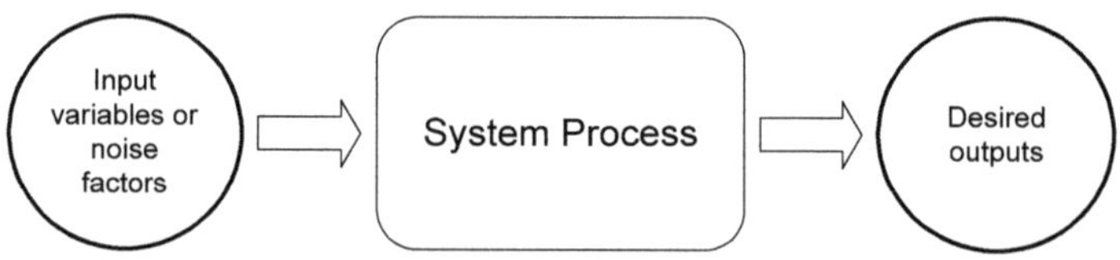

Figure 17.7. Relationship between background noise and desired outputs

A main objective of statistical process control is to quickly detect the occurrence of assignable causes or process shifts so that investigation of the process and corrective action may be undertaken before non-conforming units are manufactured. The process capability that is typical of a process is due to random or common causes of variability. Special or assignable causes are unpredictable and appear as unusual patterns of variability on a control chart. The latter part of variation is mainly attributable to:

- change in raw material;
- change in the position of equipment or part of the equipment, translated to different procedures and overlapping processes for service providers;
- change in the equipment setting, and/or over-estimated projections (financial, schedule-related, operational, *etc.*);
- damaged, or erratically operating equipment;
- operator errors.

Common or random causes are a regular feature of the process, as they are predictable and will remain so unless action is taken to eliminate them. These causes produce a natural variability pattern that is typical to a process when it is in a state of statistical control. Authors on quality technique tools, such like Bentley [17.22] and Oakland and Dale [17.23], agree that common causes are difficult to identify and correct and may be attributed among others to the following parameters:

- badly maintained equipment;
- equipment requiring refurbishment or replacement;
- inadequate operating instructions;
- the quality of raw material on which the process is performed;
- poor operating environment.

The aforementioned list is not exhaustive. It rather highlights the main influential noise factors. Performed research shows that within the area of statistical process control, companies place an emphasis on accepting the validity of capability indices. The latter are use as a key indicator to predict reject. Their implementation can:

- establish the process capability of work methods in order to ascertain whether a task can be accomplished satisfactorily;
- establish the process capability of an existing plant and, where necessary, satisfy the need to bring it up to specified requirements; and
- control and monitor process capability on a continuous basis to detect and eliminate potential causes of non-conformance and variation.

To enable the achievement of the aforementioned points, the capability is simplified by formalising a relationship between the random component of the process variability and the tolerances. The combination of the capability *Cp* and the capability index Cp_k provides a complete definition of the status of the process. *Cp* is most closely related to process capability potential and can be found below.

$$C_p = \frac{\text{USL} - \text{LSL}}{6\sigma} \tag{17.1}$$

At any value greater than 1, a process has the potential for meeting specifications if held in control at an X_o of (USL + LSL)/2. Conducted industry-specific research reveals that a minimum target value of *Cp* = 1.33 is used as the fundamental level of quality standard. The implementation of modern manufacturing tools, company-wide training and awareness, and the imposition of higher standards amongst others, have raised the capability index to $Cp_k = 1.67$ and $Cp_k = 2.0$, respectively.

The application of quality technique tools, such as the capability index in a variety of areas has, however, led to difficulties. The performed research work has shown that these constraints often arise in non-manufacturing areas, such as in the agricultural or the educational sector. In other words, Cp_k is a more concise measure of its ability to conform to the specification limits [17.24]. Such an index is the Cp_k, which is rapidly becoming industry's means of communicating process [17.25].

However, Motorola's six-sigma approach has focused the aims at reaching capability indices of at least 2.0, if reject levels are to be accepted by all involved [17.26]. Ford Motor Company requires their suppliers to achieve long-term capability indices of at least 1.67. This in its own right is a difficult task and might not be possible if tolerances are set at unachievable levels. Taguchi has discussed the subject of tolerance design but companies are more interested in the robustness of a product than individual elements [17.27].

Performed research has revealed that the following two points are important amongst other factors when defining capability indices and alike.

- Capability has been defined on the basis that *Cp* must be greater than 1. This is a minimum requirement. Some organisations change the minimum over time as SPC experience develops and processes improve. These external requirements force never-ending improvement by concentrating on reducing the variation, keeping the specification limits unchanged. If Deming's philosophy has not been understood, there can then be a danger of using *Cp* and Cp_k indices as targets measuring progress against time.
- Both *Cp* and Cp_k indices are required if we are to define a process completely. Unfortunately, some software manufacturers only provide Cp_k values. This can be misleading. For example, a Cp_k value of 2.3 does indicate that at least the process is capable, but it tells the customer nothing about the process setting. Hence, if a *Cp* value of 3.9 was also provided, it would be possible to assess the process better, recognising that the setting was unsatisfactory.

Dr K. Ishikawa in his book 'Introduction to Quality Control' [17.28] sets the following categories (Table 17.2) into which the process capability index is divided, arguing that a $Cp > 1.67$ is 'too high for general purposes'. Performed company-specific research has indicated that industry is aiming at setting standards. However, this statement falls short, due to the comparatively high number of rejected parts per million (PPM).

Table 17.2. Categories of process capability indices

$Cp>1.67$	Special class	$Cp>1.67$ should be targeted when aiming at PPM control, or extra reliability
$1.67 \geq Cp \geq 1.33$	Class A	Very good quality. Inspection can be reduced
$1.33 \geq Cp \geq 1.0$	Class B	Quite good quality. Sampling inspection is sufficient
$1.0 \geq Cp \geq 0.67$	Class C	Some defectives will be produced. *Cp* should be raised to 1.0 or above
$0.67 \geq Cp$	Class D	Very bad

17.4 Research Methodology

This work is based on preliminary studies at Olympic Airlines' Headquarters in Athens, Greece. The performed research revealed the need for facilitating fleet optimisation by acknowledging the following influential parameters:

- the acquisition of Olympic Airlines by MIG (as of 01/10/2009);
- the implementation of EEC 95/93 with regards to slot allocation; and
- the effects of the two aforementioned parameters in fleet scheduling and utilisation.

For this to be achieved, information and data were gathered from the following sources: (1) Olympic Airlines, and (2) Hellenic National Statistical Society.

Olympic Airlines kindly provided data concerning their strategic planning and maintenance of hubs and slots. In particular, company-specific policies were taken into consideration to assist a smooth transition to the new scheme Olympic Air, which at the point of writing will succeed Olympic Airlines.

The Hellenic National Statistical Society assisted the work with data regarding passenger traffic during the decade of 1997–2007, as well as showing passenger trends of Olympic Airlines during the very same period.

Performed research at Athens International Airport (AIA) indicated that the latter details and categorises the airport-specific data into the following groups:

- passenger traffic;
- cargo traffic;
- delays of inbound and outbound flights;
- destinations;
- number of aircrafts;
- new airline companies;
- AIA's performance in relation to other national airports. The latter is indicative of AIA's role as major hub-airport in Southeastern Europe.

The analysis and evaluation of the listed data will serve as an indication

- to establish the importance of maintaining a hub-and-spoke system; and
- to use gathered data as input to the development of a factorial experiment.

Literature review has shown that published papers refer to airports as entities that inadvertently affect the socio-economical life of the community. This is further enhanced by the findings of Frank *et al.* [17.29] who evaluated air traffic systems by stochastic depeaking techniques and economic optimisation methods. They argue that for effectively addressing block time distributions as opposed to the probability of a flight arriving late, the resulting plot resembles a normally distributed population. The outcome thereof represents a bottleneck. As such, the resulting graph is similar in structure to Taguchi's robustness graph. However, the study of their work suggests that two contradicting outcomes ought to be matched. They are:

- over-demand for arrival runway capacity; and
- air-traffic related control policies, *e.g.* start-up delay and holding patterns.

It is noteworthy of mentioning that full service airlines (FSAs) focus on the creation and further development of hub-and-spoke networks. The latter places an emphasis on the continuous feed of the spokes. Business travel and a high level of seat availability may contribute to a large profit margin, under the provision that the costs of the offered product and/or service are high. Therefore, a hub-and-spoke system may add congestion to an airport, due to the small time-distances between arriving aircrafts. As such, stated congestion jeopardises overall airport-specific capacity limits.

It is the author's view that new market entrants on the other hand aim at developing low-cost point-to-point connections. However, based on performed research, an FSA policy is deemed by the author as being a high-cost strategy due to the following reasons:

- higher costs owing to internal operational procedures that are company-specific and entail sensitive operating sections, and cannot, therefore, be sub-contracted or assigned to third parties; and
- the structure of a hub-and-spoke network calls for a relatively low level of reproducibility per capita, *i.e.* aircraft (including flight personnel).

In view of the above, the proposed model for bridging the contradicting outcomes between signal-to-noise ratios and capability indices will be subject to further analyses and evaluations. Furthermore, the outcomes of the model will be implemented in a number of case-studies pertaining to a variety of market sectors.

Airport hub management adheres to IATA imposed rules and regulations. These are briefly highlighted below"

- *grandfathering right* – if an airline company owed the right for the specific slot in previous seasons, it may keep it during the succeeding one;
- *use-it or lose-it* – if an airline company owed the slot during the preceding period, but made inadequate use of it compared to the allowable time, then the slot may be given to another carrier;
- *priority for regular service* – under competition laws the available slot will be given to the company/flight that explains itself ready of making the most use of it; and
- *directed discretion* – this category is enacted, if the previous rule does not produce any results.

Directed discretion assigns priority to the following factors:

- re-design of specific flight route, to allow for a different arrival time at the destination airport;
- re-design of specific flight route with a larger aircraft;
- development of a more realistic flight schedule;
- update and upgrade of existing flight route on an annual basis.

The analysis and evaluation of gathered data allows for an input of influencing parameters to form the basis of a preliminary model introducing factorial designs.

17.4.1 Areas of Further Improvement between Cp_k and SNRs

This research so far has shown that airport-related management and co-ordination issues may result in contradicting outcomes that may inadvertently affect financial and operating performance. Airport management techniques outlined in the previous sections have indicated that Dr Taguchi's approach is focusing on the target value. Indeed, it may be argued that although target values are met, no real reduction in variation is achieved.

Sullivan [17.30] points out that the use of signal-to-noise ratios and capability indices contribute to achieving good quality levels.

Although in Figure 17.8 the ends of the bell-shaped curve fall beyond the upper specification limit and the lower specification limit, respectively, followers of Taguchi's philosophy would accept the process as it meets the target value.

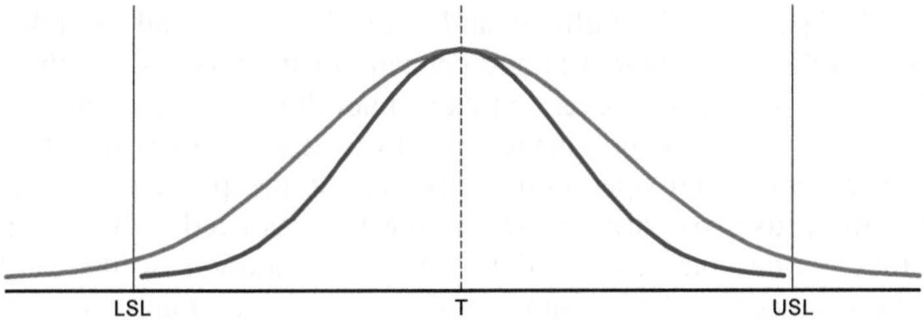

Figure 17.8. Comparison of process improvements

The red bell-shaped curve in Figure 17.8 shows an improved process, as the ends of the bell-shaped curve fall within the specification limits, albeit showing a considerable variation. At this point, Sullivan brings the *process capability ratios* in play. The brown bell-shaped curve shown in Figure 17.9 reveals a process of a capability index of Cp_k = 1, whereas the red curve in the same figure shows a capability index of Cp_k = 1.67, respectively.

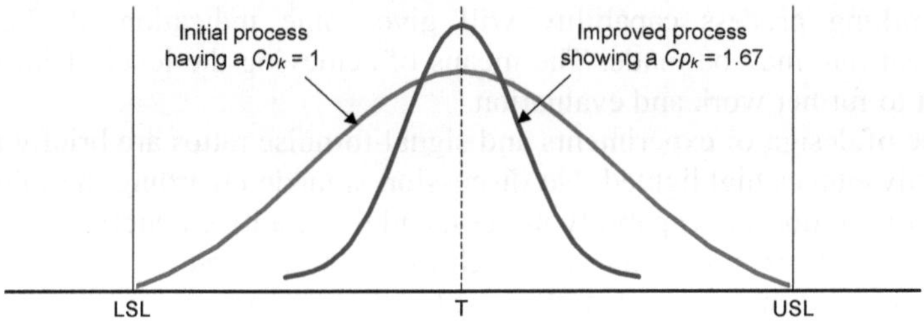

Figure 17.9. Comparison of processes with different capability indices

Although these ratios vary significantly, the difference in terms of variation by comparison to the SNRs becomes even greater for higher capability indices values, as shown in Figure 17.10. The two curves in the figure demonstrate the superiority (tightness) of the higher capability indices values as opposed to the same process demonstrated in Figures 17.8 and 17.9, respectively.

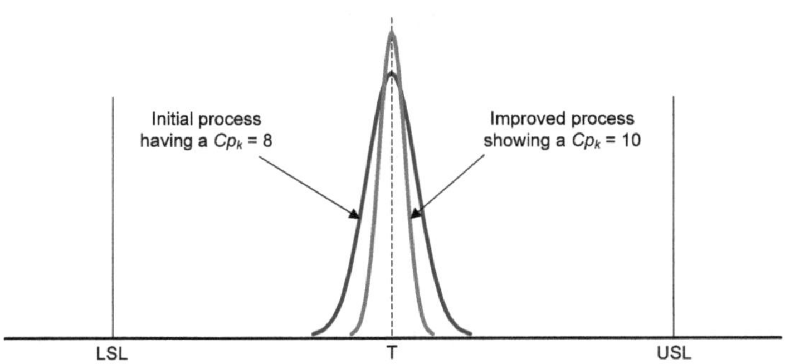

Figure 17.10. Comparison of processes with different capability indices

Straker [17.31] agrees with Sullivan and states that '...*Cp* and Cp_k taken together give a measure of both the potential and centring of the process distribution within the specification limits'. However, Straker underlines the fact that '...Process capability is more than just measuring *Cp* and Cp_k; it involves understanding the statistical performance and operational working of the process'. In addition, he believes that the causes of variation within the process need to be understood. He concludes that '...the conditions under which the variation occurs, and how the variables interact need to be examined. The purpose of doing this is to enable confident process improvement that steadily reduces variation'. He thus, recognises the need for further improvement. However, rather than proposing a new method, he suggests the development of an optimum environment. The latter includes the realisation of the following measures:

- it ought to proceed the measurement of the process capability; and
- it includes the acquisition of new tools, graded materials, the employment of more skilled people, the materialisation of slower execution times, *etc.*

The outcome of the aforementioned measures will result in the evaluation of the potential of the process. The difference between this and the measure taken from the normal working process capability will give some indication of the possible improvement that may be made. The means of achieving the level of improvement are subject to further work and evaluation.

The use of design of experiments and signal-to-noise ratios are briefly addressed and their advantages highlighted. No discussion is made regarding the joined use of signal-to-noise ratios and capability indices, and the resulting benefits.

Taylor [17.32] uses a similar approach as Straker. It can be said that Taylor is working on the same wave-length as Straker. Taylor places an emphasis on understanding the causes of variation, as well as the associated candidate input variables. Candidate input variables (or CIVs) are defined by Taylor as those inputs that might affect the system. Among many variables, Taylor mentions material selection and properties, tooling and process parameters, working methods, operator skills and training, manufacturing and usage environment, *etc.* He continues in characterising the most important input variables that affect the system as key input variables (or KIVs). KIVs are those inputs that can affect the output, either by affecting the average, or by contributing to the variation. Taylor uses Pareto's principle in describing transmitted variation. He states that 80% of the transmitted variation is the result of 20% of the KIVs. Thus, '...the elimination of the variation transmitted from the other 80% of the key inputs does little to reduce the total variation. This is the result of the non-additivity of variation'. He furthermore outlines three basic approaches to reducing the variation transmitted by an input variable. These are briefly summarised below:

- reducing the variation of the input variable;
- making the system less sensitive to input variable, *i.e.* making the system robust;
- changing the relationship between the inputs and the outputs. To achieve the latter one requires fundamental changes in the design or materials. Indeed, it is the author's view that such changes are best applied early in the design stage of the product, and/or service.

To reduce variation, Taylor proposes a set of strategies for system and/or process optimisation. These strategies affect system, parameter and tolerance design, respectively. Among some of the proposed strategies, he recommends to changing the relationship between the input and outputs to a more favourable one, and adjusting the targets of the key input variables move to the average closer to target.

However, changing the relationships would mean to alter the true characteristics of the system, and/or the process. This in turn would change the characteristic of the desired output.

Taylor uses a similar approach to Straker. The former proposes a set of guidelines to incorporate changes to reduce variation. These sets of guidelines include amongst others changes in design specification, tooling and machine equipment, as well as parameter and system design changes, respectively. Some of Taylor's main points are worth mentioning, and are briefly outlined below.

- During parameter design:
  - set key input variables to get the average as close as possible to ideal;
  - use interactions among materials and manufacturing conditions to reduce transmitted variation;
  - use interactions among design parameters and materials to widen material specifications.
- During system design:
  - select a design concept that is not prone, *i.e.* is robust, to wear and deterioration;
  - select materials that are not prone to deterioration and wear;
  - select a design concept that is robust to the manner and conditions of use.

In addition to the aforementioned set of guidelines, Taylor emphasises the objectives during the manufacturing stage of the process. He stresses that '…once the product has been designed, the materials selected, and the process developed, it is manufacturing's job to produce product in which the average is as close to ideal as possible, and the variation is at a minimum'. Taylor concludes that manufacturing cannot generally change the system design of the product or process. He emphasises the fact that '…the stages and strategies for optimising the average are system design – change the relationship between the inputs and outputs to a more favourable one; parameter design – adjust the targets of the key input variables move to the average closer to target; and tolerance design – the average is not affected by tolerance design'.

It is the author's view that the implementation of robust management techniques and reliability tools may contribute to a company's overall operational effectiveness and improved performance. As such, the changes of a system design's product or process are indeed very difficult and expensive to incorporate. It is rather a matter of finding methods and techniques to improve quality by using the existing infrastructure. As such, the identification of factors that are associated with good performance by neglecting the remaining ones only partly addresses the problem. If the remaining factors are primary factors influencing at a high degree the process, the resulting error will equally be high. Moreover, it is a case of improving the totality of factors affecting the process for a company to reach good performance levels.

It is worth mentioning at this point the fact that capability indices on their own do not determine the degree of influence and interactions of variables; this is rather the scope of *design of experiments* and *signal-to-noise ratios*. The latter will be introduced so as to suggest a different approach in effectively dealing with airport-related contradicting outcomes.

17.4.2 Summary of Most Commonly Used Approaches

This research shows that the most commonly used approaches may be categorised in the following two groups:

The first group includes capability indices and signal-to-noise ratios. Advocates of this group emphasise the importance of capability indices and signal-to-noise ratios and the resulting benefits in processes alike, in particular in manufacturing. Most of the authors focus on the study of the conditions under which the variation occurs, and the interaction of variables. Some other authors use Pareto's principle in describing variation. They state that 80% of the transmitted variation is the result of 20% of KIVs [17.32]. However, most authors propose the development of an optimum environment. This includes the purchase and use of new tools and equipment, graded materials, manufacturing and usage environment.

Operator skills and training play an ever-important role in achieving – in the optimum manufacturing environment – an improved quality level. These input variables will give a measure of the potential of the process. The difference between this and the measure taken from the normal working process capability will give some indication of the possible improvement that may be made. However, as sound as this may be, it is associated with high expenditures. The company will need to heavily invest in high-tech machinery and tooling equipment. Training of skilled workforce and the continuous update on modern methods and techniques is yet another financially demanding area. At last, the aforementioned strategy albeit from requiring a sound financial basis, is mostly viable in the mid- to long-term period.

As already mentioned, capability indices and signal-to-noise ratios are widely used in industry. It is common to measure process capability in units of process standard deviations [17.33]. In particular, it is common to look at the relationship between the process standard deviation and the range between the upper and the lower specification limits (see Equation (17.1)). The minimum acceptable value for Cp is considered to be $Cp = 1$.

Many companies and research institutions use the capability index, Cp_k [17.34]. Cp_k relates the process mean to the nominal value of the specification. Capability indices compare the match between the process capability and the product specifications. They are as much a measure of the manufacturability of the product as of the ability of the process to produce the product.

Alternate capability indices exist for the cases of upper and lower specification limits, only. Capability indices also exist for considering the average along with the variation. Regardless of how the capability indices are calculated, they are interpreted similarly.

The performed literature review has shown that a consistent strive for achieving capability indices of Cp_k =2 or higher for every process. This is called six sigma capability, which assures that product with 15 or fewer key characteristics have less than 50 defects per million. It furthermore verifies that a product with 1,000 or fewer parts are perfect at least 99.66% of the time.

The second group includes design of experiments, or DoE. Performed research has indicated that a serious shortcoming of past approaches has been the inability to rationally deal with the quality issue in the early stages of the product and process development life cycle. Taguchi, however, has directed attention towards parameter selection at the early stages of design. This can be enhanced by measuring quality by functional variation during use and by using DoE methods.

Figure 17.11 illustrates Taguchi's concept regarding robust design. It is the author's view that although the terms 'testing' and 'experimentation' have their rightful place in industry, one should not serve as an alternative for the other. Indeed, Japanese companies have used DoE for parameter selection at the product and process design stage. In this case, the aim is to experiment with various combinations of the important design parameters for the purpose of identifying the particular combination(s), the latter of which optimise certain design criteria or performance measures. Western companies [17.35] have placed a great deal of time, money and emphasis on life testing of components. In this area, many identical units are subject to field conditions for the purpose of determining the life expectancy of performance.

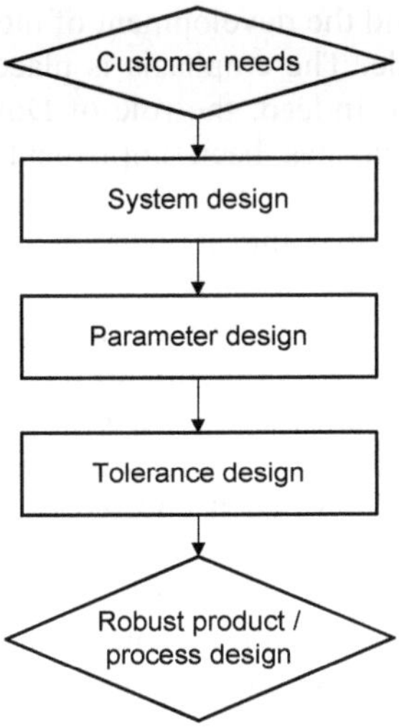

Figure 17.11. Taguchi's approach pertaining to the design process

By design failure, changes are made to the system and/or the component, which is again re-tested. These changes are identified through a deliberate experimental approach or through more *ad hoc* procedures. Life testing of product performance is important but is not a substitute for experimentation to determine what ought to be tested. Published research work by scientists and engineers [17.36] in the field of air-transport management are commonly involved in experimentation as a means to describe, predict and control phenomena of interest. An emphasis is placed on optimising slot controls and the resulting parameters.

In view of the above, the collection of data is a fundamental activity toward the building and verification of mathematical models, whether such models are derived from first principles or are purely empirical in nature. Comparative experiments are an important means to discern differences in the behaviour of processes, products, and other physical phenomena as various factors are altered in the environment.

Too often, data analysis, modelling and inference are stressed at the expense of the activities that enhance the planning and the execution of experiments. Performed literature review indicates that valid and meaningful data are available either from passive observation of the process or from purposeful experiments, and that statistical methods embrace the analysis of such data. The same authors state that it is the planning or design stage leading toward the collection of data that is critical and needs to receive more attention. Indeed, this point makes a statistical approach to the design of experiments important.

The purpose of most experimental work is to discover the direction(s) of change, which may lead to improvements in both the quality and the productivity of a product or process. Performed literature review has indicated that in the past, there has been a tendency to conduct studies farther downstream at the process.

Since the beginnings of quality, the engineering community had been less commonly embracing the use of DoE concerning product design purposes. The imposition of the QS 9000 and the ISO 9000 series of quality standards that have led to the merger between these and the development of the TS 14949 [17.37] has begun to change over the past decade. The emphasis is placed by Taguchi and others on using DoE for product design. Indeed, the role of DoE in the earlier stages of the product development life cycle has been implemented in a growing number of industries and service providers.

Performed research shows that the application of concurrent or simultaneous engineering methods concerning products and their manufacturing processes is receiving widespread attention. Mathematical modelling, computer simulation, and the associated use of design of experiments are all playing a central role in this activity. In investigating the variation in performance of a given process, most authors distinguish between qualitative and quantitative factors or variables. Both qualitative and quantitative factors, when allowed to vary cause performance, vary in the same way.

Qualitative factors are also called categorical variables, while quantitative factors possess an inherent continuity of change. Kounis and Panagopoulos [17.38] have shown that there is indeed a growing need for the precise recognition of the relative roles that factors play in governing the nature of product and process performance. Some factors are external and environmental but nonetheless having an important impact on performance. Some other factors have a strong influence on performance on average, while others tend to influence the level of variation in performance.

17.5 Analysis of Noteworthy Approaches

In order to address the increasing number of aircraft movements at an airport and to effectively deal with the phenomenon of congested airways, the following strategies have been proposed and implied:

- ground delay program suggested by the Federal Aviation Administration;
- slot controls primarily introduced by the Eurocontrol authorities in Brussels.

The latter strategy affects rendered services that are subject to the following influencing parameters:

- no changes at the level of offered services;
- changes that affect the timing of the flights concerned;
- reduction in the final number of offered flights;
- elimination of all different service levels.

Abeyatne [17.39] states that the slot control system in Europe was developed during the transition phase of public airline companies, including then-national airports. He argues that national air carriers were actually competing against one another, rather than co-operating within the principles of an equal and unified financial union.

Another noteworthy case study is introduced by Le Loan *et al.* [17.40]. The authors argue that slot controls are not effectively utilised by smaller aircrafts. Hartsfield Atlanta International Airport (ATL) is utilised by aircrafts larger than 210 seats, *i.e.* B747, B777, L10. These aircrafts only make up 4% of seat share and 1.7% of flight share, respectively, at ATL airport. In retrospect, 75.1% of the flights range between 97 and 210 seats that are available in jets, such as B767, B757, MD80, represent 87.7% of the total seats. Additionally, 21.7% of the total flights have less than 70 seats (ATR, CRJ, *etc.*) and only contribute 8.3% of the overall available passenger capacity, with the cargo flights representing 1.5% of the slots.

In order to address the aforementioned parameters, Diana [17.41] focuses his research in determining whether time delays are more apparent in market-concentrated airports, or in less concentrated ones. He introduces the Herfindahl-Hirschmann Index (HHI) and applies this to the schedule of seller carriers. The HHI is a measure of the size of companies operating in the market in relation to the industry. It serves as an indicator that determines the amount and level of competition amongst these. In his study, the measurement of a time delay is conceived as a propagating signal that is characterised by its magnitude (amplitude), cycles (referred to as phases) and speed. The application of a Fourier transform results in preliminary findings that delay ratios occur. The use of non-parametric tests further enhanced the first findings and indeed, no time delays occur and they are irrespective of the nature of an airport. Unanswered questions include the identification of factors that may impact the variation of the magnitudes of stated delays.

Gillen [17.42] argues that the trend in the airline business follows the following two different operational strategies:

- full-service airlines (FSA) focus on creating hub-and-spoke networks; and
- new entrants aim at creating low-cost services that will connect and establish a point-to-point diagram.

FSA's strategy is characterised as a high-cost one for the following reasons:

- higher expenses due to operational procedures, and the latter cannot be sub-contracted; and

- a hub-and-spoke network structure is associated with low re-producibility per capita (aircraft, flight crew, ground handling personnel, *etc.*).

Indeed, a hub-and-spoke system fosters spoke-related air traffic. Such a system, although highly profitable, results in congestion, owing to the small time blocks between incoming and outgoing aircraft(s). Thus, the congestion inadvertently affects the operational performance of an airport and places a burden on the network controlled by the tower. This issue becomes of paramount importance when the airport experiences severe weather conditions. In this case, FAA has introduced the collaborative decision making (CDM) procedure [17.43] in order to effectively meet operational requirements due to bad weather.

The establishment of slot controls in accordance to a CDM model applied at airports that are subject to severe weather conditions is a procedure that is not market-oriented. In the spring of 1995, FAA introduced a scheme called FAA Airline Data Exchange (FADE) in order to analyse and evaluate whether the updated information pertaining to weather conditions provided by the airliners could improve decision making [17.44]. As such, FAA assigned slot controls to aircrafts according to a priority mechanism. The latter included taking into consideration the following factors:

- weather scheduling procedures; and
- distance between flights. In this context, longer flights are issued with a priority number.

The parameters listed below are deemed by the author as important in the introduction of a slot control mechanism:

- changes in flight scheduling;
- reduction of the number of offered flights;
- no changes regarding flights and existing service levels;
- abolition of flight diagram and route.

Bard and Mohan [17.45] introduce a dynamic programming method that incorporates algorithms to re-allocate arrival slots during a GDP at a specific airport. In this model the capacity of an airport was regarded as being limited. Additionally, potential delays and associated costs were deemed as following a linear representation. In their study, they argue that '...the results were good for relatively small instances', but '...as more flights were included, computation times grew exponentially.' They propose a split-down of the system into sub-systems and groups. In doing so, the suggested computational experiments manage to effectively address real and randomly generated data. As such, the flight-parameters linked to a slot are

- cabin crew and passenger delays;
- additional fuel costs; and
- passenger compensation owing to delayed flights and missed connections.

The last parameter is important, as only single connection was taken into consideration. Thus, any given flight can neither have more than one preceding, nor more than one succeeding. However, it is the author's view that given continuous growth in air traffic a dynamic model incorporating potential changes in the impact of outer and/or inner noise variables ought to be formulated.

Martin and Voltes-Dorta [17.46] introduce a method for dealing with problems linked to spatial concentration indices. These form part of a hub-and-spoke network, as connecting passengers are channelled through a hub to their final destination. They differentiate between airports that concentrate a high number of passengers, which use the spokes of the network to arrive at their final destination, from the ones that have no further connections. The proposed new hubbing concentration index is addressing standard spatial concentration indices in a more effective manner. However, the effect of network configuration on airline-specific costs is yet to be determined.

Rietvelt and Brons [17.47] suggest a '...measure for the quality of the co-ordination of timetables by carriers in hub airports.' They concluded the following.

- The average re-scheduling time was found to be inversely proportional to the lower frequency F_1 of the two legs of the journey;
- The average waiting time at the hub is inversely proportional to the higher frequency F_2 of the two legs; and
- Smaller waiting times at hub airports do not result from a high frequency timetable.

Indeed, the aforementioned authors stress the fact that the '...reliability of hub operations is a candidate for further research', as this topic has not been effectively addressed.

17.6 Discussions on Current Techniques

Based on the outcome of 'The Centre of Politics and Economics' as received by the author, the aforementioned organisation states that a number of hub airports expand by adding new runways, as well as new terminals, so as to cope with increased demand. Slack [17.48] indeed confirms this by mentioning that airport managing authorities suggest the possibility of developing and/or creating new satellite terminals. This measure was deemed necessary so as to address increased air-traffic, whilst reducing exterior influencing parameters.

This act, however, is regarded by the author as not being completely financially viable, as hub airports would not create stand-alone satellite terminals – albeit of a lesser magnitude, but would be in charge of managing a limited amount of the main airport's operations. As such, the areas of further improvement are listed below:

- optimum distance between nearest city and main terminal;
- cost/benefit analysis pertaining to proposed satellite operation;
- comparative analysis between existing infrastructure and provided services;
- statement of problem/bottleneck areas as experienced by airport users, such as airliners, passengers, transport and logistics companies, *etc.*

In order to effectively address on the one hand the continuous increase in air traffic and on the other the resulting bottlenecks, Bard amongst others [17.45] introduced a model of dynamic programmable algorithm. This model regards an airport's capacity as given and static. Additionally, potential delays and arising costs were deemed as following a linear behaviour. Research so far has indicated that air-traffic from and to international airports is increasing.

The suggested model incorporated only single air-route, *i.e.* any given flight can neither have more than one preceding it, nor can it have more than one succeeding it. Hence, by increasing flight numbers by a factor of say 100, the need for developing a dynamic and reliable model capable of managing increased air traffic arises. This model ought to take into consideration influencing parameters.

17.6.1 Development of New Hubs: Strategic Uses and Applied Policies

The Liberalisation of the European Skies in conjunction with the Open Skies policy resulted in the market entry of low-cost airline companies. The factors that assisted in such a move are summarised below:

- financial market trends;
- development of Pan-European high-speed rail network;
- globalisation;
- total utilisation of under-utilised airports due to lower taxes.

Research work by Lijesen *et al.* [17.49] verifies that hub-premiums may be characterised as a strategic market power, subject to product diversification and economies of density. Economies of density justify a hub-and-spoke system, as the resulting cost advantages are associated with a more dense air route. In addition to the above, economies of density are associated with lesser expenses per passenger.

The aforementioned authors suggest a mixed management of the hub-and-spoke networks between airliners and high-speed rail providers. Rail-alliances, such as 'Railjet' and the high-speed rail products Inter City Express (ICE), Train a Grande Vitesse (TGV), Thalys and Eurostar, to name but a few, are competing, and/or co-operating with the relevant hub airports. It is worth mentioning at this point that the German Lufthansa has introduced a code-share agreement with Deutsche Bahn, DB AG on the ICE feeder services to Frankfurt airport.

Economies of density, globalisation, and the liberalisation of European Skies have resulted in the following.

- Continuous fleet renewal by low-cost airline companies. This leads to low maintainability costs, as new generation aircrafts require lesser maintenance levels.
- In order to eliminate the initial purchase costs, airline companies tend to lease aircrafts.
- Owing to the higher average speeds by modern tractive power units, new generation railways offer services that are either similar or supersede the equivalent ones delivered by airliners. Another added advantage is the fact that railways connect large European cities. It is noteworthy mentioning here that the Basel-Paris route has been dropped by EasyJet due to competition from the French TGV.
- Low-cost airliners have managed to eliminate maintenance-specific and ground-handling costs. The former is due to national airliners performing all

service and maintenance on a sub-contracting basis, whereas the latter occurs due to continuous elimination of the company's financial basis.

The complete utilisation of secondary airports resulted, amongst others, in the financial development of the region. In particular, low-cost airliners regarded this as a profitable move due to the development of secondary airports in major hubs. In addition to the above, this turn led to a number of new jobs being generated.

Dennis [17.50] confirms this statement by highlighting Kuala Lumpur's major stakeholder's airport management. The latter introduced an investment plan of an estimated US$550 million to be completed by the end of 2011. In order to face increasing air-traffic needs in this particular region, the following aims are to be materialised:

- construction of a third runway at Kuala Lumpur International Airport;
- construction of four parallel taxiways for allowing a smooth operation for the independently handled runways 2 and 3;
- extension by 1.5 km of the existing railway line;
- complete upgrade of the airport's overall handling capability so as to allow for an increase in passenger traffic from 25 to 30 million passengers.

Low-cost airliners opt for airports that have low-costs, low taxes, fast turnaround times, low handling costs and low aircraft parking fees [17.51].

17.6.2 Proposed Model by Martin and Roman

Martin and Roman [17.52] suggested a theoretical model to address and evaluate congestion phenomena pertaining to increased competition by airline companies. This model addresses the optimum placement of an airport to act as a hub airport, taking into consideration issues such as globalisation and liberalisation of skies.

The entire process may be regarded as being dynamic and has the form of a mathematical game, which is materialised in two steps:

- positioning of the new airport; and
- competition by other airliners.

The first step fosters on the location analysis and optimum positioning of the airport; an act that is left to the airliners to decide subject to a detailed analysis and evaluation procedure. The second step pertaining to competition includes the assessment and evaluation of 'worst-case' scenarios. This point equally considers all potential outcomes that will be used as inputs in the location analysis procedure. Factors that are included in the suggested model are listed below:

- potential passenger and freight traffic between cities and hub airports;
- central geographical location in relation to market served, so as to minimise associated flight costs for a specific market requirement;
- good airport infrastructure, associated with optimum flight co-ordination and minimum change-over time(s);
- satisfactory weather conditions to allow for a smooth operation and flight handling; and
- location of hub-airports in relation to competition.

17.6.3 Proposed Model by Rietveld and Brons

Hub-and-spoke networks provide passengers with a wide range of destinations in frequent intervals at an adequate price. In some cases though, a passenger may be confronted with a further in-between stop, so as to reach the final destination. Diagrams with a high frequency do not necessarily refer to high transportation cost.

In their study, Rietveld and Brons [17.53] observed that more often than not, a high frequency diagram was showing ineffective co-ordination. As a result thereof, passengers were required to wait longer at connecting/transfer points.

The authors observed the paradox that higher frequency diagrams did not necessarily result in smaller transfer times. This is attributable to a reduced effectiveness, the latter of which is dependent on the lower frequency leg of the diagram. On the other hand, the waiting time is defined by the higher frequency leg of the same diagram.

The authors conclude that congested airports face a bottleneck in adhering to published arrival and departure times. Preliminary applications indicate that the implication of Taguchi's *design of experiments* and *robustness* techniques confirms this very statement. The authors break-down the total travel time from their starting point to their final destination into incremental smaller time-legs, by assigning the average incurred waiting costs to the given total journey time. They introduced the following equation

$$rc = \frac{eT}{2F_1} \tag{17.2}$$

where rc = waiting cost, e = cost due to waiting time at hub airport, T = period, and F_1 = lower frequency leg.

This equation is used as a benchmark, so as to optimise flight-related parameters, such as waiting time, flight time, associated costs, *etc.*

In order to effectively address the resulting quality issue at major hub airports, a so-called quality index A is introduced by Rietvelt and Brons. This index takes three values, namely 0, 0.5, and 1. Although the proposed quality index relates the frequency of scheduled flights with an optimum co-ordination, the open-ended question gives rise to the introduction of design of experiments and capability indices.

17.6.4 Evaluation of Hub-influential Parameters

Milgram [17.54] argued that a network's characteristic quality is the possibility and the amount of 'easiness' between connecting nodes that represent departure and destination airports, respectively. Non-stop flights offer the shortest path.

As already stated, Martin and Voltes-Dorta [17.46] evaluate hub-specific indices that may affect an airline company's network. The implementation of the Herfindahl-Hirschman index, as well as the normalised GINI index leads to a graphical representation of hubbing points. The outcome of their research verifies the hypothesis that all measurements taken are highly correlated with one another.

Malighetti *et al.* [17.55] discussed the connectivity of the European air transport network. The *between-ness* and *essential between-ness* pertaining to the European intra-connectivity is presented by means of tables and scatter diagrams. In particular, special notion is done on the amount (in percentage) of the waiting time spent at airports. The implication of 'optimum departure time' may be used as the target value for calculating the associated robustness in a DoE case. Although the term 'robustness' is mentioned albeit to measure the fastest travel leg between two destinations, no further analysis is performed. In addition, the aforementioned study does not model a traveller's choices regarding flight leg(s) and airports. It is the authors' view that the following variables ought to be taken into consideration:

- offers and prices;
- level of services provided;
- membership card(s);
- aircraft type;
- level of services rendered at in-between airports, *etc.*

The proposed methodology is based on a 1960s' connection time.

17.7 Preliminary Model

The term 'robustness' is more often than not referring to the measurement of fastest journey times between airports. It is worth mentioning at this point that London's Stansted Airport serves 130 direct destinations, but only 16 are the fastest legs. This is due to the fact that Stansted Airport does not form a major hub of airliners' strategic alliances.

As already mentioned, Frank *et al.* [17.29] place an emphasis on the financial optimisation of airport systems. Hub-and-spoke systems offer travellers a continuous flow to and from airports. As a result, the relevant reliability degree of such a system is regarded as important.

QS 9000 [17.56] defines *reliability* as 'the probability that an item will continue to function at customer expectation levels at a measurement point, under specified environmental and duty cycle conditions.' Indeed, the term 'reliability' may be substituted by the terms 'possibility' or 'level of success'.

Additionally, the term 'item' may be replaced by the terms 'service' or 'system'. Parameters such as weather conditions, aircraft maintenance, and crew availability may inadvertently affect an airliner's overall reliability. Any changes or fluctuations of the parameters may result in a random variation between the true and the slot co-ordinated demand.

The analysis and evaluation of the scientific data leads to the following conclusions.

- The development of a hub is of paramount and strategic importance to the operational effectiveness and fleet planning of an airline company; and
- The factors that affect amongst other the development of a hub may be summarised as: (1) airport infrastructure and position, (2) airline company and market position, and (3) customer.

17.7.1 Input Parameters for Development of a Factorial Experiment

The analysis and evaluation of gathered data in conjunction with open-ended scientific questions leads to the introduction and development of the DoE method. The focal aim of Taguchi's method is to minimise the 'loss to society' by eliminating all potential error causes pertaining to fault design, improper services, product recalls, *etc.* For a design to be effective, it has to be robust. The latter is achieved by making the former insensitive to inner and outer noise factors.

Table 17.3 shows a $L_{12}2^8$ factorial experiment and the factor settings with their respective outcome(s). It, furthermore, encapsulates the factor settings with their respective outcome(s). Preliminary research has revealed eight parameters to be important at their minimum and maximum levels. For such an experiment to materialise, Taguchi requires twelve (12) experimental runs.

Table 17.3. Factorial experiment $L_{12}2^8$

Abbreviation	Factor	Min.	Max.	Outcome
SA	Seat availability	105	231	Aircraft utilisation
AA	Aircraft availability	1	3	Effective turnover
DAD	Delays at departure	5'	120'	'Domino' effect
DDTTR	Delays due to technical reasons	10'	60'	Customer satisfaction
DAC	Distance between airport-city	15'	75'	Upgraded accessibility
TT	Turnover time	27'	70'	Aircraft utilisation-connectivity
FS	Fare structure	80€	300€	Competitive pricing
R-PL	Runway parking lots	1/5	1/50	Effective airport space utilisation

Based on performed research, the factor 'seat availability' has been defined as 70% of an aircraft's overall passenger capacity and represents a 105 minimum seat capacity and a 231 maximum seat capacity, respectively. At any time, there are up to three aircrafts serving a certain travel leg. This data is shown in the second row of columns 3 and 4, respectively. The factors 'delays at departure' and 'delays due to technical reasons', as indicated in rows 3 and 4, are taken from the data contained in Annual Handbook of Athens International Airport [17.57]. It is worth mentioning that as per IATA Regulations, a delay is defined as an aircraft arriving at the destination airport more than 15 minutes late. The distance between an airport and the nearest city has been found to be in the range of 15'–75' (measured in minutes). Similarly, the turnover time was established by taking into consideration both statistical and empirical data and lies within 27–70 minutes. Fare price refers to the average ticket pricing from domestic and European flights, excluding all offers.

The last factor addressing the availability of runways in relation to the available parking lots is the result of the minimum runway at small airports. The latter equally includes the minimum number of small aircraft parking lots. Finally, the last column depicts a runway that acts as a feeder line to a parking lot of 50 aircrafts.

It is noteworthy that the factor presented in the 4th column is currently updated, so as to include AIA-specific data from the year end of 2009.

The evaluation of the listed data served as input numbers to DoE-specific software. The implied confidence interval (CI) is 95%.

As already mentioned, Taguchi primarily uses three different ratios, namely:

- larger-the-better;
- smaller-the-better; and
- nominal-the-best.

'Customer satisfaction' is a key variable and regarded as being of paramount importance for airline companies. In DoE terms, this variable is perceived as a *quality characteristic* (QC). Figure 17.12 and Table 17.4 show the linear graph of the factorial experiment for *larger-the-better*.

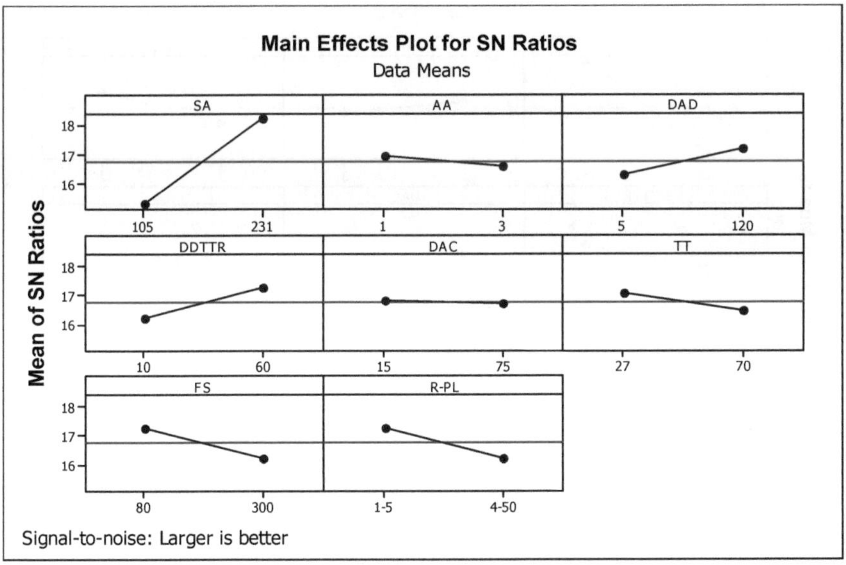

Figure 17.12. Factorial experiment $L_{12}2^8$ for *larger-the-better*

Table 17.4. SNR response for *larger-the-better*

Level	SA	AA	DAD	DDTTR	DAC	TT	FS	R-PL
1	15.29	16.93	16.32	16.22	16.80	17.06	17.28	17.27
2	18.22	16.57	17.19	17.29	16.71	16.45	16.23	16.24
Delta	2.93	0.36	0.87	1.07	0.08	0.60	1.05	1.03
Rank	1	7	5	2	8	6	3	4

The analysis of the linear graph results in the following observations.

- Seat availability (SA) shows the highest variation at 2.93 and has, thus, the biggest Loss-to-Society;

- Fare structure (FS) has the third highest variation of 1.05; and
- The distance between airport and a city reveals the smallest variation at 0.08. Hence, its effects to society are negligible.

Figure 17.13 and Table 17.5 show the linear graphs for the means and their outcomes. It reveals the following:

- SA has the largest variation in-between average values at 2.250;
- DAD shows a variation in-between average values of 0.750; and
- Finally, the distance between airport and the city verifies the outcomes of the main effects plot for larger-the-better with the smallest variation of 0.083. It indicates that its effect to society is minimal.

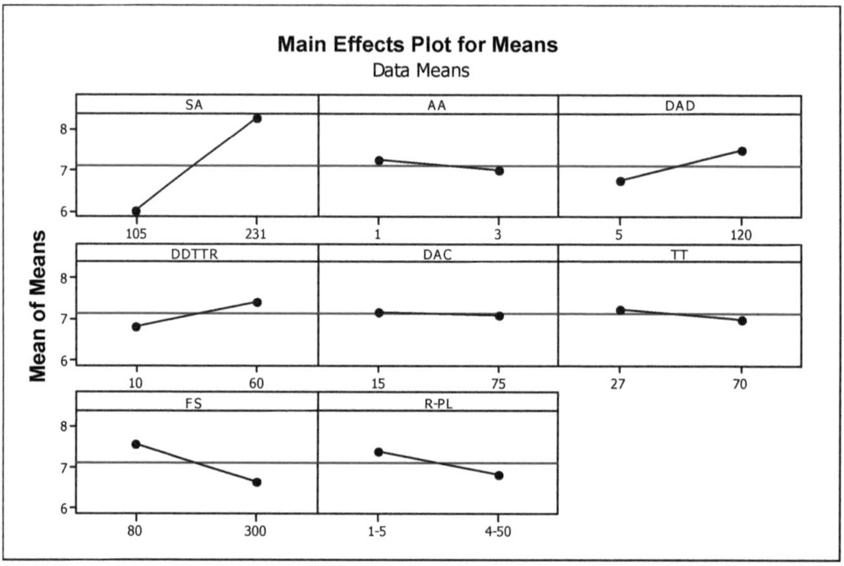

Figure 17.13. Factorial experiment $L_{12}2^8$ for the means concerning *larger-the-better*

Table 17.5. Response of means for *larger-the-better*

Level	SA	AA	DAD	DDTTR	DAC	TT	FS	R-PL
1	6.000	7.250	6.750	6.833	7.167	7.250	7.583	7.417
2	8.250	7.000	7.500	7.417	7.083	7.000	6.667	6.833
Delta	2.250	0.250	0.750	0.583	0.083	0.250	0.917	0.583
Rank	1	6.5	3	4.5	8	6.5	2	4.5

In view of the above and in order to minimise the negative effect to society, *i.e.* the impact on the airline company and its affiliates, the factor 'seat availability' ought to be improved. The assessment of Taguchi's factorial design implies that distance between an airport and a city is nominal and has the least effect on society. This is the outcome of both the application of signal-to-noise ratio and the analysis of means for *larger-the-better*. The last parameter equally satisfies initial research hypothesis and empirical observations that European airports are on average within acceptable vicinity of a city.

17.7.2 Factorial Experiment for Smaller-the-Better

Punctuality is another quality characteristic that affects the operational efficiency of an airline company. Taguchi's *smaller-the-better* is used so as to minimise potential variation to target and keep scheduling robust and insensitive from inner, and/or outer noises. Figure 17.14 shows the linear graphs for the SNR case of smaller-the-better. The evaluation of the linear graphs and the corresponding response given in Table 17.6 reveal the following:

- DAD is associated with the highest variation of 8.12. In this case, this parameter has the biggest influence to the airline company and the society;
- DDTTR is the third most influential parameter showing a variation of 4.28;
- AA indicates the least variation at 0.48 and is, thus, negligible in terms of loss-to-society.

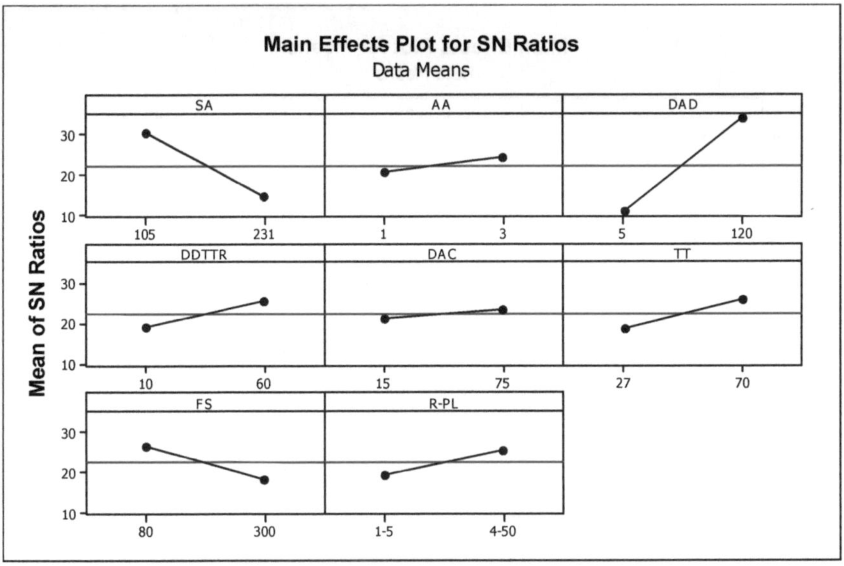

Figure 17.14. Linear graphs for the SNR case of *smaller-the-better*

Table 17.6. SNR response for *smaller-the-better*

Level	SA	AA	DAD	DDTTR	DAC	TT	FS	R-PL
1	−26.92	−24.02	−20.19	−22.11	−25.15	−23.07	−26.17	−22.13
2	−21.59	−24.50	−28.32	−26.40	−23.36	−25.44	−22.34	−26.38
Delta	5.33	0.48	8.12	4.28	1.79	2.38	3.83	4.25
Rank	2	8	1	3	7	6	5	4

Figure 17.15 shows the linear graphs and Table 17.7 shows the equivalent responses for the means for the case of smaller-the-better. The results indicate the following:

- The delays at departure show the highest in-between average values variation of 22.83, and therefore, DAD is the most influential parameter with a considerable effect in loss-to-society;
- Fare structure is the third most influential parameter, having a variation of 8.33; and
- The distance between an airport and a city has the least variation with a factor of 2.33.

Its effect to the loss-to-society is negligible. By comparing the means of the outcome of larger-the-better regarding DAC, one might observe that both cases verify the minimum influence and effect of this factor to the society and airline company. For a company applying the case of smaller-the-better, the evaluation of the SNR and the associated responses indicate that an emphasis ought to be placed at reducing and eliminating delayed departures.

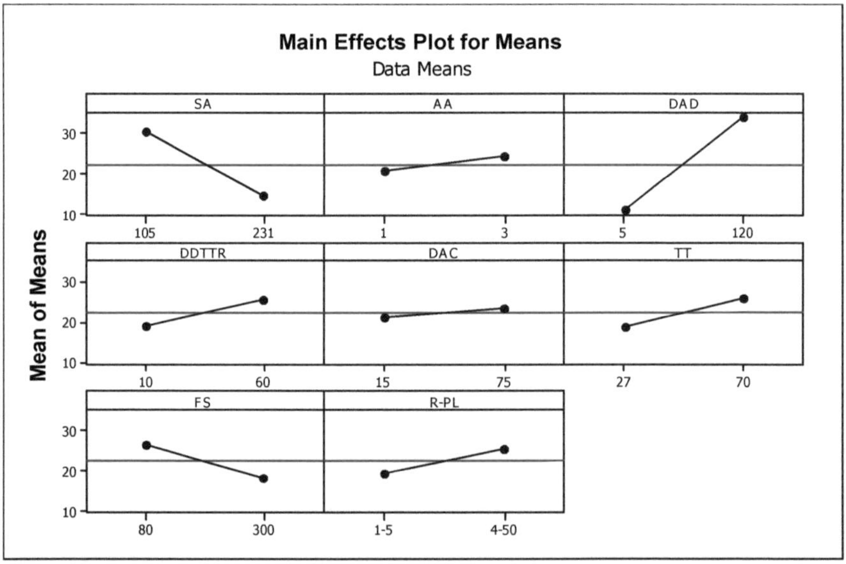

Figure 17.15. Factorial experiment for means regarding *smaller-the-better*

Table 17.7. Response of means for *smaller-the-better*

Level	SA	AA	DAD	DDTTR	DAC	TT	FS	R-PL
1	30.42	20.58	11.00	19.17	21.25	18.83	26.58	19.25
2	14.42	24.25	33.83	25.67	23.58	26.00	18.25	25.58
Delta	16.00	3.67	22.83	6.50	2.33	7.17	8.33	6.33
Rank	2	7	1	5	8	4	3	6

17.8 Conclusions and Future Work

The study presented in this chapter highlights the importance of bridging the contradicting outcomes between capability indices and signal-to-noise ratios. Based on continuous on-going research, the focus will be on applying a 2D model on a number of real-life cases.

Within the field of aviation engineering, this work has shown the criticality of maintaining hubs-and-spokes. The evaluation and assessment of a number of scientific papers has indicated that the utilisation of critical hubs is subject to further studies and research. A variety of cases that require further research and analysis evolve, not least due to the very fact that airport allocated space, during the maiden years of aviation, was not regarded as being at a premium.

The continuous growth in passenger travel and aircraft movements has resulted in a gradual congestion over major European airports. The latter contributed to an incremental level of delays. In order to minimise the effects of delays, the tendency is to follow one of the following two strategies:

- *Expansion of already existing and vital hubs*. International practices focus on the operational readiness, functionality and utilisation of satellite terminals. Although this may satisfy an even higher passenger demand, it is the author's view that the bottleneck will be re-positioned, as larger number of passengers allow for an increased number of associated aircraft movements.
- *Creation of a new hub outside the main zone of the initial hub*. Low-cost carriers play an important role in accelerating the finances of the region.

The establishment and introduction of Airport Scheduling Committees was equally aimed at dealing with means of reducing congestion at airports. Athens International Airport will do so during 2010.

Overall, operational practices include the following:

- grandfathering rights;
- use-it or lose-it policy;
- priority for regular services;
- directed discretion; and
- code sharing.

Although a considerable amount of research has been made in reducing airport-related congestion, the problem still exists. The imposition of quality standards and the endeavour for continuous improvement both in the manufacturing industry and in service providers has resulted in eliminating failures. Capability indices are a measure of variation reduction of a process. Signal-to-noise ratios, although aiming at set target values do show a larger percent of variation, also referred to by Dr Taguchi as noise. When both quality technique tools are used, more often than not, contradicting outcomes may result. Taguchi's design of experiments is a means to indicate potential bottleneck areas and assign these to an associated loss-to-society.

Performed research at Olympic Airlines' headquarters has revealed that the preliminary implication of the propose model does indeed show some promising results. So far, the proposed model has shown itself capable of bridging the gap between capability indices and signal-to-noise ratios. In addition, the suggested model offers a set of guidelines that meet quality requirements when followed, as set by the ISO 9000 and QS 9000 quality standards, respectively.

Initial practical testing has shown that the preliminary model equally describes the procedure required for improving set quality levels, by minimising variation, whilst maintaining robustness. It finally addresses quality-specific dilemmas at the beginning, rather than during the process.

In view of the above, conducted analysis indicates that the application of the model may assist aviation planners to effectively co-ordinate their flights by bridging the contradicting outcomes between capability indices and robustness. Additionally, further evaluation and continuous updates incorporating real-life cases will foster the model's applicability in a variety of cases. As such, additional influencing parameters and factors will be included.

Cases pertaining to the environmental footprint generated by aircraft emissions in conjunction with an optimum flight level, including a comparative analysis with global practices, will assist in indicating areas of further improvement and optimisation. The outcome of the latter will aid the proposed model to equally contain management-related procedures, in order to minimise and eliminate the detrimental effect of inadequate implied practices.

In addition to the above, a viable, sustainable, dynamic and flexible model acknowledging all potential influencing parameters in a hub-and-spoke system may be developed to maximise an airline's financial benefits.

Further work may be performed by taking into consideration principles of location analysis practices, which may be further enhanced by the implementation of ANOVA and capability indices. Future studies will be performed in establishing a hand-shaking model incorporating management-related procedures to optimise company-specific practices.

References

[17.1] EEC, 1993, *Commission Regulation No 1617/93*, 25/06/1993.

[17.2] EEC, 1993, "Common rules for the allocation of slots at community airports," *Council Regulation 95/93.*

[17.3] European Commission, 2001, "European transport policy for 2010: time to decide," *White Paper.*

[17.4] Rassenti, S.J., Smith, V.L. and Bulfin, R.L., 1982, "A combinational auction mechanism for airport time slot allocation," *Bell Journal of Economics*, **13**(2), pp. 402–417.

[17.5] Kounis, L.D., O'Sullivan, J.M. and McAndrew, I.R., 1999, "Preliminary experiments on a model to relate Cp_k and SNR indices," In *Proceedings of the 15th National Conference on Manufacturing Research*, Advances in Manufacturing Technology-XIII, pp. 409–414.

[17.6] Kounis, L.D., O'Sullivan, J.M. and McAndrew, I.R., 2001, "Modelling of parametric relationships in jet fans," *International Journal of Condition Monitoring and Diagnostic Engineering Management*, **4**(1), p. 34.

[17.7] Kounis, L.D., O'Sullivan, J.M. and McAndrew, I.R., 1999, "Geometrical evaluations for balancing robustness and capability indices," In *Proceedings of the 15th*

International Conference on Production Research, Manufacturing for a Global Market, **2**, pp. 1115–1118.
[17.8] Juran, J.M. and Gryna, F.M. (eds.), 1988, *Juran's Quality Control Handbook*, McGraw-Hill, New York.
[17.9] Kounis, L.D., O'Sullivan, J.M. and McAndrew, I.R., 1998, "Capability indices in a robust quality environment," In *Proceedings of the 15th Conference of the Irish Manufacturing Committee*, pp. 545–553.
[17.10] Besterfield, D., 1990, *Quality Control*, 3rd edn., Prentice-Hall, Inc., New Jersey, pp. 264–390.
[17.11] Bergman, B. and Klefsjo, B., 1994, *Quality: From Customer Needs to Customer Satisfaction*, McGraw-Hill, Columbus, OH.
[17.12] Montgomery, D.C., 1997, *Introduction to Statistical Quality Control*, 3rd edn., John Wiley and Sons Publishers, Hoboken, NJ.
[17.13] Taguchi, G., Elsayed, A.E. and Hsiang, T., 1989, *Quality Engineering in Production Systems*, McGraw-Hill, Columbus, OH.
[17.14] O'Conor, P.D.T., 1991, *Practical Reliability Engineering*, John Wiley and Sons Publishers, Hoboken, NJ.
[17.15] Kounis, L.D., O'Sullivan, J.M. and McAndrew, I.R., 1998, "Relationships between capability indices and signal-to-noise ratios for use in the manufacturing industry," In *Proceedings of the 14th National Conference on Manufacturing Research, Advances in Manufacturing Technology-XII*, pp. 785–790.
[17.16] Moen, R.D., Nolan, T.W. and Provost, L.P., 1999, *Quality Improvement through Planned Experimentation*, 2nd edn., McGraw-Hill, Columbus, OH.
[17.17] Peace, G., 1996, *Taguchi Methods*, Addison Wesley Publishing Company, Boston, MA.
[17.18] McAndrew, I.R. and O'Sullivan, J.M., 1995, "Designed experiments for assembling CPR," In *Advances in Manufacturing Technology: the 11th National Conference on Manufacturing Research*, Professional Engineering Publishers, pp. 65–71.
[17.19] Garwin, D.A, 1988, *Managing Quality: the Strategic and Competitive Edge*, The Free Press Publishers, pp. 54–55 and p. 266.
[17.20] Jones, H.J., 1996, "Developing designed experiments," *International Journal of Quality and Reliability*, p. 25 and pp. 87–95.
[17.21] SMMT, 1986, *Guidelines to SPC*, Society of Motor Manufacturers and Traders, London.
[17.22] Bentley, J.P., 1993, *An Introduction to Reliability and Quality Engineering*, Longman Scientific and Technical, Harlow, Essex.
[17.23] Oakland, J. and Dale, B., 1991, *Quality Improvement through Standards, in Association with BSi*, Stanley Thorns Publishers, Cheltenham.
[17.24] Grant, L. and Leavenworth, R.S., 1996, *International Edition, Statistical Quality Control*, McGraw-Hill, Columbus, OH, pp. 324–326.
[17.25] Bissell, D., 1994, *Statistical Methods for SPC and TQM: Tests in Statistical Science*, Chapman and Hall, London, pp. 236–239.
[17.26] Oakland, J.S. and Followell, R.F., 1990, *Statistical Process Control: A Practical Guide*, 2nd edn., Butterworth-Heinemann, Oxford, pp. 133–150.
[17.27] Fox, M. J., 1993, *Quality Assurance Management*, Chapman and Hall, London, pp. 107–109, 226–231, 247, 255, 262.
[17.28] Ishikawa, K., 2003, *Introduction to Quality Control*, Taylor & Francis, London.
[17.29] Frank, M., Mederer, M., Stolz, B. and Hanschke, T., 2005, "Depeaking-economic optimisation of air traffic systems," *Aerospace Science and Technology*, **9**, pp. 738–744.
[17.30] Sullivan, L.P., 1984, "Reducing variability: a new approach to quality," *Quality Progress*, American Society for Quality Control.

[17.31] Straker, D., 1995, *A Toolbook for Quality Improvement and Problem Solving*, Prentice Hall, Inc., New Jersey, USA, pp. 167, 172–177, 310–319.

[17.32] Taylor, W.A., 1991, *Optimisation and Variation Reduction in Quality*, McGraw-Hill, Sponsored by the American Society for Quality Control, pp. 99–117, 119–129, 163–178 and p. 180.

[17.33] Barker, T.B., 1994, *Quality by Experimental Design*, 2nd edn., Marcel Dekker, Inc., New York, p. 229.

[17.34] DeVor, E.R., Chang, T.-H. and Sutherland, J.W., 2006, *Statistical Quality Design and Control*, Pearson Education, Upper Saddle River, NJ, pp. 19–24, 264–267, 326–327, 336–337, 504–511.

[17.35] Anonymous, "The cracks in quality," *Economist*, **323**(7755), pp. 85–86.

[17.36] Tsandilis, P., Tzimas, G. and Kounis, L.D., 2009, "Optimising Olympic Airlines' fleet by implementing Design of Experiments," *International Journal of Aviation Management*, (accepted).

[17.37] Kounis, A.B., O'Sullivan, J.M. and McAndrew, I.R., 2000, "Determining capability indices with small samples," *In Proceedings of the 16th National Conference on Manufacturing Research*, pp. 449–453.

[17.38] Kounis, L.D. and Panagopoulos, N., 2007, "Total quality management and benchmarking: bridging the gap in the public sector," *International Journal of Services and Operations Management*, **3**(2), pp. 245–259.

[17.39] Abeyratne, R.I.R., 2000, "Management of airport congestion through slot allocation," *Journal of Air Transport Management*, **6**, pp. 29–41.

[17.40] Le Loan, Donohue, G., Chen, C.-H., 2005, "Using auction-based slot allocation for traffic demand management at Hartsfield Atlanta International Airport: a case study," *Journal of the Transportation Research Board*, (1888), pp. 50–58.

[17.41] Diana, T., 2009, "Do market-concentrated airports propagate more delays than less concentrated ones? A case study of selected U.S. airports," *Journal of Air Transport Management*, pp. 1–7, (in press).

[17.42] Gillen, D., 2003, "Airline strategies, competition and network evolution in aviation: how important are slots?" In *GARS Workshop: How to Make Slots Work*, University of Applied Sciences, Bremen.

[17.43] Zografos, K.G. and Madas, M.A., 2003, "A critical assessment of airport demand management strategies in Europe & U.S.: a comparative perspective," *TRB 2003 Annual Meeting*, CD-ROM.

[17.44] Mehndirata, S.R. and Kiefer, M., 2003, "Impact of slot controls with a market-based allocation mechanism at San Francisco International Airport," *Transportation Research Part A*, **37**, pp. 555–578.

[17.45] Bard, J.F. and Mohan, D.N., 2008, "Reallocating arrival slots during a ground delay program," *Transportation Research Part B*, **42**, pp. 113–134.

[17.46] Martin, J.C. and Voltes-Dorta, A., 2009, "A note on how to measure hubbing practices in airline networks," *Transportation Research Part* E, **45**, pp. 250–254.

[17.47] Rietveld, P. and Brons, M., 2001, "Quality of hub-and-spoke networks: the effects of timetable co-ordination on waiting time and re-scheduling time," *Journal of Air Transport Management*, **7**, pp. 241–249.

[17.48] Slack, B., 1999, "Satellite terminals: a local solution to hub congestion?" *Journal of Transport Geography*, **7**, pp. 241–246.

[17.49] Lijesen, M.G., Rietveld, P. and Nijkamp, P., 2001, "Hub premiums in European civil aviation," *Transport Policy*, **8**, pp. 193–199.

[17.50] Dennis, W., 2009, "Malaysia aims to become aviation hub," *Journal of Engineering and Technology*, **4**(6), p. 13.

[17.51] Barrett, S.D., 1999, "Peripheral market entry product differentiation, supplier rents and sustainability: a case study," *Journal of Air Transport Management*, **5**, pp. 21–30.

[17.52] Martin, J.C. and Roman, C., 2004, "Analysing competition for hub location, in intercontinental aviation markets," *Transportation Research Part E*, **40**, pp. 135–150.

[17.53] Rietveld, P. and Brons, M., 2001, "Quality of hub-and-spoke networks; the effects of timetable co-ordination on waiting time and re-scheduling time," *Journal of Air Transport Management*, **7**, pp. 241–249.

[17.54] Milgram, S., 1977, "The small world problem," *The Individual in a Social World: Essays and Experiments*, Addison-Wesley, Reading, MA.

[17.55] Malighetti, P., Paleari, S. and Redondi, S., 2008, "Connectivity of the European airport network: self-help hubbing and business implications," *Journal of Air Transport Management*, **14**, pp. 53–65.

[17.56] Chrysler, Ford, General Motors, 1998, *QS 9000 Quality System Requirements*, 3rd edn., Michigan.

[17.57] AIA, 2008, *Annual Handbook of Athens International Airport.*

Index

MIX
Papier aus verantwortungsvollen Quellen
Paper from responsible sources
FSC® C105338
FSC
www.fsc.org

If you have any concerns about our products,
you can contact us on
ProductSafety@springernature.com

In case Publisher is established outside the EU,
the EU authorized representative is:
Springer Nature Customer Service Center GmbH
Europaplatz 3, 69115 Heidelberg, Germany

Printed by Libri Plureos GmbH
in Hamburg, Germany